MYSTERY CATS OF THE WORLD REVISITED:

Blue Tigers, King Cheetahs, Black Cougars, Spotted Lions, and More

By Dr. Karl P.N. Shuker

Anomalist Books
*San Antonio * Charlottesville*

An Original Publication of ANOMALIST BOOKS
Mystery Cats of the World Revisited

ISBN: 978-1-949501-17-9

Cover: Mystery cat art images provided by William M. Rebsamen
Book Design: Seale Studios

For information about the publisher, go to AnomalistBooks.com, or write to:
Anomalist Books, PO Box 6807, Charlottesville, Va 22906

CONTENTS

In fondest memory
of my mother Mary Shuker, and all of my family
—for your love and unwavering support.
I wish that you were all still here with me.

NEW PREFACE (2020)

I am too much of a sceptic to deny
the possibility of anything.
— Thomas Henry Huxley, Letter written in 1886

My first book, *Mystery Cats of the World: From Blue Tigers to Exmoor Beasts,* was published in 1989. Little did I think back then that just over 30 years later, I would have 30 books to my credit, plus over a dozen more featuring me as consultant and/or contributor, having succeeded in my ambition to become a full-time freelance zoological author and researcher specializing in cryptozoology, animal anomalies, and zoomythology. Given such a long passage of time, it was inevitable that a fair few of my books, particularly the earlier ones, would eventually go out of print. Happily, however, at one time or another and in one form or another they have all come back into print, either as straightforward reprints or as expanded, updated new editions—all but one, that is.

That lone exception was—until now—*Mystery Cats of the World.* It proved very popular both within and far beyond the cryptozoological community, attracting positive reviews from such diverse figures as Nobel Prize-winning authoress Doris Lessing *(Daily Telegraph,* 25 November 1989) and world-renowned felid geneticist Roy Robinson (*Cats Magazine*, December 1989). This in turn crucially provided me with the necessary encouragement and impetus to continue my cryptozoological pursuits. Moreover, having swiftly become the standard work on mystery cats (even today, it remains the only book to survey such cryptids on a worldwide basis), it became so sought-after and collectible that copies regularly sold for up to £200 on various online auction and bookselling sites, and I regularly received communications from readers and fellow researchers alike beseeching me to prepare a second edition. Yet it had been such an inordinately intensive project to research and prepare that the thought of one day producing an expanded, updated version was one that had always daunted me. And so the years passed by with no such edition coming into being.

By 2011, however, I had begun work on what seemed to me to be a satisfactory compromise—namely, a second book on mysterious, controversial cats, which would include certain updates of cryptozoological cases previously documented in *Mystery Cats of the World* but whose remit would go far beyond that earlier book of mine, by also including reports of fascinating felids outside the realm of cryptozoology. These included cats that were ostensibly paranormal (e.g. the Indian devil cat and the demonic Big Ears), or undeniably legendary (e.g. Japanese vampire cats and feline deities), or teratological (e.g. winged cats and Janus cats), or exhibiting unexplained abilities (e.g. psi-trailing cats or cats that can sense impending death), or even of exclusively literary origin (e.g. Lewis Carroll's Cheshire cat and Edward Lear's runcible cat). Quite a weighty tome when complete, and lavishly illustrated throughout in full color, it was entitled *Cats of Magic, Mythology, and Mystery: A Feline Phantasmagoria*, and was published in 2012.

This too was well received, yet still the requests continued for a new version of *Mystery Cats of the World*. Finally, a couple of years ago I decided that the time had come to accede in some form or another to popular demand. Initially, I had planned simply to release what to all intents and purposes would be a simple reprint, containing only essential updates or amendments required in order to maintain the book contents' accuracy. Yet once I began working upon it, I soon realized that to do nothing more adventurous than this would be to waste a golden opportunity to incorporate the exciting, compelling new data on mystery cats (not to mention some of the many superior illustrations) that I had uncovered or received during the intervening three decades since 1989. Equally, however, it swiftly became clear that any attempt to incorporate all such information would result in a book so voluminous that publishing it would likely not be a financially viable prospect. Somehow, therefore, once again a compromise had to be achieved, and after a period of reflection I finally conceived one.

From 1989 to 2019, in certain areas of the world reports of mystery cats had continued to emerge on a regular basis but were of an entirely repetitious nature, i.e. the types of cat reported, their behavior, theories concerning their identities, etc, remained the same, year after year. This was particularly true with regard to those variously black pantheresque, puma-resembling, and lynx-like felids reported from Britain and mainland Europe, plus the pantheresque and puma-like felids being sighted in North America (in out-of-place localities there with regard to puma-like cats) and Australia. Consequently, I saw little sense in devoting numerous additional pages to such repetition when the many cases concerning such felids that I had already documented leading up to 1989, as well as the various theories put forward to explain them, in my original *Mystery Cats of the World* book were still wholly representative of such animals 30 years later.

Conversely, in certain other areas of the world, notably tropical

Africa, Asia, and South America, many reports of entirely new mystery cats, i.e. feline cryptids totally different from those documented in my 1989 book, had come to light and therefore demanded preferential treatment, so this is what I decided to give them. Consequently, as this new book documents a sizable number of novel, additional mystery cat forms and reappraises various previously documented ones, I have duly entitled it *Mystery Cats of the World Revisited.*

I had anticipated that this book would be published in 2019, thereby commemorating exactly 30 years since its predecessor appeared in print, but fate decreed otherwise, inasmuch as my health took a rather worrying turn that year, which need not be dwelt upon here other than to say that having a medical Sword of Damocles suspended over one's head for several months does not exactly focus the mind upon writing and research, and therefore severely impacted my ability to prepare this book within my originally conceived timeframe.

Happily, however, that ominous Sword eventually dematerialized, and despite two further major delays (caused by a couple of very serious computer meltdowns), I have worked unstintingly since then with dogged (or should that be catted?) determination in an attempt to make up as much lost time as possible, and I am delighted at last to see this long-awaited, long-requested volume finally in print. I also wish to take this opportunity to thank my publisher and editor Patrick Huyghe most sincerely and gratefully for his kindness and understanding shown to me during my indisposition last year, which provided me with immense encouragement at a very difficult time in my life.

Looking back, it seems like only yesterday that my little box of six complimentary copies of *Mystery Cats of the World,* sent to me by its publisher, Robert Hale Limited of London (now no longer in existence), arrived through the post one sunny morning in June 1989, and how excited my mother Mary Shuker and I were to see them, hold them, browse through them, and later see copies in the shops—my very first book! How proud once again she would be now to know that it has risen like a veritable feline phoenix 31 years later, and that it will, I hope, be enjoyed by a whole new generation of readers, not to mention my loyal, ever-expanding, ever-faithful following, some of whom date back with me right to the early 1980s and my debut within the cryptozoological community.

How I wish that you were still here too, Mom, to be a part of it all, and how truly, eternally grateful I am to you for all of the love and support that you gave to me throughout my life and zoological career. I personally signed to you the very first copy of *Mystery Cats of the World* that I took out of that box of complimentary copies, and I still have it here today. And when I receive the copies of *Mystery Cats of the World Revisited*, I shall sign the very first one of those to you too, because no-one in this entire world could ever deserve it more than you. God bless you Mom, with all my love, always.

Original Preface (1989)

I saw a proud, mysterious cat.
— Vachel Lindsay, "The Mysterious Cat"

Once all-too-commonly derided and dismissed as a subject for dreams and for dreamers, cryptozoology—the scientific investigation of animals whose existence or identity has yet to be formally ascertained—nowadays attracts ever-increasing international interest. So much so in fact that in 1982 the world's first scientific Society devoted to mystery creatures was established—the International Society of Cryptozoology (ISC), based at Tucson, Arizona. Its Board of Directors comprises some of the world's most distinguished professional zoologists, and its President is French zoologist Bernard Heuvelmans—a renowned cryptozoological researcher and the person responsible for first popularizing the term "cryptozoology."*

It is not difficult to understand their interest and belief in the existence of such creatures. All of the following large-sized animals, for example, were unknown to science until the 20th century: the okapi (relative of the giraffe, discovered in 1901); mountain gorilla (world's largest ape, 1901); giant forest hog (over 8 ft long, 1904); Komodo dragon (world's largest lizard at up to 10 ft in total length, 1912); kouprey (new species of wild ox, 1937); Shepherd's beaked whale (radically new species, 1937); coelacanth (large fish belonging to a group hitherto believed extinct for over 64 million years, 1938); cochito (new species of porpoise, 1959); Chacoan peccary (pig-like creature believed extinct since the Ice Age, 1975); megamouth (world's third-largest shark species, 1976); bilkis gazelle (alive and well in Qatar, 1986). And these are only a few such discoveries!

Huge expanses of forested, mountainous, desert, and polar areas still await scientific exploration, and are currently known only from aerial photographs. The zoological content of much of the inland waters of the world is very incompletely known too, and that of the immense oceans is shrouded in mystery. In short, the potential for new and significant zoological discoveries is tremendous, a fact that cryptozoology's critics attempt lamely to deny by dismissing that century's major animal finds as irritating exceptions, and by conveniently overlooking our world's vast untraversed, unexplored regions—an attitude

* When I wrote this preface, i.e. 1989, the ISC was still a very active, thriving scientific society; sadly, however, by the end of the 1990s it had ceased to exist and has never been reestablished.

that is happily and ever increasingly ridiculed by every new animal discovery that takes place.

Astute students of cryptozoology will note readily that a considerable number of mystery animals fall into one of three main categories—and for good reason. Namely, that these animals have the greatest chance of remaining undetected. They are:

> Category 1 - aquatic creatures, whose natural habitat is not conducive to meticulous, sustained searches (especially the larger inland stretches of water and the voluminous oceans).
>
> Category 2 - higher apes and primitive humanoid forms, whose high intelligence would keep them well away from modern humanity.
>
> Category 3 - mystery cats, whose inherent elusiveness is a primary requirement for their entire way of life.

Several books devoted to aquatic "monsters" and unknown primates have already been published, and more continue to emerge. Never before, however, has a book appeared that is devoted to mystery cats of the world. I have been interested in such animals for many years, and my personal researches have yielded a considerable archive of relevant material—a notable proportion of which has not received serious cryptozoological attention before. Additionally, many items have not been previously published, whereas others have never before appeared in an English-language book. In recent years, mystery cats from many lands have gained very notable attention—e.g. the notorious Beast of Exmoor, the cheetah-like onza shot in Mexico, the resplendently-striped king cheetah filmed in southern Africa, the procurement in Scotland of specimens of a black gracile cat hitherto unrecorded by science, the sleek panther-like creatures that roam almost every state not only of the USA but also of Australia. Periodicals specializing in mysterious phenomena are devoting ever-increasing amounts of space to reports of unidentified and strange cat forms too.

Consequently it seemed an appropriate time for me to prepare an entire book dealing with mystery cats worldwide—but not an exhaustive, rigidly-formal work. For this subject is supported by such an immense body of material—documented and undocumented—that, quite frankly, a really detailed account would require several hefty volumes! Indeed, as we shall see, entire books have already been devoted to certain specific examples. In actual fact, as my cryptozoological researches progress, I do plan to produce a specialized series of scientific treatises on selected mystery felids. At the present time, however, I feel that something rather different is needed first of all.

For it is only too clear that the phenomenon of mystery cats worldwide is a subject that currently fascinates people worldwide—and with an appeal not limited to any single type of person. Specialists and nonspecialists, scientists and the general reading public; all are very greatly intrigued by the numerous reports of a vast variety of cat-like creatures that continue to appear and which continue to baffle, involving feline fauna being sighted in regions where such animals are not supposed to exist, or fitting the description of no species known to exist anywhere.

With this in mind, I have therefore sought to produce a concise (but not superficial!) and relatively informal guide to these animals, which brings together all of the principal mystery cat forms of the world within a single volume, and which defines and displays the scope and essential components of the subject. Hence my book has been designed to fulfill the following functions: to demonstrate the number and diversity of mystery cat forms; to provide for each of these a resume of its basic history (which concentrates upon the more lucid and informative of reports currently available, rather than attempting an unselective blanket coverage); to offer an analysis whenever possible of likely zoological identity involved; and to produce a comprehensive reference list of all sources used in this book, which can be consulted by readers wishing to pursue any aspect further.

Incidentally, felids once mysterious but nowadays relatively well-known and well-studied (e.g. Rewa-type white tigers, Timbavati-type white lions) are only treated briefly here. Additionally, unusual and out-of-place forms of domestic cat (e.g. sphynx, "winged" cats, West Australian ferals) are not referred to at all. Instead, these are fully documented in a subsequent book of mine [a very early forerunner of what ultimately became *Cats of Magic, Mythology, and Mystery*, 2012].

One further point: it has been frequently suggested that certain mystery cat forms (notably the black pantheresque felids of Britain and continental Europe, Australia, and North America) are not corporeal beasts at all, but are paranormal entities instead. A variety of identities of this type have been put forward for such cats, ranging from were-panthers, thought-forms (tulpas), and specters, to UFO occupants, and denizens of a subterranean world at the Earth's core.

As far as I am concerned, however, all of the mystery cat reports contained in this book are ones that I firmly believe can be satisfactorily explained by straightforward cryptozoological reasoning, following traditional scientific principles. Having said that, I am nevertheless aware of certain additional reports describing cats that do appear to be more akin to the phantasmal Black Dogs of folklore than to any corporeal cryptozoological cat.

Such creatures include, for example: the eerie devil cat of India, exhibiting a propensity for eldritch shrieks and a notable fear of holy water (as documented by J. Morrison in *Fate*, February 1955); a ghostly black cat said to haunt Northern Ireland's Mount Pellier (noted by J.J. Dunne in *Haunted Ireland*, 1977), and a comparable creature allegedly frequenting the ruins of Suffolk's Borley Rectory (documented by Graham McEwan in *Mystery Animals of Britain and Ireland*, 1986); plus China's deadly devil tiger—more like an amorphous animated cloud than any form of animal (E.G. Bentley reported it in *Fate*, July 1960). Even if such bizarre entities as these really do exist, their origins and identities are evidently subjects that pass far beyond the realms of cryptozoology. Consequently they are not dealt with in this book.

The past few years have witnessed the publication of cryptozoological books on a hitherto unprecedented scale. Worthy of especial note here is that a sizable proportion of these are still in print, demonstrating that there is indeed an appreciable public interest and demand for books dealing with cryptozoology. Similarly, the two-day ISC Annual Members Meeting held in July 1987 at Edinburgh's Royal Museum of Scotland attracted three times the audience present at previous years' ISC Meetings. Long may this upward trend continue, so that cryptozoology itself can flourish and thereby enable its researchers to succeed in carrying out those activities common and intrinsic to all scientific disciplines—transforming the myths and superstitions of the past into the facts and realities of the future.

Original Foreword (1989) By Paul And Lena Godsall Bottriell

This thoughtful work of scholarship by Karl Shuker is an ideal book with which to enter the 1990s; he has performed an admirable service. As the century pitches inexorably towards the year 2000, humankind's awareness of the world's environment has taken a dramatic leap forward. Significantly, an increased appreciation of nature's limitless ability to bring forth life has accompanied it—a new comprehension of evolution happening around us. Suddenly those vestiges of wilderness that remain are being viewed as more than just the marvels of nature, without which our spirits would be the poorer. It's not only the untold benefits the preservation of our tropical forests offer, in cures for terminal diseases, that's coming home to humanity: rocketed into the spotlight by environmental events and resultant pronouncements on changes in weather patterns and the ozone layer, it is the compelling significance of what preserving the symbiotic processes of nature, with its remarkable interdependence of species, represents for the modern world.

Amidst this new thinking can be glimpsed the budding of a universal realization that the saving of individual species is not a romantic, Franciscan whim reserved only for impractical eccentrics, but the concern of everyone; that it's not only possible to discover new species of flora and fauna, but animals once taken to be just attractive aberrations or compelling myths may be new races evolving—and new races evolving to survive, for reasons linked directly to the changing fortunes of the Earth's environment. Legendary animals, like the king cheetahs, for example—for the "re-discovery" of which we're privileged to be deemed responsible. It's here this exemplary treatise best sits; an intelligent study of a neglected subject that doesn't sensationalize information to fit personal theories. With an investigative, systematic flow of research details, Karl Shuker reviews his data critically and constructively, leaving room for readers to add their own analysis, and by extension, follow-up research and fieldwork. Equally worthy is an attractive, uncomplicated style, which makes it an enjoyable read, and thus as accessible to laymen as to zoologists.

In *Mystery Cats of the World*, Karl Shuker treads worthily in the footsteps of Bernard Heuvelmans. As a reliable, up-to-date reference, this is a book that both professionals and amateurs should carry in their hip pockets. Perhaps a sequel on another group of mysterious animals may be forthcoming from the author in the not too-distant future. We avidly await it.

L.G.B. & P.B., England, March 1989

Introduction

Cats, no less liquid than their shadows,
Offer no angles to the wind.
They slip, diminished, neat, through loopholes
Less than themselves.
— A.S.J. Tessimond, "Cats"

Cats are among the most elegant, aloof, and awe-inspiring of all animals—they are at once ubiquitous and evanescent, docile and deadly, familiar and mysterious. Their ensemble includes the majestically-maned lion, black-and-gold tiger, richly-rosetted leopard and jaguar, gracile cheetah, densely-furred snow leopard or ounce, mountain-haunting puma, keen-eyed lynx, and a host of smaller species, not least of which is the tabby tiger that lies curled up in front of many a hearth-side fire.

A total of 40 wild modern-day species are currently recognized by science (41 if we accept the delineation of the Sunda leopard cat as valid). However, a vast body of information exists—ranging from anecdotal reports and native lore to unidentified paw tracks (spoor) and even the very occasional preserved pelt or specimen—which implies very strongly that additional cat forms also inhabit our world. Some seem to be forms undescribed by science (at least in the living state); others appear to be felids already known to science, but occurring in unexpected, non-native localities or of aberrant, non-standard appearance. All are awaiting investigation and possibly even actual discovery by anyone with sufficient interest and persistence to traverse through little-explored terrain and peruse through dusty, obscure publications in pursuit of conclusive evidence for their existence—evidence that can then at last be submitted for formal scientific analysis to determine these animals' identities.

Such cats constitute the subject of this book, and include among their number: Tanzania donkey-sized, brindled mngwa (nunda); Colombia's striped saber-tooth lookalike; Fujian's blue tiger; East Africa's spotted lions; Australia's Emmaville panther and Queensland tiger; southern Africa's elegant king cheetah; Mexico's equally gracile, puma-like onza; the Ile du Levant's fierce wildcat; Angola's water lion and Patagonia's water tiger; North America's mysterious maned lions; Guyana's white-chested jaguarete; Ethiopia's tigerine wobo; England's Surrey puma and Exmoor Beast; and many others too.

In order to set the stage for such animals and appreciate their especial significance, however, we should firstly familiarize ourselves with the known cats, via a concise consideration of cat characteristics, form, evolution, and present-day contingent, emphasizing those aspects particularly relevant to the investigation of mystery cat types.

Characteristics and Form

Taxonomy is the scientific classification of organisms. Cats belong to the taxonomic order of mammals known as Carnivora, and its extant families are: Felidae (cats), Nandiniidae (nandinia or African palm civet), Prionodontidae (Asian linsangs), Eupleridae (Madagascan mongooses, fossa), Herpestidae (true mongooses, meerkats), Hyaenidae (hyaenas, aardwolf), Viverridae (civets, genets, binturong), Canidae (dogs, foxes), Ursidae (bears, giant panda), Ailuridae (lesser panda), Mephitidae (skunks, stink badgers), Mustelidae (weasels, martens, otters, wolverine, badgers), Procyoniidae (raccoons, coatis, cacomistles, kinkajou, olingos), Odobenidae (walruses), Otariidae (eared seals), and Phocidae (earless seals).

Incidentally, although "feline" is often used colloquially as a noun, in strict zoological terminology it functions exclusively as an adjective. Consequently, phrases such as "the mystery feline" and "unknown felines" that frequently crop up in newspaper reports and other popular-format media coverages are technically incorrect. Grammatically speaking, the cat is a felid, its behavior is feline.

When dealing with mystery cats, it is essential to recognize that felids are undoubtedly the most efficiently evolved of all mammalian carnivores for predation and flesh eating, in terms of hunting behavior and anatomical specialization. Indeed, their prowess as predators and their inherent elusiveness are legendary. Hence we should not be surprised to discover that mystery cats possess these same qualities.

Except for certain absolute and relative size differences, all modern-day cat species are remarkably similar structurally, as the following blueprint of the felid hunting machine demonstrates.

The felid's head comes complete with large, highly-developed eyes; mobile, acute ears; a sensitive, albeit smallish nose; and touch-responsive whiskers (vibrissae). Coupled with a brain that possesses large cerebral hemispheres—responsible for coordinated, intelligent behavior (a vital requirement for a successful hunter)—all of these contribute significantly to very efficient prey detection.

Its jaws are relatively short, and contain 30 teeth that constitute a dentition specialized to accomplish the two principal functions required of a successful predator's teeth—seizing and tearing the living prey, then cutting its carcass into manageable pieces for swallowing. Consequently, a cat's canines are long, strong, and sharply-pointed to carry out the first of these functions. The second is served by specialized teeth that are characteristic of mammalian carnivores—the carnassials. These large teeth consist of the last pair of upper premolars and the only pair of lower molars, and are modified by the possession of sharp blade-like edges that slice through meat like shears when brought alongside each other by the closing of the jaws. The remaining teeth are reduced in size and used mostly for holding the food in position (the incisors), and chewing (those premolars and molars not adapted as carnassials).

In felids, the tongue is long, and bears many horny, backward-pointing projections called papillae. These are used by the cat to prise meat away from bones (felids do not normally break the bones of their prey while eating), and to wash itself effectively.

The felid body is both flexible and strong—the vertebral column is exceedingly so, especially in the lumbar (waist) region—and its intricate muscle system in combination with its powerful and supple limbs is coordinated superbly to yield the graceful fluid gait and movements epitomizing all types of cat and integral to their hunting technique.

The mode of hunting carried out by most felids consists of a short series of creeping runs (referred to as stalking runs) that bring the cat ever closer to its prey, interspersed by pauses for renewed observation of the latter from behind cover (the watching posture). When the cat ultimately reaches its intended victim, it then pounces with forelimbs outstretched to strike.

Although the cat's typically long tail can certainly assist balance while jumping and climbing, it is clearly not indispensable—the short tails of lynxes do not seem to impair their agility in any way. In fact, probably its most important function is as a signaling device to other cat individuals; it is also an indicator of felid behavior—its movements and position vary according to the cat's mood and intentions.

Spoor—Paws and Claws

Cat spoor, i.e. the tracks of their paw prints, have certain characteristics that enable them to be differentiated in many instances from those of dogs and other mammals. Thus if an unidentified mystery carnivore leaves behind spoor, these provide the investigator with important clues for determining the creature's basic identity. Such evidence has been obtained in relation to a number of beasts currently deemed to be mystery cats. Hence a short guide comparing felid and canid spoor will follow shortly here.

First of all, however, a few words concerning feet are warranted. Mammals such as ourselves, as well as bears and monkeys, walk on the soles of our feet, and are therefore termed plantigrade. Conversely, dogs and cats walk only on their toes (digits), and are therefore termed digitigrade. This latter mode of locomotion is of benefit to those animals to which running and springing are of such great importance, e.g. hunters, because the act of moving on one's toes actually increases the length of the limbs, and hence the length of each stride taken.

Comparison of cat and dog spoor (plus alleged Queensland tiger spoor) (Karl Shuker)

Cats and dogs possess five toes on each front foot (forefoot), but the fifth toe or pollex (corresponding to our thumb) is extremely small and is positioned well above ground level, so these animals only walk on four toes on each of their forefeet. The same is true for their hind feet, although with these the fifth toe or hallux (corresponding to our big toe) is missing altogether.

The spoor of dogs, foxes, and cats all yield paw prints that each contain four toe pads and one posterior plantar (palm) pad, but certain differences also exist.

With foxes, the distinctly-perceived toe pads are elongate, all four produce claw impressions, and the two front toe pads are positioned very close to one another. Moreover, the plantar pad is two-lobed, with its rear edge concave on the forefeet, convex on the hind feet; and the overall spoor outline is elongate.

Dog spoor, regardless of breed, is very similar to that of foxes, except of course for overall size, which varies according to the breed in question. In addition, the overall outline in dog spoor is rounder than in that of foxes, and the two central toe pads are not positioned so far forward.

Although varying between species in terms of its size, cat spoor otherwise remains much the same in appearance (although spoor of larger species tends to be slightly more asymmetrical than spoor of smaller species), but is recognizably different from that of canids. Its overall outline is almost circular, enclosing four clearly separate, oval toe pads that point straight ahead plus a large three-lobed plantar pad with a convex rear edge on all four paws.

Additionally, the pads' alignment in cat spoor is more asymmetrical than the very symmetrical pad alignment in dog spoor. This can be readily demonstrated by what is commonly termed the X test. Simply place a pencil, straight stick, drinking straw or some other similar item so that it lies directly below the lower edge of the left front toe pad and the lower edge of the right rear toe pad, and a second such item directly below the lower edge of the right front toe pad and the lower edge of the left rear toe pad, so that the two items form an X. In a dog spoor, the X usually will not touch any edge of the plantar pad, but in a cat spoor the X will (due to the cat spoor's asymmetrical shape).

Furthermore, cat spoor does not contain claw marks under normal conditions. There is very good reason for emphasizing those three words. The terminal bone of each of a cat's 18 toes bears a sharp claw, which can be fully retracted by all known felids except for the cheetah and serval. Numerous books and articles contain the statement that cats retract their claws when moving and thus do not leave paw prints possessing claw marks. This is at best ambiguous and at worst suggests a lack of understanding concerning not only felid behavior but also felid anatomy. For such a statement implies that claw retraction is an active process, i.e. one requiring effort on the cat's part, without which its claws would automatically extend. In fact, the exact reverse of this is the true situation.

Paws of a black panther (melanistic leopard) showing their pads (Karl Shuker)

This is because when its paws are in a relaxed, passive state (i.e. unexerted, with no current use for their claws), the claw-bearing terminal bone of each toe is positioned completely off the ground, held on top of the bone immediately behind it by a retractor ligament, so that each claw is retracted. Conversely, when the claws are needed, a flexor muscle actively contracts, pulling upon a tendon that in turn lowers the claw-bearing bone of each toe, so that the claws then become fully protracted and in contact with the ground, i.e. they are extended.

Although it is certainly true that a cat's claws remain retracted under normal conditions when it moves (thereby keeping them razor-sharp for predatory purposes), they do not stay retracted under all conditions while the cat is moving. On the contrary, claw extension will occur if conditions necessitate.

For example: as any cat owner will verify, when a cat leaps from an above-ground surface, e.g. a wall, a tree branch, even a chair, it will extend its claws in order to secure a firm grip upon the ground when it lands, thereby ensuring its continued stability. Similarly, if a cat is moving across a hazardously uneven or slippery surface, it will once again extend its claws, to obtain a better grip on that surface and thus maintain its stability. In short, a cat's claws are used not only for seizing prey but also for coming to the rescue of its stability when moving. Needless to say, in a situation where claws have been employed for the latter purpose, any spoor left behind will contain clear claw marks.

Consequently, the frequent tendency of mystery beast investigators to eliminate immediately a felid as a possible identity in a case involving clawed paw prints is evidently unwise and incorrect. Instead, whenever clawed paw prints are discovered in such a situation, an assessment of the ground's surface—is it slippery, rough, at a steep gradient?—should always be conducted before any thought of dismissing a felid identity for the mystery beast in question is aired.

Evolution

As will be seen in the following chapters, certain mystery cats apparently resemble various felids officially believed to be long extinct. If these mystery cats are ultimately discovered and are indeed found to constitute supposedly demised forms or hitherto-unknown modern-day descendants of them, it would be exceedingly significant scientifically. Yet it would not be unprecedented because quite a number of creatures officially long extinct have indeed been discovered alive and well or represented by living descendants in remote regions—like the two known modern-day species of coelacanth *Latimeria chalumnae* and *L. menadoensis*, the Chacoan peccary *Catagonus wagneri*, and the mountain pygmy possum *Burramys parvus*, to name but four extremely important rediscoveries of this type made in modern times. Consequently, in view of this exciting possibility, the following brief history of felid evolution is presented to illustrate and identify the major cat types of the past (and present?).

All modern-day members of the taxonomic order Carnivora (technically referred to as carnivorans) are descended from a taxonomic family of weasel-like mammals called miacids, of which the typical genus, *Miacis*, lived about 50 million years ago during the Eocene epoch. Felid development from this humble ancestor then occurred very rapidly.

By the late Eocene/early Oligocene (approximately 36 million years ago), the first recognizable representatives of distinct felid taxonomic tribes had already appeared, beginning with the nimravids or false saber-tooths. They were so named because their upper canine teeth were somewhat larger than their fairly normal lower canines. Examples included the puma-sized *Dinictis* and the gracile *Nimravus*, both from the North American Oligocene.

Early on in nimravid evolution, a side-branch diverged, giving rise during the Eocene to a second taxonomic tribe—the hoplophoneids. These were the first true saber-toothed cats—characterized by upper canines that were very greatly enlarged in size and lower canines that were very greatly reduced in size. This highly-specialized dental arrangement meant that the jaws and teeth of the hoplophoneids functioned by stabbing (although the precise mechanism utilized varied from species to species), their enlarged upper canines being used to puncture the lower neck and throat of their prey. Conversely, the jaws and more comparably-sized upper and lower canines of the nimravids functioned not by stabbing but instead by biting. Hoplophoneids included *Hoplophoneus* itself of the North American Oligocene, and the initially Eurasian *Eusmilus*.

By the Pliocene (5.3-2.6 million years ago), all of the nimravids had died out. Before their demise, however, they had given rise to two further taxa—the machairodontid subfamily and the true felid subfamily, collectively constituting the taxonomic family Felidae (although some felid specialists categorize both of these subfamilies as families).

The machairodontids constitute those familiar but formidable, large-fanged saber-tooths popularly nicknamed "saber-toothed tigers" (even though they are only very remotely related to true tigers). Like the hoplophoneids, they possessed greatly enlarged upper canines and very reduced lower ones, yielding a specialized stabbing apparatus (which once again varied between species with regard to certain morphological modifications). Indeed, it was once believed (incorrectly) that the machairodontids had descended directly from the hoplophoneids. However, subsequent researches revealed that this was not so. Instead, the hoplophoneids apparently died out during the late Miocene (*Barbourofelis* was their last representative), whereas the machairodontids emerged as a wholly independent side-branch of the nimravids—in short, the second time that these latter carnivorans had given rise to a taxonomic group of saber-toothed cats.

Machairodus was a machairodontid from the late Miocene, of near-ubiquitous distribution, and of leonine stature. By around the same time, the dirk-tooth *Megantereon* had also appeared. This was a panther-sized form, initially inhabiting Africa but extending its range into the New World and ultimately Eurasia too. But it was in North America where the most massive of all machairodontids evolved. This was the enormous, awe-inspiring *Smilodon*, which was larger than a lion and sported colossal upper canines that measured 6 inches long! An interesting machairodontid side-line was the scimitar cat *Homotherium*, which will feature in Chapter 2. Despite their variety and impressive forms, however, with the demise of *Smilodon* in North America about 12,000 years ago the machairodontids officially died out. Note well for further reference the word "officially"!

The second subfamily to emerge from the nimravids just before they became extinct constituted the true felids—which include all of the cat species known to be alive today. Such emergence must have involved secondary reduction of the nimravids' rather large upper canines. Some authorities do not believe that this reduction occurred and suggest that the true felids arose instead from some other, separate, obscure cat group not currently known from the fossil record. Yet regardless of their origin, by the Pleistocene many true felids existed, including not only modern-day species but also a host of exotic forms no longer with us (officially!), such as the giant American lion *Panthera (leo) atrox*, Owen's puma *Puma pardoides* (found in Europe), Eurasia's lion-sized giant cheetah *Acinonyx pardinensis*, and a huge form of tiger from China. Whereas the hoplophoneids and machairodontids were stabbing cats, the true felids, like the nimravids, were and still are biting cats, with normal-sized lower canines.

Modern-Day Cats—Big and Small

The modern-day species of true felid are divided into two subfamilies. One of these, Pantherinae, houses the five so-called big cats (lion, tiger, leopard, jaguar, and snow leopard), all belonging to the genus *Panthera*, plus the two species of clouded leopard, belonging to the genus *Neofelis*. The second subfamily, Felinae, houses all of the other cat species, most of which were once housed together within the single genus *Felis* but are now split up into several different genera, plus the very distinctive cheetah, sole member of the genus *Acinonyx*. Excluding the cheetah, all of this second subfamily's species are known collectively as small cats (even ones as physically sizable as the puma). To prevent any possible confusion between using "small cat" as a size indicator and as a taxonomic indicator, when using it in the latter capacity I have capitalized it, i.e. Small Cat, throughout this book.

In popular-format literature, the term "big cat" is used indiscriminately to describe any cat of large size. Accounts appertaining to mystery cats often include phrases such as "mystery big cat," or "Surrey's big cat." Unless the taxonomic identity of the cat in question is known conclusively to be that of a *Panthera* species, however, it would be most desirable not to use the term 'big cat' in this manner, confining it instead to its correct, scientific definition (as I have done throughout this book, and capitalizing it, i.e. Big Cat), thus preventing ambiguity.

For in actual fact, body size is not the feature that differentiates Big Cats (genus *Panthera*) from Small Cats (subfamily Felinae) and clouded leopards (genus *Neofelis*). Even though it is true that all five species of Big Cat are indeed big, not all Small Cats are small (as already noted regarding the extremely large puma, for instance). Instead, the character responsible for splitting felids into these two basic taxonomic groups is the composition of a structure called the hyoid apparatus, located in the throat region and incorporating a bony arch or suspension.

In the Small Cats (and also in the cheetah and clouded leopards), this arch is composed wholly of bone (i.e. fully ossified), thereby restricting the movement of the larynx (voice-box), retaining it close to the skull, and in turn yielding a high-pitched voice. Consequently, although they are capable of purring continuously and emitting various yowls and screams, Small Cats cannot produce deeper sounds; in short, they cannot roar.

Conversely, in the Big Cats, one segment of the arch constitutes an elastic ligament, so it is not made entirely of bone (i.e. it is incompletely ossified). In these species, therefore, the larynx is held less rigidly and thus has greater mobility, yielding more resonant, deeper voices, exemplified by the Big Cats' diagnostic and most familiar vocalization—roaring. They also purr on occasions, but only between breaths, not in the normal continuous manner typical of the Small Cats.

The structure of the hyoid apparatus should be of interest not only to felid anatomists and taxonomists but also to mystery cat investigators because it enables deductions to be made regarding the fundamental taxonomic identity of a mystery cat in the field—even when the cat in question cannot be seen. Any felid heard purring continuously can be confidently identified as a Small Cat, whereas any felid giving vent to full-throated roars can be identified with equal confidence as a Big Cat.

Another fundamental difference between these two basic felid groups concerns their eyes. Namely, the eye pupils of Small Cats can be contracted into vertical slits, whereas those of Big Cats (and the cheetah too in this particular instance) remain round in shape.

In this book, usage of the terms "big cats" and "small cats" will be restricted to their respective scientific meanings whenever possible (and capitalized). Correspondingly, terms such as "large," "large-sized," "sizable," and "small-sized" will be used when simply describing a cat's size.

Morphs, Subspecies, and Coat Color Genetics

Some animal species exhibit several strikingly different, separate morphological types all within a single interbreeding population. For example: in addition to its normal spotted form, the leopard *Panthera pardus* can produce individuals with black coats (melanistic), white coats (albinistic), red coats (erythristic), and several other coat color types too. Such morphological types as these are termed morphs; but as they are merely genetic variants, they have no taxonomic significance, and do not receive formal scientific names (although in earlier days some morphs were mistakenly assumed to constitute full species in their own right). Species such as the leopard that exhibit several different morphs are said to be polymorphic; species that only exhibit two different morphs are termed dimorphic, the jaguarundi *Puma yagouaroundi* being one such species among felids.

The greater the frequency of a particular morph's occurrence within a population, plus the greater the number and the wider the distribution of populations containing that morph, the more common it will be. Some morphs of a given species may occur over most of that species' distribution range. Thus morphs represent variation within individual populations, regardless of geographical locality.

Consequently, morphs should not be confused with subspecies (although they often are, especially in popular media accounts) because subspecies represent variation between collections of populations occurring in specified geographical localities. A subspecies can be defined as a collection of morphologically similar populations of a species, inhabiting a geographical subdivision of that species' total distribution range, but differing taxonomically from other populations of that species. Subspecies, therefore, do have taxonomic significance, so they receive formal scientific names (trinomials). Thus the leopard's Arabian subspecies is called *Panthera pardus nimr*, its African subspecies is *P. p. pardus*, its Sri Lankan subspecies is *P. p. kotiya*, and so on.

As will be seen, some mystery cats may be undiscovered subspecies of known species. Equally, certain other mystery cats may be undocumented morphs of known species.

Back in 1989 when this book's original edition was published, it was widely believed that there were only six principal genes responsible for felid coat coloration across the entire spectrum of felid species. Today, conversely, over 30 years later, the situation is known to be far more complex, and to explain all of the intervening genetic discoveries that have been made is far beyond this new edition's remit. Wherever possible, therefore, I shall limit coverage of felid genetics to the bare minimum required in order to comprehend fully each mystery cat case in which such matters arise.

For its historical significance, however, I am still including at the end of this introduction the gene table from the original edition, which summarizes much of what was known back in the 1980s concerning the genetics of coat coloration in wild species of felid. But as will be noted in the respective relevant coverages elsewhere in this book, some of its information is now outdated, having been superseded by newer findings. In particular, the postulated chinchilla mutant allele of the Full Color gene appears not to exist; similar chinchilla-type coat coloration in different species seems to be caused by totally different mutant alleles, so nowadays the term "chinchilla" is used simply as a descriptive, non-genetic term for a particular coat color phenotype (appearance). Consequently, please bear all of that in mind.

Meanwhile, a few words concerning the basic principles of genetics as relevant to felids and felid morphology. In most cases, each of a cat's genes is represented per cell nucleus by two copies—one on each member of the specific pair of chromosomes possessing that gene.

However, these two copies are not always identical to one another, i.e. a given gene can have more than one form (allele). This can have notable consequences in relation to an animal's external appearance, as demonstrated with the following examples using leopards and the Agouti gene.

The coat of a leopard individual possessing two copies of the Agouti gene's normal (wild-type) allele exhibits the typical yellow-brown background coloration for this species, which serves to enhance its coat's contrasting black rosettes. Conversely, the background coloration of the coat of a leopard individual possessing two copies of the Agouti gene's mutant non-agouti allele is jet black or very dark brown. This is because the non-agouti allele stimulates an over-production of the black pigment eumelanin. Consequently, a black leopard's rosettes are virtually hidden amidst its coat's abnormally dark background coloration, so that the animal initially appears to be totally black; such individuals are termed melanistic. Nevertheless, their cryptic rosettes are still present and can often be discerned if observed closely and at certain angles. Melanistic leopards are popularly called black panthers (a term also used in the New World for melanistic jaguars).

Of great interest and genetic significance is the fact that the coat of a leopard possessing one copy of the wild-type allele of the Full Color gene plus one copy of this gene's mutant non-agouti allele exhibits wild-type (normal) coat coloration. This is because the wild-type allele inhibits the effect of the mutant non-agouti allele. The former allele is therefore said to be dominant, and the latter allele is said to be recessive.

Felid individuals with two copies of the same allele for a given gene are termed homozygous for that particular allele. In contrast, felid individuals with one copy of one allele and one copy of another allele for a given gene are termed heterozygous for both alleles.

Consequently, a leopard that is either homozygous or heterozygous for the Agouti gene's dominant wild-type allele will be spotted, whereas only leopards homozygous for its recessive non-agouti allele will be black. In other words, no leopard heterozygous for non-agouti can be black.

Moreover, as long as spontaneous mutations do not occur, spotted leopards homozygous for the Agouti gene's wild-type allele bred with other spotted leopards also homozygous for this allele will always produce spotted cubs. Similarly, black leopards (i.e. always homozygous for non-agouti) bred amongst themselves will always produce black cubs (as long as, once again, further mutations do not spontaneously occur).

However, spotted leopards that are heterozygous for the Agouti gene's wild-type allele will, when bred amongst themselves, produce some spotted cubs and some black cubs, in an approximate 3:1 (spotted:black) ratio. The black cubs are those that have inherited a non-agouti allele from both parents, thereby lacking a wild-type allele for this gene.

Undoubtedly, felid genetics is of considerable importance relative to mystery cat identification. Yet before we put too much store in this, we would do well to recall the following lines from Patrick Chalmers's poem "The Tortoiseshell Cat":

> Every cat in the twilight's grey,
> Every possible cat.

Cryptozoological cats are certainly no exception to this!

Genetics of Cat Coloration [1980s]

As described in full by felid geneticist Roy Robinson in his many papers, genetically-induced mutant color forms (morphs) in felids involve the mutant alleles of six major genes. These genes, their normal (wild-type) and mutant alleles, and the appearance of felids expressing these alleles, are as follows:

GENE	ALLELE	APPEARANCE OF FELID EXPRESSING ALLELE	
Agouti	1)	wild-type	Hairs of fur each banded with yellow and black stripes, yielding grey- or gold-colored fur - locally or widely distributed over animal.
	2)	non-agouti (melanistic) mutant	Yellow stripes on fur hairs absent - entire animal black.
	3)	black-and-tan mutant	Dorsal body regions black, ventral body regions pale.
Black pigmentation	1)	wild-type	Animal possesses various black body regions.
	2)	brown mutant	In combination with non-agouti, these body regions are brown instead of black.
Full color	1)	wild-type	Animal possesses black (eumelanin) and yellow (phaeomelanin) pigments.
	2)	chinchilla mutant	Eumelanin present, phaeomelanin absent—animal's background body color pale, body markings still present, eyes often blue.
	3)	acromelanic mutant	Phaeomelanin absent, eumelanin limited to body extremities - animal very pale all over except at extremities, eyes often pink.
	4)	complete albino mutant	Phaeomelanin and eumelanin both absent - animal totally white, eyes pink.
Dense pigmentation	1)	wild-type	Normal degree of coat color density for species concerned.
	2)	dilute mutant	Coat color pale - appears blue {blue dilute) when in combination with non-agouti.
Extension of black	1)	wild-type	Black pigment distributed to normal degree in hairs, yielding normal fur color for species.
	2)	non-extension of black mutant	Black pigment's distribution restricted in hairs - fur appears red or yellow.
	3)	supra-extension of black mutant	Black pigment excessively distributed in hairs - fur predominantly/entirely black.
Dark-eye	1)	wild-type	Eye color dark, coat color normal.
	2)	pink-eye mutant	Eye color pink/red, coat color variable but generally pale.

Other genes exist whose mutant alleles also affect felid coat color, and are responsible for silvering, piebald spotting, roan coloration, etc.

Chapter 1
Great Britain: Surrey Pumas and Exmoor Beasts

But cats to me are strange, so strange
I never sleep if one is near;
And though I'm sure I see those eyes,
I'm not so sure a body's there.
— W.H. Davies, "The Cat"

During the Pleistocene epoch (approximately 2.6 million to 11,700 years ago), the British Isles (Great Britain and Ireland) still possessed a diverse felid fauna. This included such examples as the cave lion *Panthera (leo) spelaea* and fossil lion *P. (l.) fossilis,* European Ice Age leopard *Panthera pardus spelaea*, Owen's puma *Puma pardoides* (previously thought to have been a relative of the leopard and therefore originally dubbed Owen's panther *Panthera pardoides*), European jaguar *Panthera gombaszoegensis*, European wildcat *F. silvestris*, Eurasian (northern) lynx *Lynx lynx*, and scimitar cat *Homotherium latidens*. Out of all of these species, however, only one has survived into the present day here—the European wildcat, and even that one is now confined entirely to Scotland. In addition, populations of feral (run-wild) domestic cats *F. catus* exist throughout the British Isles. This is the official history of their felids. The unofficial version is rather different.

Many descriptions and sightings have been reported from all parts of the island of Great Britain concerning a wide variety (judging at least from the diversity of eyewitness accounts) of felids alive and well in its countryside but allegedly much larger than any wildcat. Since the beginning of the 1960s, Britain's multitude of mystery cats has attracted ever-increasing attention from the media and cryptozoological researchers, but let us begin our examination of this subject by looking at various accounts that pre-date (sometimes very considerably) the 1960s.

Pre-1960s Evidence

Although it was a Pleistocene inhabitant of England and Wales, the cave lion seemingly never reached Scotland, and in any event it died out entirely in the British Isles approximately 14,000 years ago according to fossil evidence. Hence it must be somewhat disconcerting to paleontologists to learn that the 16th-century chronicler Ralph Holinshed wrote: "Lions we have had many in the north parts of Scotland, and those with manes of no less force than those of Mauretania; but how and when they were destroyed as yet I do not read."

As recognized by writers Anthony Dent and Michael Goss, worthy of mention in connection with British mystery cats is the 18th-century poem "Black Annis." Its mythical subject, an ill-defined witch-like creature, possessed notable talons and a tendency to leap down from the overhanging branches of an oak tree onto victims passing beneath and suck their blood. As Dent and Goss have suggested, this could perhaps be taken as an anthropomorphic description of some felid form; worth noting is that in later legends, Black Annis became known as Cat Annis.

Although superficially dubious, that particular example acquires greater potential as a result of this next one, documented by Chambers's Journal in 1904. During the spring of 1810, flocks of sheep all over the Ennerdale valley in what was then the county of Cumberland (now part of Cumbria) were being slaughtered by an elusive but ferocious beast that became known as the Ennerdale or Girt Dog. Yet rather surprisingly (in view of these names), it was never heard to bark. Moreover, it exhibited the rather macabre, non-canid behavior of biting into its prey's jugular vein and sucking the blood (comparable with another mystery sanguinivore, this time reported from Cavan, Ireland, in 1784, and included in Charles Fort's books).

> The Girt Dog was sighted many times throughout the spring and summer of 1810, and was hunted frequently but without success—until a dog supposedly identified as this bizarre beast was cornered and shot dead on 12 September. Its body was preserved and displayed for a time at a museum in Keswick but has since been lost. Although it was generally classed as a dog, i.e. a canid, eyewitness Will Rotherby—who was actually knocked over by this creature while it was escaping from a hunting party, and who therefore probably saw it at closer range than anyone else—remained convinced that the animal he observed was not a dog at all but was instead a lion-like cat.

In *Rural Rides, Vol. 1* (1830), the eminent English Member of Parliament and journalist/pamphleteer William Cobbett described a visit that he made with his son on 27 October 1825 or thereabouts to Waverley Abbey in Cobbett's home town of Farnham, Surrey. While there, he showed his son:

> ...a very old elm tree, which was hollow even then, into which I, when a little boy [he was born in 1763] once saw a cat go that was as big as a middle-sized spaniel dog, for relating which I got a great scolding, for standing to which I at last got a beating; but stand to which I still did...When in New Brunswick [Canada] I saw the great wild grey cat which is there called a Lucifee; and it seemed to me to be just such a cat as I had seen at Waverley.

With Waverley being at the very centre of Surrey, which would eventually become a notable hub for alleged puma *Puma concolor* sightings in England during the 1960s (see later), it would be very tempting to presume that the lucifee is indeed the puma, which in Cobbett's time was still well-established in New Brunswick. In truth, however, "lucifee" is apparently a corruption of "loup cervier," a term used throughout New Brunswick and also the USA's New England states from the late 18th century onwards for the Canadian (aka North American) lynx *Lynx canadensis*. Nevertheless, if Cobbett's identification of the sizable mystery cat that he saw was correct, a lynx sighted in Surrey approximately 250 years ago is still very worthy of note.

One of the most remarkable pre-1960s mystery cat episodes on record from Great Britain took place in 1927 and was featured in a detailed account published by London's *Daily Express* newspaper (14 January 1927). Due to a series of sheep kills in Inverness, northern Scotland, plus sightings made there of an animal "...like a leopard, but without spots on its coat," and the discovery of "...tracks like those of a gigantic cat," a steel trap had been set by one alarmed farmer. Upon checking it the next morning, he was stunned to discover that it had snared a "...large fierce yellow animal," which he promptly shot and sent to London Zoo for identification. It was found to be a lynx!

Was Cobbett's Waverley lucifee a lynx, like this felid? (public domain)

Moreover, according to the newspaper report, two other lynxes had been shot in this same area during this same period. No clue was given as to the whereabouts of these specimens, and with regard to their equally perplexing origin the report merely offered (but without any corroborating evidence) the opinion that "...they must have escaped from some traveling menagerie." One further twist to this already very tantalizing tale is that according to James Kirkwood, who was the London Zoo's Senior Veterinary Officer when I contacted him during the late 1980s, the zoo had no record of ever having received such a creature at that time (the late 1920s).

In 2010, Bristol University zoology undergraduate Max Blake made a remarkable discovery in the archives of the Bristol City Museum and Art Gallery—a hitherto uninvestigated stuffed lynx whose mounted skin and skeleton are from a specimen that had been shot dead in Newton Abbot, Devon, southwest England, in or prior to 1903. When subjected to a comprehensive anatomical analysis via a research team that included Blake and British paleontologist Darren Naish, it was found to be almost certainly a Canadian lynx *Lynx canadensis*, but from where it originally escaped (or was released) remains a mystery (*Historical Biology*, 2014).

On 19 March 1938, *The Field*, a then-weekly (now-monthly) British countryside-related periodical, published a letter written by Irene Roberts of Lightwater, Surrey, in which she recalled having frequently heard some very strange cries that she was unable to associate with any known type of animal in the area. She described them when heard at 2:00 a.m. in July 1937 as being "...of peculiar intensity, expressing, it seemed, mortal fear and physical pain," continuing for about 10 minutes and "...punctuated by a queer half-snoring, half-purring sound, produced by something probably considerably larger than a cat." On a later date, but this time at 1:30 a.m., "...they seemed to be something between a bark and a scream, and did not appear to be produced from anything of the shape of a beak. They seemed to begin with a B. 'B'yow! B'yow!' long drawn-out, and penetrating." It lasted for about five minutes, periodically yielding a crescendo.

Roberts's letter stimulated much ensuing correspondence in *The Field*, and included offered identities for the unseen vocalist as diverse as badgers, foxes, otters, owls, domestic cats, and even lovestruck hedgehogs. Little did anyone realize at that time, of course, that a mere two decades later this very same location would become the centre of a prolonged media-fueled hunt for a certain felid species whose screaming cries correspond very closely indeed with Roberts's quoted verbal description—namely, the puma. Just a coincidence? Equally interesting is the fact that puma sightings were being reported from Surrey during the 1940s, i.e. almost 20 years before the Surrey puma saga hit the news headlines, but they received little attention due to the much weightier events of World War II. By the end of the 1960s, however, the Surrey puma had become one of the most famous British mystery beasts.

Please note that in all of the following accounts from the 1960s onwards of so-called "puma," "cheetahs," "panthers," "lionesses," "lynxes," etc, the presence of quotation marks should be understood because in most instances no conclusive evidence to confirm the taxonomic status of these animals has been obtained. In addition, so many mystery cat cases had emerged from Great Britain up to the publication of this book's original edition in 1989 (let alone during the intervening 31 years between then and now) that it would have been impossible to provide anything remotely approaching a comprehensive coverage within an entire book, let alone within a single chapter of one.

Instead, therefore, in that original edition (as well as again here) I sought to provide a selected coverage concentrating upon those pre-1990s cases that had proved to be of greatest importance and interest, and which had offered the most significant clues regarding the identities of the animals featured in them. To supplement this coverage and also to bring it fully up to date, I heartily recommend the excellent ongoing series of British mystery cat reviews by Paul Sieveking, Bob Rickard, and others that regularly appear in the British monthly periodical *Fortean Times* (originally titled *The News*, and which I drew upon when compiling the following survey). Other publications containing valuable coverages of British mystery cat reports from this same time period include the books of Janet and Colin Bord, as well as Graham McEwan.

Shooters Hill Cheetah and Surrey Puma

The first British mystery cat to attract widespread media attention during the early 1960s was a long-limbed cheetah-like felid with a lengthy upward-curing tail, sighted on the roadside in the area of Shooters Hill, Woolwich, in southeast London, by lorry driver David Back at 1:00 a.m. on 18 July 1963. It moved off into the nearby woods as he approached, and despite a well-manned police search of the region following later sightings, it evaded all attempts to capture it—although it made its presence felt very effectively by audaciously leaping over the hood of one of the police cars! In addition, tracks were discovered in the soft mud leading down the path of a dried-up woodland stream. These measured several inches across and were recognizably different from those of the police Alsatians. Furthermore, they bore the clear impressions of claws. Cheetahs habitually leave clawed prints, which would therefore support the cheetah identity and moniker popularly applied to this mystery cat by the press—except, that is, for the fact that, notwithstanding their difference from those of the Alsatians, clawed prints measuring several inches across are much more suggestive of a canine than a feline identity.

Following the unsuccessful cheetah hunt, some snarls and sightings were reported during the next few days, but by August 1963 Shooters

Hill's strange visitor had apparently moved on because it was not reported again. However, sightings of shorter-legged, puma-like beasts would be documented from various Kentish localities in future years.

Moreover, on the media front the Shooters Hill cheetah's disappearance was itself eclipsed by a puma of sorts; namely, an unidentified felid destined to become known worldwide as the Surrey puma.

At about 7:45 a.m. on 16 July 1962, while cycling down a wooded path leading to Heathy Park Reservoir, Mid-Wessex Water Board, official Ernest Jellett observed a rabbit being stalked by a most unusual animal—one that he felt certain was neither fox nor dog. Instead, he considered it not to be unlike a young lion, with a flat face, large paws, pale-brown pelage, and a height of 18-24 inches. As he watched, it pounced upon the rabbit but missed, and then bounded up towards Jellett himself—who, understandably apprehensive, shouted loudly at it and succeeded in scaring it away.

During the winter of 1962-63, a variety of odd cat-like beasts of predominantly nocturnal nature made several visits to Bushylease Farm, sited between Crondall and Ewshot in Hampshire, close to the Surrey border. According to manager Edward Blanks, such visits were usually accompanied by a strong ammonia-like smell and yowling cries, and left the farm dogs terrified. He described one such felid as being Alsatian-sized and sandy-brown in color, with a cat's head and long tail.

Eighteen months later, events here took a more sinister turn. On 30 August 1964, a 4-cwt Friesian bullock on the farm was discovered badly mauled, although still alive, bearing severe lacerations upon its flanks, shoulders, and neck. Ministry of Agriculture officials who later examined this animal claimed that these wounds had been made by barbed wire and identified strange red pellet-like objects found in the vicinity of the bullock as clots of congealed blood. Conversely, a veterinarian called in to see the bullock favored an animal assailant as the originator of the wounds, but not of any species native to Britain. Furthermore, while lying in wait for the mystery animal on subsequent nights, Blanks found more "blood clots" on the ground, and this time there was no victim from which they could have originated. The elusive predator itself, meanwhile, maintained its low profile.

During the remainder of 1964, puma pandemonium not only filled Surrey but also spilled over into neighboring counties, with dozens of reports emerging during the last months. Notable among these were: a short-legged beast with a cat-like face, dirty-brown or golden pelage, large paws, a height of three feet, and a total length exceeding five feet, sighted by blackberry picker George Wisdom near Munstead's water tower in Surrey (4 September); a large unidentified creature that left behind gigantic clawed spoor measuring 5.25 inches across near to some riding stables at Munstead (7 September); a dead roe deer with broken neck and wounds compatible with a felid killing discovered at Cranleigh, Surrey (23 September); a puma-like cat sighted by two policemen at the Thomas Grey memorial at Stoke Poges, Buckinghamshire (12 November); and a similar cat sighted at Nettlebed in Oxfordshire (20 November) as well as at Ewhurst in Surrey (15 December). Moreover, frequent reports of spine-chilling screams and growls were documented from Surrey and its environs during periods of alleged puma activity.

By now, using Godalming's station as headquarters, the Surrey police were compiling a puma sightings book, carrying out frequent searches, and consulting with numerous wildlife experts in an all-out attempt to solve the mystery of their county's cryptic cougar. Victor Manton, curator of the Whipsnade Zoo at that time, made several forays into Surrey puma country armed with a tranquilizer-bearing rifle, and although he never cornered his quarry he did come across a roe deer whose neck had been broken—an act requiring a very sizable, powerful predator. He also collected hairs from a barbed wire fence through which a puma-like animal had allegedly passed; the hairs were later found to compare closely with those from the tail-tip of a puma (*Radio Times*, 17 May 1986).

Little of note happened during 1965, but sightings returned with a vengeance in 1966, of which the following were especially significant. A classic mystery cat encounter took place on 4 July at Worplesdon, Surrey, when a puma-like felid stalked and killed a rabbit in full view of police and villager eyewitnesses—watching from only 100 yards away, and for a full 20 minutes. It was described as being ginger-brown in color, and the size of a Labrador dog but with a cat-like face. Curiously, its tail was tipped with white, whereas a puma's is normally black-tipped, like the hairs that had been discovered by Victor Manton. Worth noting, however, is that black hairs can sometimes be replaced by white ones following an injury—such as a tail-tip being trapped on barbed wire, for instance?

In early August 1966, ex-police photographer Ian Pert actually succeeded in taking a photograph of a strange cat at a distance of only 35 yards, again sighted at Worplesdon. This animal had a small feline

Ian Pert's photograph of Worplesdon mystery cat (Fortean Picture Library)

head, a muscular body (especially the hindquarters), a rather thick tail measuring an estimated 13 inches, a sandy-colored pelage, and a total length of about 3 feet (*Sunday People*, 14 August 1966). Two months later, another photo was taken of a mystery cat, this time at Upper Hale, but it appears to have been lost, with no details concerning it having been recorded.

Encouraged by these recent sightings, providing positive proof of the existence of at least one strange Surrey felid, further police searches were mounted at the end of August and October, but once again all in vain. Meanwhile, E.J. Rogers had noted that puma sightings had actually been reported on both sides of the River Thames at much the same time, thereby suggesting the existence of at least two such animals, but one was more than enough for most investigators! Although Surrey's mystery cat (or cats?) showed an apparent dietary fondness for deer and rabbits, no evidence had been collected to indicate that it posed any threat to humans—quite the reverse in fact, judging from its evident desire and ability to elude them.

Consequently, in mid-August 1967 the Surrey police officially terminated their investigation. The great Surrey Puma Hunt—which according to British zoologist Dr Maurice Burton had given rise between September 1964 and August 1966 to 362 formally-listed sightings and many more claimed but not officially declared (*Animals*, December 1966)—was over, which in turn led to a fall in this mystery beast's public profile. Indeed, that may well have fueled the rumor prevalent at the close of the 1960s that the Surrey puma had been shot by a farmer and then buried without publicity on his lands because he did not have a license for his gun. Similarly, a fox shot in daylight near one of the mystery cat's known haunts in Surrey's Ash Ranges during February 1970 was "identified" as the puma. The recurrence of puma sightings in the latter locality and in others within (and beyond) Surrey's county borders soon afterwards, however, quelled both of these attempts to end the mystery.

Indeed, sightings have continued right up to the present day here. A significant footnote: while compiling an impressive dossier of records concerning a tawny puma-like beast frequently reported from various Cornish localities during the 1980s, Exeter-based zoologist Frank Turk learnt of a strange feline animal sighted in August 1984 close to the Surrey village of Peaslake. Hair samples had been collected and were passed on to him. After subjecting them to exhaustive microscopical analyses to discover their originator's identity, Turk announced to the media in January 1985 that they were from a puma. He also stated that he was "absolutely certain these animals are living wild in Britain," and his find was considered by London's Natural History Museum to be "very exciting" (*Western Morning News*, 14 January 1985). Respectability at last for the Surrey puma?

Pumas (and Black Panthers) Elsewhere in Southern England

The Surrey puma's disappearance from national headlines in 1967 created a British mystery cat hiatus that was to last for almost a decade. By this, I do not mean that mystery cat sightings were not being made in Britain—far from it!—it was simply that their media profile was less overt. Hertfordshire and Hampshire, for example, played host to many reports of such animals between 1971 and 1973.

These included: a dog-sized, long-tailed creature with large paws, grey-brown pelage, and (according to one eyewitness) "an ugly cat-like face" that leapt out of a tree to chase a dog in Farnborough, Hampshire (25 May 1971); a grey cat roughly two feet high and described as puma-like by its police eyewitnesses when seen at St. Albans, Hertfordshire (15 July 1971); a sturdy brown cat lacking markings, standing two feet high, with large paws and a long tail curving upwards at its tip, spied in Hampshire by Aldershot's Fleet Station (August-September 1972; amazingly, this felid was ultimately "explained" as a pet Siamese cat!); a tawny-brown cat larger than an Alsatian, with a sizable head and pricked ears, encountered in Hampshire's New Forest and compared with a puma illustration by its boy eyewitnesses (September 1972); and an animal seen at Farnborough that was described by its police observers who spied it at less than 30 yards away as being "the same size and colour as a puma," with pointed ears and long tail (19 June 1973).

Breaking the monopoly (and monotony) of puma sightings, a large all-black feline beast was sighted one evening in late April 1971 near to Surrey's Road Research Laboratory at Crowthorne; the tail of a huge but similarly pantheresque cat with short head and long but low-slung body was reputedly hit by a car when the cat ran in front of the vehicle along the A35 road from Christchurch to Lyndhurst, Hampshire, in January 1973; and a large black felid also ran in front of a car on the Rake-Harting Combe road in Hampshire on 3 November 1973.

Yet although such reports were of local press interest, the national British media were not impressed—until the coming of the Nottingham lioness in 1976. Noteworthy is that this year also saw the beginning of a marked increase in the number, diversity, and geographical distribution of British mystery cat reports—a trend sustained in subsequent years.

Nottingham Lioness, Glenfarg Lynx, and Others

In late July 1976, two milkmen were rather taken aback to see a large beast that they unhesitatingly identified as a lioness (they specifically recalled that the tip of its tail was tufted—a diagnostic feature of *Panthera leo*) walking round the edge of a field at Tollerton, near Nottingham in Nottinghamshire. Other sightings followed (documented in a series of *Nottingham Evening Post* reports), large paw prints were reportedly spied in various local inhabitants' gardens, and on 31 July the mysterious felid visited the garden of Nottingham's deputy coroner, John Chisholm, of Normanton-on-the-Wold. Like the milkmen, he noted in particular that its tail was identical to that of a lioness. The police were called at once, but as their cars drew near, Chisholm's wife saw the animal move away, probably frightened by the vehicles' headlights.

By 6 August, the number of sightings had risen sharply, reaching a very noteworthy tally of 65, but still unaccompanied by any clues regarding the creature's origin or current location. Animals that were tracked down turned out to be big dogs—and in one decidedly surreal incident, a particularly sizable, dark-colored carrier bag! Finally, the police made a formal statement to the effect that the Nottingham lioness was truly nothing more than a large dog, and the case was officially closed. The fact that no large dog (and certainly no large carrier bag!) resembles a lioness did not appear to attract much attention.

Moving from England to Scotland and from the leonine to the lynx-like: On the evening of 9 August 1976, an inhabitant of Glenfarg, Perthshire, walked into her garden to find out why her terrier was barking—and discovered that the reason was a three-foot-high feline beast crouching on top of her garden wall, spitting and snarling at her! The two features that caught her attention most of all were its "burning orange" eyes, and its ears, which were long and pointed with tufts at their tips. As she cautiously retraced her steps, with terrier carried aloft, towards her open door, her unexpected visitor leapt down onto the ground beyond her garden and made its exit through the bordering fields.

Mystery cats sighted elsewhere in Britain that year included a lioness-like beast at Skegness in Lincolnshire, England, a puma-like creature at Nuneaton in Warwickshire, England, and an elusive felid at Blantyre in South Lanarkshire, Scotland. Intermittently reported through 1976 and 1977 were the Bettyhill/Skerray beast in Sutherland, Scotland, and the North Berwick puma in East Lothian, Scotland. These in turn were supplanted during 1978 by the Scottish Highlands' Conon Bridge lioness (so-called because eyewitness Helen Fitch of Bishop Kinkell considered it to be very similar to animals that she had seen while living in Africa).

In 1979, a photograph was taken of a black panther-like felid seen in Ayrshire, Scotland. However, Glasgow Zoo officials felt that it was merely a large domestic cat.

Pumas were rife throughout 1979 and 1980, and provided some significant reports. In a letter to *Mystery Animals of Britain and Ireland* author Graham J. McEwan, J.M. Jenkins recalled that in the summer of 1979 he had observed a felid in Midlothian, Scotland, at a range of six feet, which had a black bar crossing its face, pelage the color of golden Labrador, and appeared to be pregnant. Yet it was not aggressive; in contrast, it seemed rather curious concerning Jenkins's activities—he was using a metal detector at the time. Jenkins affirmed that it was identical in appearance to the pumas at Edinburgh Zoo.

During the summer of 1980, urban Wolverhampton in the West Midlands, England, became the unlikely venue for a puma hunt, as documented in a series of reports by Wolverhampton's *Express and Star* newspaper. A number of closely-corresponding eyewitness descriptions confirmed that the mystery cat in question was a brown unspotted felid the size of a large dog, with a round head, pricked ears, long tail, and shortish limbs, closely resembling a puma. Yet despite repeated searches by police and Dudley Zoo curator Mike Williams, who had caught sight of it himself by 25 July, this Midlands mystery cat remained uncaptured, and in the best crypto-felid traditions it eventually vanished without trace.

A puma was also the star of one of the most sensational of all episodes in the long-running saga of Britain's unofficial feline fauna, charted in several fascinating articles by Glasgow's *Daily Record* newspaper. Namely, the actual, highly-publicized capture of a genuine large-sized non-native cat species in the British countryside.

The Cannich Puma and the Powys Beast

On 27 October 1979, farmer Edward ("Ted") Noble saw a lioness-like animal stalking ponies on his Highlands farm, near Cannich in Inverness-shire, Scotland. He reported this to the police, also mentioning that livestock had been disappearing inexplicably from his farm for more than a year. The police organized a search of the area but found nothing. Over the next 12 months, Noble continued to see large cat-like beasts, and lose livestock, especially sheep, but even though other locals by now were reporting strange beasts too, he failed to engender sufficient interest for another search.

Finally, in desperation, Noble decided to solve the mystery himself—by erecting a specially-constructed steel cage near his farm and baiting it each day with fresh meat. And when he inspected the cage with his son on 29 October 1980, he discovered that he had at last obtained the evidence that he had sought for so long—five feet of furry evidence that snarled and spat at him. For his cage contained an adult female puma! Noble's claims had been vindicated—or had they?

For when the puma was examined by wildlife experts (including Highland Wildlife Park director Edward R.J. Orbell and veterinarian George Rafferty), she was found to be aged, arthritic, and atrociously tame—even allowing her visitors to tickle her ears. Hardly the ferocious sheep-slaughterer that everyone had envisaged! More like a pet recently released into the wild, in fact—which was also the opinion of a number of wildlife authorities. Had Noble been the victim of a hoax—perhaps of deliberate intent by person(s) unknown to make him appear foolish, or simply through some opportunist using Noble's cage and reports of strange cats in the area as a perfect cover for letting loose an unwanted exotic pet?

In favor of Noble's belief that the puma had genuinely been in the wild for some considerable time, however, was Hans Kruuk from the Institute of Terrestrial Ecology, in Banchory, Aberdeenshire, Scotland, who based his belief upon analyses that he had carried out on two of the puma's fecal droppings produced within hours of her capture. One of these was composed almost entirely of deer remains, the other of sheep, which is exactly what one would have expected from a puma living in such surroundings for an extensive period.

Nevertheless, some authorities remained—and remain—skeptical. As for the puma herself—duly dubbed Felicity—she spent the remainder of her life comfortably at the Highland Wildlife Park, dying peacefully of old age in February 1985, and leaving behind the riddle of her origin still unsolved (*Scotsman*, 7 February 1985). Curiously, sightings of a puma on the loose in the Cannich area continued to be reported long after Felicity's capture (there were even claims, reported in the *Sunday Express*, 27 January 1985, that an actual puma colony existed in the remote Guisachan Forest). Perhaps more than one mystery cat had been responsible for the livestock farm disappearance. At one stage, the police were investigating a rumor that a prisoner at Winchester Prison had released two pumas into the Highlands before being jailed in 1979, but nothing substantial emerged from this (*Daily Record*, 10 November 1980). Alternatively, as Felicity's critics would aver, possibly the real feline culprit had yet to be captured.

Felicity the Cannich puma, preserved for posterity at Inverness Museum, Scotland (Geni/Wikipedia)

While Scotland was being confused by the Cannich puma, Wales was being perplexed by the Powys Beast. On 23 October 1980, just a month after a lynx-like beast had been sighted near to Churchstone, Powys, by a district nurse, farmer Michael Nash of Llangurig, Powys, heard strange snoring sounds emanating from some bales of straw in his barn. Being mindful of the fact that four of his sheep had recently been killed in a manner unlike earlier dog kills, and having observed a large spoor just outside the barn, he contacted the police. They surrounded the barn and kept watch all night.

Yet when they entered the barn the next morning, they found nothing. Nevertheless, even though the beast seemed to have dematerialized, the spoor had not. Almost five inches long, it possessed a relatively small plantar pad and four elongate toe pads (each of which terminated in a claw), yielding in overall shape a long rather than rounded footprint—totally unlike that of a lynx or indeed any other felid.

Isle of Wight Mystery Cats

Situated off the southern England county of Hampshire, the Isle of Wight (I.O.W.) is a county in its own right and is England's largest island. It has also hosted a surprising number and diversity of mystery cat reports down through the decades. Having said that, its most famous pre-1960s case was something of a non-starter, inasmuch as when a supposed lion on the loose there, nicknamed the Island Monster by the press and claimed to be maned but otherwise virtually hairless, was finally shot in 1940, it proved to be an old fox in an advanced state of mange. Almost all of its fur had been lost, except for some still covering its neck, creating the illusion of a mane (*Isle of Wight County Press*, 24 February 1940).

During 1983 and 1984, a resurgence of I.O.W. mystery cat reports occurred, but now they appeared to feature unequivocally feline creatures, as chronicled via a series of articles in Portsmouth's newspaper *The News*.

They began in January 1983, but the first notable sighting, reported by *The News* in August, featured what was described by one eyewitness, Buckinghamshire schoolteacher Alan Goodwin, who had been holidaying on the island earlier in the year with his wife, as a sandy-colored cat measuring approximately 3.5 feet long and 2.5 feet high. After studying reference books in search of an identity for it, he considered it to have been a young adult puma. At that time, there was a zoo on the I.O.W., at Sandown, which owned a pair of pumas, but both of them were adult and fully accounted for.

By mid-September, seven separate mystery cat sightings had been reported, in two different areas, and most of them describing a black and grey panther-like felid as big as a Labrador or Alsatian. On 15 September, *The News* reported the discovery of what may have been this animal's first victim—a half-eaten, disemboweled, day-old Friesian calf, found close to East Upton Manor Farm, near Ryde, with its back legs stripped to the bone. After examining its carcass, bearing 4-inch claw marks on its body, and taking into account the amount of meat eaten, Sandown Zoo boss Jack Corney could not rule out that it was the work of a big-sized cat, possibly a puma. By late October, at least 15 sightings had been reported.

After an uneventful beginning, 1984 saw its first mystery cat action on the I.O.W. when, following reports of puma-like and black panther-like felids in 1983, March highlighted a third version, an apparent lynx. On the evening in question, Hilding Ecklund and Stephen

Matthews had parked their car near to Newport's Lynnbottom rubbish tip in Newport, and were listening to CB radio, when they suddenly heard some very loud wailing sounds reminiscent of a baby crying, coming from the direction of a small copse close by. When Eklund shone his torch there, its beam hit an Alsatian-sized cat standing about 6 feet away, which screamed and growled. The beam revealed that it sported a mottled tan or sandy-colored coat, a flat face, and pointed, tufted ears.

A month later, several similar sightings had been reported, including one featuring a dramatic role-reversal in which a dog-sized felid with pointed ears played eyewitness to an apprehensive Elizabeth Dowers for a time before slowly advancing towards her (fortunately, her car was close at hand!). In a *Sunday Express* article of 29 April, John Roberts, owner of the Robin Hill Country Park on the island, was quoted as stating in relation to the I.O.W. mystery cat: "I think there is a real possibility that it could be a lynx. I had a visit about 10 days ago from a man who said he had been only yards away from this animal. He gave me an absolutely perfect description of a lynx." This may be so, but as can be appreciated by the range of descriptions given by eyewitnesses, it would seem that a lynx was not the only cat type being reported on the island.

By the end of 1984, about 80 sightings of mystery cats of varied appearance had been reported, and further sightings were made in 1985 too. Moreover, the lynx phase of the I.O.W. felid phenomenon was giving way to a panther resurgence.

A dark-colored cat of Alsatian size, with pricked ears and bright eyes, was spied near Ryde in early January 1985, and a fortnight later a jet-black felid with a long tail and small erect ears was seen by a policeman at Merstone. A charcoal-colored cat with a round head, very long tail, and in overall size about four times that of a domestic cat, was observed in early February by Fred Kershaw and wife as it ran across the entrance to Beapers Farm, south of Ryde, vanishing into the woods beyond, and a black felid with a somewhat thicker tail and of Labrador size but leopard-like outline was reported from Cowes on 13 March. Interspersed between these were further accounts of sandy-colored, long-tailed puma-like cats. Most startling of all, however, was the discovery of large and small feline spoor together, at Oakfield, near Ryde, suggesting the presence of an adult mystery cat with young.

Further kills were also recorded. One was of a three-month-old, 60-pound lamb that was almost completely eaten, and whose owner, farmer Roy Kingswell, discounted foxes, badgers, and dogs as culprits. The killings of several other sheep, including a rare Manx Loaghtan ewe, were also attributed to a mystery cat, but none was ever captured or, to my knowledge, even photographed during this 1980s I.O.W. "cat flap"—to quote a media-popular term for an outbreak of mystery cat reports in a given area.

What makes this particular "flap" even more perplexing than normal is not only the presence of at least three very different cat types (black panther, puma, lynx) but also where on an island of only 147 square miles they could have come from. If they had been maintained privately in captivity, surely neighbors or friends would have known about it?

Having said that, an Asian leopard cat *Prionailurus bengalensis* (thus unquestionably an escapee/release, yet of unconfirmed origin) was shot dead on the I.O.W. in 1997 by teenager Stuart Skinner. He and friend Jason Ward had found it trapped in a snare that they had set near Yaverland Manor, Sandown, in order to catch a fox that had been worrying their ducks. Alarmed at killing what they thought may be a valuable, significant animal, however, after taking photographs of its body the two teenagers buried it and maintained a discreet silence for seven years—before learning that they had not committed a crime by shooting it after all. This persuaded them to release their photos and information to the media in early 1994 (*Isle of Wight County Press*, 14 January 1994).

An Asian leopard cat (Karl Shuker)

Other English mystery cat reports during the early 1980s included a puma at Fobbing in Essex, a cheetah at Brighton, East Sussex, and another puma at Ilkeston, Derbyshire.

The Beasts of Exmoor

Apparently incorporating quite a variety of different carnivore types, the saga of southwest England's formidable but highly elusive Exmoor Beast(s) is without doubt the most complex and significant of all British mystery cat cases. And since the end of 1986, very important new evidence from several different sources has been obtained regarding Beast identities. Yet prior to the publication of this book's original edition in 1989, much of this information had never been included in any cryptozoological tome. Consequently, in order to document it while concomitantly preventing this section from becoming disproportionately long, I chose then (and choose still) to represent here the earlier exploits of the Beasts (already extensively chronicled in a number of previous publications) via a concise summary of those most integral incidents. (For further details concerning early Beast history, I strongly recommend the accounts provided by Graham McEwan and Bob Rickard respectively, the lengthy series of newspaper reports appearing in Portsmouth's *The News*, Barnstaple's *North Devon Journal-Herald*, and Plymouth's *Western Morning News*, plus the respective books of on-site Beast experts Trevor Beer and Nigel Brierly, all as listed in the bibliography for this chapter.)

Sightings of jet-black panther-like animals on Dartmoor in Devon, and at Tedburn St Mary, sited between Dartmoor and Exmoor (the latter massive expanse of moorland straddling the border between Devon and Somerset), had been occurring for some considerable time, and had been reported spasmodically by the media from the beginning of the 1980s onwards. Conversely, the unexplained losses of several lambs during spring 1982 from Drewstone Farm, owned by Eric Ley of South Molton, Exmoor, and rumors of a similarly pantheresque creature having been seen in the area, had somehow attracted little attention. Consequently, when news broke regarding the dramatic events taking place in the very same location but this time during April 1983, as far as the media were concerned the Exmoor Beast saga appeared to have sprung into existence in best Athena style—i.e. fully-formed and fully-armed.

Local and national newspapers revealed that between the beginning of the 1983 lambing season and 15 April of that year, Ley had lost no fewer than 30 lambs (one almost every evening) to some unknown predator, and that other farmers had also suffered losses. Searches by the police and locals failed to discover the culprit; several stray dogs suspected of the killings were shot, but the killings continued. The method that Exmoor's mysterious slaughterer—soon nicknamed the Beast—adopted was to crush the skull of its lamb victims between its jaws or to break the neck and then eviscerate the victim from the neck downwards, leaving its skeleton virtually untouched.

By May, the situation had become so serious that the Royal Marines stationed at Lympstone were called in, bringing along their infrared searching equipment, and digging camouflage trenches for all-night monitoring of the South Molton region. At 5:35 a.m. on 4 May, Marine sniper John Holden saw a very large and powerful all-black creature cross a railway line some distance away. Holden declined to shoot as a farmhouse was sited directly behind it; also, the Marines' primary instruction was to capture the Beast alive if possible.

Throughout the Marines' stay on Exmoor, a very interesting, paradoxical dichotomy prevailed. Whereas Beast sightings made by local inhabitants generally suggested a large felid (supported by the frequent sounding out of high-pitched cat-like screams in the area at night), the Marines and police pursued the belief that it was a dog—and offered dog fur found at kills plus spoor identified as canid in support of this. So who was right? Quite possibly everyone—because Major John Watkins, who was in charge of the Marines' endeavors, ultimately stated that there may be two animals at large and commented that he had been informed of rumors concerning a possible puma in the area.

A reward of £1000 offered on 7 May by London's *Daily Express* newspaper for the first unpublished photograph of the Beast was criticized by police and the Marines as a potential danger (encouraging people to enter an area already besieged by gun-toting officials and locals). It finally led to the Marines formally withdrawing—although by popular demand they secretly returned on 17 May for a second surveillance period.

The police continued to popularize a canid identity for the Beast, even holding a canine identity parade in order to illustrate the kind of dog (a dark-colored lurcher type) that they felt was involved, but locals continued to see both dogs and black panthers. In early July, the Marines pulled out again, confessing defeat and utter bewilderment at the uncanny cunning that their quarry had shown, eluding their every move with consummate, calculated ease. Such extreme elusiveness seemed to favor a feline rather than a canine identity. During the remainder of July, reports of sheep killings seemed to be on the wane, and on 29 July the police announced that the Beast was dead—a proclamation hastily forgotten when killings bearing the hallmarks of

Dog spoor (left) and cat-like, alleged Exmoor Beast spoor (right) (Trevor Beer)

its mode of operation began again in August. By now, ideas concerning its identity had greatly proliferated—from a rogue dog, wolf, or puma, to a black panther, eagle, or even a laboratory-created escapee! A radio listener known only as "Bob" claimed that the Beast was a black bear that he had killed three weeks earlier and had buried on Molland Common near South Molton. The police were reported to be following up this story, but it was ultimately denounced as a hoax.

Two local naturalists who had been investigating the Exmoor Beast case from its very beginning had their own hypotheses. Nigel Brierly favored a mutant race of feral domestic cats of abnormally large size. In contrast, Trevor Beer (d. 2017) believed that although large-sized cats of some type were indeed being sighted, they were not responsible for the actual killings, suggesting instead that these were the work of big dogs. Although differing from one another on the subject of the Beast's identity, Trevor Beer and Nigel Brierly were united in their conviction that more than one specimen of large cat was present, both men having filed eyewitness accounts of such beasts that were made far apart in location but virtually simultaneously. Moreover, during 1984 Trevor had revealed that not all mystery cat sightings on Exmoor were of black animals; as many as 20 percent of the reports that he had gathered described fawn-coated felids resembling pumas.

During this same year, Trevor announced that he had learned of the existence of a cache of several deer carcasses present within a broad-leaved wood on Exmoor; one carcass that he had examined and which had not fully decomposed sported a gaping hole at the neck base but bore no other sign of wounding. What had killed those deer? Poachers would not have left venison to rot, and water samples taken for analysis were shown not to contain toxins. Trevor revisited the site several times, and was intrigued to note that his dog was always agitated when there. On his sixth visit, moreover, he came face to face with what may well have been the explanation for these anomalies.

While crossing a stream, Trevor observed the head and shoulders of a large jet-black animal emerge out of some bushes nearby. He felt that its head seemed almost otter-like—sleek and broad with small ears, raised above a thickset neck, deep chest, and powerful forelegs. Suddenly it turned away and moved swiftly through the trees. Its gait reminded Trevor of a greyhound's, its forelimbs pushing through the hind limbs. Seen in its entirety, it resembled a large black panther, long in the body and tail. Trevor estimated its body length at 4.5 feet, its shoulder height at around 2 feet, and its weight at 80-120 pounds.

Also noteworthy from Exmoor in 1984 were sightings of felids resembling the European wildcat *Felis silvestris* during the spring at Holcombe near Dawlish, estimated by eyewitnesses to be twice the size of normal domestic cats. In July, while conducting an ornithological field-trip at Cloutsham on Exmoor, Trevor and several members of his tour watched a family of very large tabby-like cats (two adults, two juveniles) at Stoke Ridge. Trevor noted that their markings were all vertical, and their tails were bushier than those of normal domestics. This feline family eventually moved off into the edge of Horner Woods.

Black panther-like cat seen and pencil-sketched by Trevor Beer in summer 1984 (Trevor Beer)

In 1985, more puma and black panther reports occurred, and in November Nigel Brierly announced that fur identical to that of feral domestic cats had been found at a sheep kill from 1983 and again on barbed wire through which a possible sheep assailant had passed just prior to another killing. He had also noted the presence at some killings of a very strong odor reminiscent of Brussels sprouts. November was also the month in which the Exmoor Beast doubled its media exposure—literally—with newspapers recording the fact that sightings were now being made of two black cat-like beasts prowling the east moor together, leaving behind two sets of tracks.

December saw the Beast gain—and lose—a new identity. "Experts" identified a large spoor discovered near an Exmoor out-house as that of a wolverine *Gulo gulo*—a non-native, bear-like relative of weasels—until Trevor Beer pointed out that whereas a wolverine's spoor always possesses five distinct toe prints, the out-house example only exhibited four such prints (*Western Morning News*, 10 December 1985). Exit the wolverine!

(Remarkably, however, at the beginning of January 1994, two readily-identified wolverines were seen together, running along a disused railway line near South Molton on the edge of Exmoor, by Trevor Beer and Endymion Beer, only about 40 yards away. And a dead animal again identifiable as a wolverine was seen lying beside a road at Wembworthy, Devon, in summer 1994 by professional photographer Joanne Crowther from London as she drove by. Sadly, she didn't stop

—nor, ironically, take any photos of it—because at the time she didn't realize its significance. It was only later, when she casually described it to some naturalists, that she—and they—realized what it was. Where had these wolverines come from? No-one knows.)

Puma and black panther sightings continued to emerge all through 1986, together with some sheep kills reputedly similar to those of the nefarious 1983 outbreak, but this year was most notable for reports of yet another felid form on Exmoor—the lynx. In January, the Mugleston family spotted a creature near Muddiford whose pointed ears reminded them of a lynx; at the same time, their description of its tail as "bushy" is not suggestive of the short slim tail characterizing these felids. More substantial was the announcement in November by Nigel Brierly that hairs found at a sheep kill had been formally identified as those of a lynx (*Sun*, 6 November 1986); and in a *West Briton* newspaper article of 31 December 1986, zoologist Frank Turk openly favored the likelihood of this species' existence in Great Britain. Also noteworthy is that a brown spaniel-sized cat with a short tail and notably tufted, pointed ears was sighted at close range by 14-year-old Marcus Matthews and his mother while driving through the Mendips. Sadly, a photograph taken of the creature (which had reminded them of a North American bobcat *Lynx rufus*) as it bounded away only captured a small portion of its back (Matthews, pers. comm.).

Wolverine (public domain)

In January 1987, nine paw prints (each about 3-inches in diameter and also length) discovered by Trevor Beer at Muddiford were considered by mammalogist Stephen Harris to be definitely feline and consistent with those of a wildcat or lynx. This was soon followed by an equally significant revelation—the positive identification of a puma hair from Exmoor. After four years of anecdotal evidence, the much-needed physical proof for the existence of large cats here was finally surfacing. The lynx form continued to stay in the news too. A sighting of one such beast was reported by two eyewitnesses at Tarr Steps in late January, with further observations coming from Cloutsham and from Malmsmead on the River Oare during late spring. In one of these latter incidents, the spaniel-sized felid in question was actually sighted carrying a rabbit in its jaws.

The year 1987 also saw its fair share of puma and (especially) black panther reports—notable among the latter were sightings of a huge black felid near to Muddiford in mid-March, and a similar (or the same?) creature in the Dunster deer park area (later implicated in sheep kills within this region and hunted by a search party, though not discovered). But at the August Bank Holiday, the panther saga entered a new and most significant phase—because one such animal was not only seen by Trevor Beer and three other witnesses but also photographed!

Trevor estimated that it was about 100 yards away when he took his nine photos, using a hand-held 300 mm Pentacon lens on a Praktica MTL50. They show a large black beast with rather small head and powerful limbs. Trevor estimated its length to be 4-4.5 feet, and its shoulder height at 2 feet. Although it is not really possible to identify the creature positively from the photos themselves, Trevor and his companions were able to observe the creature in detail as it stalked and caught a rabbit, and had no doubts that it was a large panther-like cat and not a dog. One of the photos was published in a *BBC Wildlife* report; Trevor never publicly released any details of the precise locality of his sightings; hence in deference to his wishes, I shall refrain from doing so here.

Over several years, Trevor compiled a detailed map of black mystery cat sightings from Exmoor, recording all of those that were made in good viewing conditions over a decent period of observation and in which the eyewitness concerned was certain that the creature itself was a cat and not a dog. In addition, he marked most of these sightings with the month and year in which they occurred. As I discovered when he very kindly loaned it to me for inspection, the scatter of sighting points in relation to the months in which they had taken place clearly demonstrated that the cats underwent a seasonal migratory passage across Exmoor, which began and ended in the same location each year, and which appeared to follow the movements across the moors of the deer. Statistical analyses that I carried out on the scatter of the points in relation to their months confirmed a marked migratory pattern. This was the first time that a British mystery cat form's movements had been plotted in such detail, providing a very significant insight into the creatures' behavioral patterns.

The year 1988 opened spectacularly with a close-up photograph of a large black creature claimed by the person who took it—a mysterious unidentified figure known only as "The Jungle Man"—to be the infamous Exmoor Beast. The photo was published on the front page of the *Mid Devon Star* newspaper for 1 January, and had been brought to the newspaper's attention by Beast eyewitness Peter Bailey of Colliepriest Farm Bungalow. It depicted a sturdy all-black animal allegedly the size of a large dog, with quite narrow (but not long) limbs, and a relatively large head bearing pricked, pointed ears. The head's precise shape could not be discerned as it merged into the equally dark pelage of the animal's shoulders and back. Thus it was not possible to identify the animal conclusively. A panther identity cannot be ruled out, although the prominent muzzle provided the head with a strongly canine mien (but this may be due simply to the angle at which the photo was taken; the total blackness of the animal greatly hinders visual analysis). Another problem with this photo is its complete absence of background, thereby rendering impossible any attempt at identifying the location where it was snapped or estimating the creature's size.

By the end of January 1988, police were investigating a series of Shetland lamb killings at Ovis Farm, near Bratton Fleming; 30 lambs had been killed over the past two months—and all with the same pattern. One night the predator responsible would kill and eat half of the carcass, the next night it would return and consume the remainder (leaving behind only the fleece). Automatically, the Beast was blamed, but Trevor felt that a large dog rather than a large felid was most likely to be the culprit. Certainly, it is only too easy to blame a mysterious, unidentified animal for such events rather than look for a more commonplace answer.

This was demonstrated well by a later incident. At the end of January, a day-old-foal was discovered with its throat torn out in a Muddiford field. As usual, the Beast was immediately suspected (and indeed, it is known that pumas have a great liking for foals). The Ministry of Agriculture took the corpse for formal analysis and discovered that the foal had actually died of hypothermia, after which its carcass had been preyed upon by a fox (fox scats had been found nearby that contained foal hairs) and also by crows. Trevor visited the site soon after the foal had been found, and he noted a number of dog prints there too, but no cat spoor.

Also in late January, teenagers Simon Hopwood and Sebastian Carmell spied a large skull possessing very long canines, lying at the side of a hedge near the Dartmoor village of Lustleigh. In various newspaper accounts it was described as puma-like, but judging from photographic evidence available (all depicting its left-hand profile), a leopard identity is equally if not more likely—although naturally I would hesitate to offer a definite specific identity for it without having first examined the skull directly. Nevertheless, it is unequivocally from some large-sized cat species, so how can its occurrence be explained? Leopards (in the form of black panthers) and pumas were being reported from this region (and local rumor had it that a sizable cat form was released onto the moors during the 1980s). Alternatively, it may have been placed there deliberately by a hoaxer, in the express hope that someone would subsequently come along and find it. (And indeed, some years after this book's original edition was published, I learned from Jonathan Downes, the Centre for Fortean Studies' director, who lives in Devon, that in reality the skull had been found wrapped inside a cellophane bag!)

A while later, a supposed lynx skeleton was discovered at Ilfracombe (northwest of Exmoor). However, it proved merely to be that of a large domestic cat (Beer, pers. comm.).

In mid-April 1988, an Asian leopard cat *Prionailurus bengalensis* was shot by a farmer on Dartmoor. Little larger than a domestic cat, this felid was clearly not responsible for the sheep killings attributed to the Beasts of Dartmoor and Exmoor. Nonetheless, its unexpected presence here still constitutes a significant occurrence, as it provides unequivocal support for the possibility not only that exotic, non-native cat species are indeed being kept in captivity (probably as pets) in this region, but also that some are escaping (and/or being deliberately released?) onto the moors.

Incidentally, less than three months earlier, in February 1988, after suffering several pheasant kills by an unseen predator at Minto Estates near Jedburgh, just north of the England-Scotland border, head gamekeeper Willie Thomas had encountered a very strange beast running from a pen with a dead pheasant in its mouth. According to Thomas, the animal resembled a cross between a leopard and a wildcat—so, automatically, he raised his gun and shot it. Measuring just over 3 feet in total length, and weighing up to around 11 pounds, it was identified on 1 March as a leopard cat by Jerry Herman from the Royal Museum of Scotland, and investigations revealed that it had escaped from the premises of a private collector from Cumbria in the previous autumn during a burglary (London *Sunday Telegraph*, 6 March 1988).

Perhaps the most exotic species to have been added to the list of putative Exmoor Beast identities is the binturong *Arctictis binturong*, a large black-furred Asian relative of civets and genets. Rumors abound that a pair of binturongs were released onto Exmoor during the mid-1970s, and it was briefly mentioned in a Legends of Exmoor booklet entitled *The "Beast" Strikes Again*, published in 1988 by Exmoor National Park. Remarkably, moreover, this was not the first time that the binturong had featured in the history of British mystery cat reports, as will be seen later in this chapter.

Binturong (Karl Shuker)

A final point worth mentioning about the Exmoor and Dartmoor Beasts is that in addition to high-pitched screams comparable with those made by pumas (see Chapter 5), deeper, roaring sounds (some ending in a cough-like grunt) have also been heard here; these latter are more suggestive of leopards.

In summary, it would seem plausible that during the 1980s, and also, judging from many further sightings since then, right up into the present day, the moors of Somerset and North Devon contain large felids belonging to more than one species. Whether or not they will ever receive formal scientific recognition, and whether or not they are the animals responsible for any of the stock farm killings reported here and elsewhere, however, are other matters entirely.

Incidentally, during August-September 1993 I visited Trevor Beer for a week and spent each day trekking far and wide with him across Exmoor seeking its infamous inhabitant, but we spied no Beast. However, it opened my eyes to the vastness of the region's moors and forests, and to the indisputable fact that if a large-sized cat—or cats—did lurk here, one should not be remotely surprised by how easy it would be for anything choosing to stay hidden here to do so.

The author in early 1990s holding black panther model resembling eyewitness reports of Exmoor Beast (Karl Shuker)

Also during the early 1990s, a very similar situation arose on and around Bodmin Moor in Devon's western neighboring county, Cornwall, featuring sightings and even some photographs and videos of black cats sometimes referred to by eyewitnesses and media as panthers or the Bodmin Beast(s), plus various livestock kills attributed to them. In response, the British government's Ministry of Agriculture Fisheries and Food (MAFF) ordered an official investigation, conducted by ADAS Cambridge for six months during 1995.

The official MAFF report, authored by Simon J. Baker and C.J. Wilson, and published as an illustrated booklet, stated that detailed analyses were carried out on video films of alleged big cats on Bodmin Moor submitted to ADAS by members of the public. These analyses also involved team members visiting the precise locations where the cats had been and measuring background features seen in the videos/photos using fixed one-meter-long ranging poles. These revealed the cats to have displayed the same dimensions as an adult domestic black cat (rather than a black panther). Three plaster casts of footprints taken on Bodmin Moor were submitted for analysis, and it was concluded that two were from a domestic cat and the third from a dog. Only four possible livestock predations by "big cats" were reported to ADAS by members of the public during its investigation, and no evidence was found that such creatures were involved in any of them. The team concluded that although it had not been able to prove that a "big cat" was not present, it had not received any verifiable evidence that such a creature was present, and based upon its findings with the kills it did not consider there to be a significant threat to livestock in the Bodmin area from a "big cat."

Kellas Cats, Rabbit-Headed Cats, and Fairy Cats

Worthy of brief mention here is a feline ex-mystery, the Kellas cat. A gracile felid the size of a large Scottish wildcat, with prominent fangs, black bristly pelage sprinkled with long white primary guard hairs, and a white throat patch, it is represented by several specimens obtained since 1983 in parts of northern Scotland. Their striking appearance initially led to media speculation that they represented a hitherto undiscovered species. However, as I predicted in this book's 1989 edition and also in a peer-reviewed Kellas cat paper published by the International Society of Cryptozoology's scientific journal *Cryptozoology* for 1990, this eye-catching felid was ultimately revealed (via researches at London's Natural History Museum and at the Royal Museum of Scotland—now part of the National Scottish Museum) merely to be an introgressive (complex) hybrid strain of Scottish wildcat and domestic cat.

As the Kellas cat's identity was still undetermined at the time of publication of my two above-cited works, however, they contain an extensive coverage of its history and putative nature, but repetition of such coverage here is not warranted now that this cat form's identity has been confirmed.

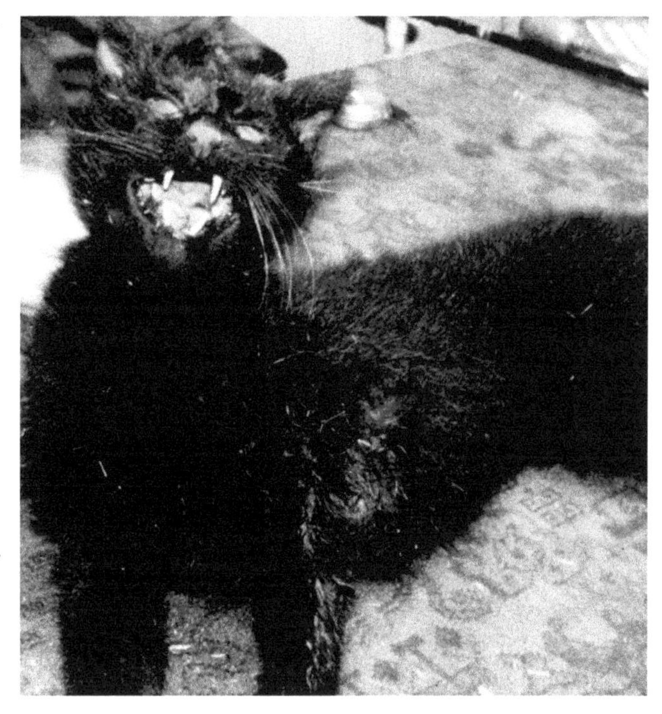

The Revack specimen—a Kellas cat, a rabbit-headed cat, or both? (Edward R.J. Orbell)

In 1988 an even more than normally eye-catching Kellas-like cat was shot by a local gamekeeper in the Dufftown area of Speyside, as later reported by Roy Kerridge (*Sunday Express*, 15 December 1996). Distinguishing this adult male example from previously publicized Kellas cat specimens was not only its alleged lack of white primary guard hairs and chest patch but also, in particular, its oddly-shaped head, reminiscent of a rabbit's, together with large ears, Roman-type nose, and notably small braincase. A second "rabbit-headed cat" was shot in December 1993 by gamekeeper Jimmy McVeigh after his dogs had flushed it out of a pond in a locality near East Kilbride where it had been swimming in pursuit of some wildfowl. This cat was an adult female but displayed much the same head shape and Roman nose as the Dufftown cat. Moreover, it has been suggested that the Revack specimen of Kellas cat from 1984, the first one to receive media coverage, may have been yet another rabbit-headed cat; but unlike claims to the contrary for the Dufftown individual, the Revack cat's pelage definitely sported long white primary guard hairs.

According to an online article on Kellas cats and rabbit-headed cats by veterinary surgeon Aron Bowers, the East Kilbride rabbit-headed cat's skull was later examined in Edinburgh at the Royal Museum of Scotland by felid expert Andrew Kitchener, who had previously examined Kellas cat material there. In his article, Bowers stated: "Kitchener's findings suggested the rabbit-headed cat skull exhibited no real anatomical differences between it and specimens of Scottish wildcats, and domestic cats and their hybrids." In her own online account of Kellas cats, Sarah Hartwell commented that the Dufftown rabbit-headed cat "...had a distinctly Siamese/Oriental profile indicating the domestic breed that had been involved with the hybridisation."

During one of the always interesting, informative telephone conversations that I enjoyed for many years with internationally-renowned felid geneticist Roy Robinson, he mentioned to me that after scrutinizing photographs and written descriptions of the Dufftown cat, his considered opinion as to its identity was that it was most probably merely a freak, teratological specimen of either a large black feral domestic cat or a Kellas cat. The second of those two options put forward by Roy is also my own personal opinion with regard to the likely identity of the rabbit-headed cats.

Lastly, while researching Scottish mythology in relation to the Kellas cat back in the early 1980s, I discovered in Katherine Briggs's many books on British folklore, and later elsewhere too, that Highland folklore includes a legendary cat form called the cait sith or fairy cat. This is a sizable creature sporting a black bristly pelage, white throat patch, and sparks or stars over its fur. If these "sparks" or "stars" refer to gleaming white primary guard hairs (and one could hardly ask for a more accurate description of them), then the cait sith bears an almost exact resemblance to the Kellas cat. Consequently, if the latter felid form is only of quite recent origin, it is truly a most remarkable coincidence that the only region of Great Britain from which specimens or sightings of Kellas cats have emerged is also the very same and only region of Great Britain that contains in its traditional folklore a creature bearing an uncanny similarity to the Kellas cat.

Over 30 years have passed since I wrote this book's first edition, but the situation regarding British mystery cats has not changed to any significant degree. Quite a few additional physical remains (carcasses, hair samples) have been obtained, their respective non-native species positively identified, and in many (although not all) cases their origins ascertained (I included a comprehensive chapter documenting these specimens in my later book *Cats of Magic, Mythology, and Mystery*, 2012). Also, a few mostly inconclusive photographs of alleged mystery cats in Britain have been snapped, and countless new unconfirmed sightings reported. Yet mainstream zoologists largely continue to deny such creatures' existence here, notwithstanding the reliability and experience in making accurate observations of quite a number of their eyewitnesses. However, if we assume that at least some such felids do indeed exist and have indeed been described accurately by their observers, what could they be?

Identifying Great Britain's Large-Sized Mystery Cats: The Composite-Identity Theory

The thought that several different regions of Great Britain harbor thriving naturalized colonies of exotic Indian ring-necked parakeets and Australian wallabies (which they do) is hardly likely to incite trepidation. Conversely, the same cannot perhaps be said in relation to images of sizable cat forms prowling its countryside (despite the incontrovertible fact that these animals are going out of their way to avoid humans at all cost). Thus it is probably for this reason more than any other that it is proving so difficult to convince non-eyewitnesses that such animals are indeed doing so too.

Moreover, although some people are willing to admit to the likelihood that large felids of one given species may exist here, the prospect that a number of different species might in fact be doing so is for the great majority of people a rather unpalatable notion. Yet in view of the notable diversity of sightings reported in this book and elsewhere, this latter solution to Great Britain's mystery cat phenomenon is the only logical one. Any attempts to identify all sightings as pumas, or panthers, or lynxes, or simply feral domestic cats, or even some hypothetical, hyper-elusive felid wholly unknown to and undiscovered by science, fail miserably to accomplish this goal in my opinion.

Having studied the British mystery cat saga for many years, I now wish to reiterate my own theory concerning the likely taxonomic status of these animals, which I first put forward in this book's 1989 edition. I have named it the Composite-Identity Theory because it comprises a series of identities that collectively account for the various forms being reported, i.e. it propounds that no single species is responsible for all such forms. There are at least four identity categories, and possibly a fifth, as now discussed.

1: Feral Domestic Cats

Scientific studies carried out with feral domestics at a number of British localities have revealed that, contrary to popular belief, they are on average no larger or heavier than their tame counterparts (oversized specimens are occasionally reported but very rarely submitted for formal scientific verification). Yet despite their small size, it seems likely that certain mystery cat incidents have feral domestics as their basis.

On Exmoor, for example, Nigel Brierly found positively-identified feral domestic cat hairs at certain sheep kills and noted a strong odor reminiscent of Brussels sprouts. Veteran British zoologist Maurice Burton reported that this odor is commonly associated with ferals, and also foxes (but not with larger felids such as pumas). Yet ferals are neither large nor powerful enough to kill sheep, so how can the presence of their fur and odor at such kills be explained? The most plausible answer is that they have scavenged meat from kills already made by some much large predator.

In contrast, Brierly suggested that these latter larger predators are themselves feral domestics, but of a mutant strain that has attained puma-sized proportions. Yet at his death on 12 March 1986 in Cairns, Queensland, Australia, the world's largest reliably recorded domestic cat (a neutered male tabby named Himmy) only weighed just under 47 pounds (by comparison, a puma on average weighs 200 pounds), and this individual was very obese. Even though feral domestics have established themselves throughout the world, no evidence has been obtained to suggest that *F. catus* can and does attain a body size commensurate for sheep- or deer-killing. And with Exmoor already providing a wide range of prey sizes suitable for feral cat predation, it seems most unlikely that natural selection would act upon such cats here in a manner necessary to engender the dramatic evolutionary changes required for the escalation in size from normal feral to puma-sized feral.

Despite being sighted in the heart of Surrey puma country during the peak of Surrey puma sightings, even the most cursory of glances at the photo by Ian Pert of the Worplesdon cat in comparison with any photo of a puma (adult or juvenile) is sufficient to demonstrate at once that the two felid forms bear little or no resemblance to one another. It is quite hopeless to reconcile as one and the same species the sharp-faced, thick-tailed, short-bodied Worplesdon mystery cat with the much larger and rangier, flatter-faced, thin-tailed, long-bodied puma. Conversely, I have seen several taxidermy specimens of domestic cats that compare very closely indeed with the Worplesdon cat—one in particular (a specimen dating from Britain's Victorian age) was a near-perfect match.

Finally, it is certainly true that large creatures can be "created" out of smaller ones via optical trickery if seen at a distance and/or in poor lighting conditions. Consequently it is by no means unlikely that some British mystery cats may indeed be nothing more than poorly-spied feral domestics (remember the findings by the MAFF investigation regarding the supposed Bodmin Beast). Coupled with this is the frequent occurrence in mystery cat episodes (in Britain and elsewhere) of such excitement and interest (not to mention apprehension) that after a time almost any animal seen is automatically "identified" as the mystery cat concerned. The conversion of a feral domestic cat into a much more formidable felid in the eyes of an observer "tuned in" to reports of mysterious "big cats" in his/her vicinity may have happened many times. Indeed, Maurice Burton suggested in a December 1966 article published by *Animals* (forerunner of the earlier-cited *Wildlife* magazine) that feral cat sightings are the explanation for most (if not all) of the 1964-1966 Surrey puma sightings.

2: Wildcat

Despite being distributed widely across the British mainland (for the Irish situation, see Chapter 2) in earlier times due to relentless persecution prior to World War I by gamekeepers and hunters, the European wildcat today is officially present here only in parts of Scotland, where it is referred to as the Scottish wildcat. Nevertheless, since the 1920s its distribution within Scotland has re-expanded considerably. In addition, certain mystery cat reports suggest that its range may even have re-extended beyond Scotland.

First of all, however, here is a concise morphological description of the wildcat. The background color of a pure-bred British specimen's pelage is buff-grey (paler ventrally, and white on the throat), marked with several vertical stripes running down from the dark dorsal stripe to the belly. The limbs are also transversely striped, and the black-tipped tail is encircled by a number of rings (of which at least the last two or three are complete). Its forehead bears four to five longitudinal stripes running down to the neck's nape where they converge to form the dorsal line. The wildcat can be distinguished from most (but not all) domestic tabbies by its absence of blotched markings. Unfortunately, however, feral domestics tend to revert to a striped pelage pattern that is deceptively similar to a true wildcat's; and to make matters even worse, wildcats will mate with such ferals to yield hybrids that are again very similar to pure-bred wildcats.

Vintage photograph of Scottish wildcat that clearly shows its broad-tipped tail (public domain)

Consequently, it is exceedingly difficult to decide from pelage morphology alone, whether a given tabby-like cat sighted within the wildcat's known range is truly a wildcat or simply a hybrid or feral domestic. Generally, however, a wildcat is about a third larger in overall size than a domestic, its head is broader, its teeth larger (in some cases the upper canines protrude even when the jaws are closed), its body stouter, its limbs longer, and its tail shorter (and ending broadly and bluntly rather than tapering at the tip).

In short, any re-expansion of its range into former haunts is only one side of the wildcat distribution coin. The obverse constitutes the exciting possibility that in some of the more secluded regions of England and Wales, the wildcat never died out in the first place—or at least, prior to its extinction there, the wildcat left behind some of its genetic heritage within the local feral domestic population by having interbred with them. Worth noting here is that in the opinion of Colin Matheson from Cardiff's National Museum of Wales, a "wildcat strain" may indeed have survived among feral domestics in some areas of Wales.

These suggestions can also be applied to the English situation. Some wildcat-like felids were sighted by Trevor Beer and others in 1984 on Exmoor, which makes the following account of particular interest. In her book *Living on Exmoor* (1963), Hope Bourne recorded that Room Hill was home for a long time to some very sizable cats, reported right up to World War I. Their ferocity was renowned throughout the area—apparently not even the farm dogs would dare to attack them—and they would unhesitatingly kill any farm cats that they encountered.

Fred Milton, past President of the Exmoor Pony Society, recalled to Bourne one of his own sightings when he observed one of these cats slinking down a nearby lane. He described it as being about the size of a male fox, with a grey or tawny-grey pelage marked all over with dark stripes. Its head seemed very large in comparison with that of a domestic cat, its fangs protruded below its lip, its limbs appeared to be rather long (especially the hind pair), it walked with a slouching gait, and its tail was blunt. Could one ask for a better description of a pure-bred wildcat? Reports of these remarkable felids dried up after World War I, so it was assumed that they had died out. Perhaps instead they simply moved on to a new location, with their descendants being responsible for the 1984 sightings.

Also worthy of mention here is that during 1922, two very large cats that had been sighted on several previous occasions on the

Westmorland-Lancashire border (in northern England) near Carnforth were shot in two separate incidents. The male was shot by a poacher and presumably died of its wounds shortly afterwards as it was not seen again. In November, the female was also shot, this time by Dr. Frederick Hogarth, who fortunately had the specimen preserved. It was displayed for many years thereafter in his doctor's office in Morecambe. Upon his death, this specimen was donated to Lancaster's City Museum, where it was subsequently measured by Ron Freethy. Following comparison of its measurements with those available for known wildcats, Freethy was unable to rule out the possibility that Hogarth's exhibit was a genuine English wildcat.

3: Escaped/Released Exotics

One of the most popular suggestions put forward by skeptics of many reports concerning mystery creatures is that sightings of such beasts simply involve specimens of various foreign/exotic species that have somehow escaped (accidentally, or with assistance!) from captivity, e.g. zoos, parks, private collections, households. In some instances, this suggestion has failed dismally to explain satisfactorily a given mystery animal sighting, but in relation to certain other incidents it is quite plainly the most logical answer. The latter would seem to be particularly true relative to many mystery cat cases not only in Great Britain but elsewhere too—which is the reason why I am covering this suggestion here in particular detail—and can offer a very convincing identity for no fewer than four of the major felid forms reported from Britain.

Known scientifically as *Puma concolor*, the puma (aka cougar, mountain lion, catamount, painter, and many other colloquial names too) is the largest of the Small Cats, averaging 5 feet in head-and-body-length, with an extra 28 inches of tail, and a similar figure for shoulder height. Its head is small, its face (bearing a black bar on each side that connects nose with jaws) and its ears are short and rounded. Its neck is relatively long, leading to a noticeably elongate, narrow body borne upon strong, muscular, large-pawed limbs, and terminating in a cylindrical and proportionately long, well-furred tail with black tip. The puma has two color morphs: red and grey. The former ranges from tawny to rufous, the latter from silver-grey to slate. In both morphs, coloration is lighter on the flanks and shoulders and almost white ventrally. This description corresponds very closely with some of those given from Surrey, Midlothian, New Forest, Isle of Wight, and Exmoor, to name just a few localities from which puma-like cats have been reported. There are also the puma hairs positively identified from Surrey and Exmoor, plus the capture of Felicity in Cannich.

Although variable in color, only a single black puma has ever been officially documented from anywhere within its native distribution range—namely, from South America (see Chapters 5-6). Hence it is highly unlikely that melanistic pumas could be the identity of the large all-black cats being sighted in Great Britain. Far more likely is the possibility that they are black panthers, the melanistic morph of the leopard *Panthera pardus*. These average around 7.5 feet in total length, of which about 32 inches constitutes the tail. The panther's body is very elongate and muscular and is borne upon massive, broad-pawed limbs that are slightly longer than those of the puma. Its tail is also proportionately longer, and its head somewhat more elongate, but its ears are once again small and rounded. At anything but the shortest of viewing distances, the panther appears uniformly black. This description clearly matches very closely those of the black beasts on Exmoor and Dartmoor, plus felids from the Isle of Wight, Farnborough, Durham, etc (in Scotland, some panther reports may be due to the Kellas cat).

On average, the European or northern lynx *L. lynx* weighs 63 pounds, has a shoulder height of 27 inches, and a head-and-body length of 40 inches. Its tail length ranges from 6 to 10 inches, thus demonstrating that, contrary to popular belief, some lynxes possess quite notable tails. Other than its tail, its most striking features are its sturdy but lengthy limbs, its long black-tufted ears, and its piercing golden eyes (the iris of most other felids tends to be greenish in color). Its pelage varies from yellowish-grey to reddish-brown, with spots of differing degrees of distinctness. This description is consistent with those of certain felids from Exmoor, Powys, Glenfarg, Isle of Wight, etc; we must also remember Cobbett's Surrey lucifee, the lynxes supposedly shot in Scotland during 1927, the Canadian lynx shot in Devon in c.1903, and the positively identified lynx hairs from Exmoor.

As for the noticeable black tuft at the tip of the Nottingham mystery cat's tail, this suggests that the latter mystery cat was quite probably an escaped lioness *Panthera leo*, i.e. representing a fourth non-native cat species in Britain.

On morphological grounds, it can be seen that the existence of naturalized pumas, panthers, and lynxes within the British countryside would go a very long way towards explaining the majority of large-sized mystery felids being reported here. But there is more than mere morphology in support of this suggested solution.

Once in range of its intended prey, the puma rapidly sprints towards it and, upon the moment of contact, generally sinks its claws into the victim's head or hindquarters (rather than the flanks). Prey of ungulate size are additionally bitten in the back of the neck (sometimes the throat), and if this is not sufficient to kill the victim, the puma increases its jaws' grip in order to break the animal's neck, although in many cases the force of impact between puma and victim is enough to do this. The carcasses of uneaten puma victims are not usually wounded to any extent externally. Panthers and other large felids follow the same principal killing mode, differing mostly only in specific details. Animals killed in this way have been reported on many occasions from various locations across Britain, as testified in some instances by veterinary officers and zoologists called in to examine them. Notable consumption of the victim's abdominal organs, lack of damage to its bones, and location of its carcass in a secluded spot (remember the Exmoor deer cache?) are typical features of felid feeding behavior and have also been described from various British kills.

Although the preferred prey species of large felids are the larger ungulates, they are certainly not incapable of surviving for considerable periods on much smaller prey species if conditions necessitate, especially rodents and rabbits (which in any case constitute the preferred prey of lynxes). Consequently, it would be unwise to suggest that the British countryside could not actually provide sufficient prey for long-term survival of the larger felids. In any event, roe deer, for example, are extremely plentiful in a number of British regions, including Exmoor and Surrey.

And of course, if prey becomes scarce in one area, such cats will extend their hunting territory by very great amounts to seek out new prey sources elsewhere, often trekking hundreds of miles in times of food shortages. Indeed, one wide-ranging individual could actually be responsible over a period of time for mystery cat sightings in a number of different localities; remember the migratory Exmoor cats? Consequently, the actual number of cats on the loose may well be a lot smaller than implied by report counts (some media articles have claimed that hundreds, if not thousands, of panthers and pumas exist in the British countryside, which in my view is patently absurd). Coupled with this is the extreme elusiveness of these cats; this is to be expected when dealing with felids. The puma's ability to remain unseen in its native Americas, for example, is renowned. Deserving of mention is an anecdote recorded by Guggisberg, in which a hunter confessed that although he had shot no fewer than 70 pumas in Montana, USA, he had never once caught sight of any of them until they had been treed by his dogs!

As previously discussed, the finding of clawed spoor in relation to supposed mystery cats should not lead the discoverer to discount automatically a feline identity for their owner. The lynx in particular is a case in point. As noted by Anders Bjärvall and Staffan Ulström in their book *The Mammals of Britain and Europe* (1986), whenever the claws of a lynx are extended to give a better grip while walking, they leave behind very clear impressions. Moreover, when extended in this way, the toes themselves are also extended, greatly increasing the overall size of the spoor. A good example of a lynx trail showing claw marks was documented in 1972 by Rare Elgmork in a *Fauna* paper. Also worth noting is that a variety of different cats, including lynxes, leopards, and lions, sometimes display graded pads (i.e. the two members of a given pad pair differ in size). For an excellent example of leonine graded pads, see the spoor diagram for the lion contained in Reay Smithers's book *Land Mammals of Southern Africa* (1986). Injuries and genetic defects can also produce abnormal pads.

Critics of the escapee theory in relation to sightings of exotic species in Britain frequently argue that such animals would be unable to survive in conditions and habitats so alien from those of their native lands. A scan through any of Sir Christopher Lever's definitive volumes on naturalized animals will very swiftly disprove this! In any case, the native ranges of the puma and lynx already contain habitats and climatic conditions corresponding very closely with those found in Britain. And as far as tropical felids are concerned, we need only examine the case of the clouded leopards.

Nowadays split into two closely-related species within the genus *Neofelis*, clouded leopards are highly specialized arboreal felids indigenous to the humid tropical forests of southeast Asia. Consequently, when one escaped from what is now known as Howletts Wild Animal Park (formerly Zoo), in Kent, England, during 1975, it was not expected to survive for more than a few days in the British countryside. It was eventually shot by a farmer nine months after its escape; it was in very good health, having fed upon rabbits and lambs. As a result of this, the possibility of the much hardier and adaptable black panther surviving here is indeed good. At the same time, of course, a lot depends upon individual disposition. In 1987, two clouded leopards escaped from Howletts during very severe hurricane conditions. One was captured almost immediately; the other, named Xiang, virtually gave himself up by walking into a baited cage and curling up inside to go to sleep, awaiting collection by his keeper.

Many known examples of naturalized animals are descendants of original escapees from small private collections or escaped exotic pets, rather than from large public zoos or parks. The reasons for this trend are numerous and varied, ranging from the difficulties faced by private individuals of meeting the high level (and cost!) of security required for ensuring retention of their animals, to the equally awesome (and again costly) legal problems to be faced if such animals do escape. Consequently it would seem reasonable to suppose that in general the private sources of captive animals are those most likely to lose animals and those least likely to report such losses.

An event having great bearing upon this and upon felid escapees in particular was the passing in 1976 of the Dangerous Wild Animals Act in the U.K. This banned the keeping of various species, including the larger cats, on all premises other than those suitably licensed to do so (and the licenses in question were extremely expensive). Is it just coincidence that the last few months of 1976 were marked by a dramatic increase in mystery cat sightings, and from all over Britain? Is it unreasonable to suppose that the license requirements proved too expensive for many private owners of large cats, and that the secret release of their pets into the countryside proved too tempting a solution to avoid? And more legislation brought into force during 1981 under the Wildlife and Countryside Act in relation to exotic pets adds further food for thought in this direction.

Exotic pets may provide the answer to the mystery of why so many black panthers (i.e. melanistic leopards) are being reported in the wild compared to a very much smaller number of felids reminiscent of the normal spotted form. If someone is intent upon buying a large-sized exotic felid as a pet, then the more exotic the form the better. And we only have to take note of TV and glossy magazine adverts, films, and pop videos to discover that the black panther is by far the most popular choice of felid for introducing an exotic image. Coupling panther popularity as privately-owned pets with earlier remarks concerning the greater chances of such animals escaping from and remaining unreported by private ownership, the prevalence of panther sightings is now less of a puzzle. The keeping of large cats as pets in the U.K. can offer other pertinent links to the British mystery cat phenomenon too.

In a paper published by the ISC's journal *Cryptozoology* in 1986, Roderick Moore of the Nature Conservancy Council commented that most of Britain's first major mystery cat sightings, in the early 1960s, occurred in affluent areas of southern Britain (e.g. Surrey, Hampshire) where one would expect to find more people who could afford to maintain exotic pets such as large cats. Linking this with the inevitable escapes that must have occurred provides a very plausible solution to the Surrey puma saga.

Worthy of mention at this point is an article published in *Wide World Magazine*, entitled "A Puma Hunt in Surrey." Nothing special, you may think, in view of the many puma searches occurring during the 1960s there. Very true—except that this particular hunt occurred prior to 1903! The puma in question was a pet that had escaped; it was finally recaptured and sold to a zoo. This demonstrates well that pumas were actually being kept as pets in Surrey more than 50 years before the Surrey puma episode, providing important support for Moore's comments.

Equally noteworthy is the following 19th-century English report, from the *Blackburn Standard* newspaper of 8 June 1836, brought to my attention by British cryptozoological researcher Richard Muirhead, and which to my knowledge has never previously appeared in any cryptozoology book:

> LEOPARD HUNT — On Monday last a strange-looking animal having been seen in the fields near Wheathamstead, Herts, a small party went in search – supposing that it was a deer which had been scared out of Brocket-hall-park by the gloomy looks of its noble occupier. Great was their surprise at finding in a hedge a large leopard, which stole away followed at a respectful distance by the sportsmen, who were only armed with fowling-pieces loaded with swan shot. As it was endeavouring to escape it met a labourer at work in the fields whom it attacked and dangerously wounded, but his life was saved by a mastiff fastening on the leopard, and enabling Mr Norman Thrale to approach within a few yards, and disable it with a discharge of swan shot. It was shortly afterwards destroyed, and was found to weigh 14 stone. It had breakfasted off a dog, whose head was found. It is not known where the beast had escaped from.

Nor, tragically, is it known what happened to the carcass of this exceedingly unexpected—and hefty—assailant.

One final facet of the exotic escapee identity to be considered here is its temporal aspect. The captivity of animals (even the most ferocious ones) within the menageries and collections of living curiosities of earlier days (some dating back to medieval times) was certainly not attended by the laws and security measures operative in relation to today's zoos and parks. The potential for felid (and all manner of other) escapees must therefore have been much greater—a most interesting and thought-provoking concept, researched by mystery cat investigator Phil Bennett from Wolverhampton, England. Certainly it provides a plausible explanation for early British cases, Cobbett's lucifee, for example, and the Girt "Dog." In the same vein, Michael Goss noted that during the 19th century, a Clent inhabitant writing under the pseudonym of "Vigorn" commented upon "... the prevalence of stories concerning wild beasts that have escaped from menageries ..." and noted a rumor that "at Broadwas, a village in the Teme Valley, a lion is roaming at large ..."

Moving back a few centuries, a panther supposedly escaped from a menagerie maintained in the park at Chillington Hall, near Wolverhampton in the West Midlands, England, during the early 1500s and was about to attack a woman and baby when it was shot through the head with an arrow by the Hall's owner, Sir John Giffard (aka Gifford). A monument known as Giffard's Cross still marks the spot today where the panther was killed.

The author standing alongside Giffard's Cross (arrowed) (Karl Shuker)

4: Non-Felids

As will be seen throughout this book, it is highly likely that some mystery felid sightings plus various examples of spoor and kills attributed to such animals involve dogs rather than cats. Certainly if seen at a distance, and perhaps for only a short time, some dogs (especially those breeds with relatively small heads, short faces, rounded ears, and/or shortish limbs) can appear surprisingly catlike. (Having said that: when on 20 December 1988 a supposed lion sighted by a motorist near a road in Nazeing, Essex, southern England, was swiftly tracked down by police, they discovered that it was in fact a four-foot-tall Irish wolfhound named Finn!) And with respect to stock farm kills, the very real menace of stray dogs is only too familiar to farmers throughout Britain. On Exmoor and Dartmoor in particular, farmers are becoming increasingly convinced that many of their savaged sheep are the work of hunting dogs being deliberately set loose, and many dogs have been shot in the act of killing sheep here; Trevor Beer's belief is gaining ground all the time.

Additionally, he has seen various kills known to have been caused by dogs that appear identical to some of the South Molton examples. Another very pertinent point that he made is that although the panther-like felids have been sighted chasing deer and killing rabbits, apparently no record currently exists of anyone having seen such a felid actually killing a sheep or even devouring the flesh from an already-dead specimen—in stark contrast to dogs. Moreover, packs of stray dogs could easily leave a sheep carcass almost devoid of flesh and viscera after a single session, a feat likely to be matched only by the most voracious of solitary felids.

As already discussed, absence or presence of claw marks within paw prints should not be looked upon as a foolproof indication of whether their owner was feline or canine. In some cases, a far more reliable guide to its identity is the overall size of the prints.

Take, for example, the set of spoor discovered near some stables at Munstead, in Surrey, England, during 1964 and attributed at that time to the Surrey puma. The fact that they were clawed does not automatically eliminate a felid identity for their owner; far more significant in this respect is their size because the diameter of each one measured a massive 5.25 inches. Any felid capable of producing such enormous spoor would need to be one of two things: either a simply colossal cat notably larger than an Alsatian if its paws were in proportion to the rest of its body, or a smaller cat but with outlandishly large paws totally out of proportion to the rest of its body. As neither of these types has been reported (as yet!) from Britain, it seems safe to assume that the owner of these prints was a dog, which is also suggested by the large size and elongate shape of the toe pads relative to the plantar pad. As a postscript to this episode, it later transpired that a lady frequently took her pet mastiff for walks in this very area—and the diameter of a mastiff's paw print just so happens to be 5-5.5 inches. Equally likely to be of canine rather than feline origin are the 5-inch prints noted at Llangurig, Powys, Wales, in 1980 (sporting a notably canine shape too).

After having discussed whether certain mystery cat sightings are actually based upon dogs, we should now consider whether alleged sightings of a certain dog type are actually based upon mystery cats. The canids involved here are the phantasmal Black Dogs—those preternatural canine entities supposedly sighted many times in earlier days and even occasionally in the present day. According to tradition, they bring death to anyone who touches them and disappear into thin air in full view of their observer, but in external appearance they usually resemble an ordinary large-sized, black-colored dog such as a Labrador or hound (except for their glowing red eyes). It has sometimes been suggested that sightings of these entities were really observations of large black panther-like cats; if correct, Britain's mystery panthers would gain a centuries-old history.

In truth, however, it is a theory beset with problems. As pointed out by Andy Roberts in his book *Cat Flaps!* (1986), accounts of Black Dogs and black panthers share only a superficial similarity. Black Dogs differ markedly from these latter felids in their frequent possession of long shaggy or woolly coats, their willingness to follow or brush against humans, their ability to pass through solid objects, their unpleasant breath (sometimes sulfurous), and the sounds of water splashing or even chains rattling that often accompany their appearances. In addition, Black Dog legends are very common and widespread in Ireland, whereas reports of black panthers from this island are very rare.

Anyway, if we take heed of the widely-stated claim that Black Dogs make a habit of approaching very closely to their observers, it is most unlikely that they would be mistaken for any form of cat. Equally, in those sightings of mystery cats that have been made at very close range and often for decent periods of time (and sometimes actually by vets, zoologists, and other eyewitnesses with detailed knowledge of wildlife), it is again very difficult to believe that a confusion with dogs could have occurred. Whatever phenomenon they represent if real, it is clear that Black Dogs have little or no connection with mystery cats. Similarly, it is evident that some but by no means all reports of such cats are the result of misidentification of normal dogs.

It is worth noting that certain mystery felid cases have involved neither cat nor dog. For example, a "cat seven feet in length" discovered in July 1929 by poultry farmer Tom Cartmell inside one of his hen cotes at Preston, in Lancashire, northern England, proved to be a binturong (see earlier for Exmoor Beast claims), which had recently escaped from a Blackpool zoo (*Reynolds's Illustrated News*, 7 July 1929). In 1970, a creature described as being like a puma, but colored tan and black with yellow streaks, attacked a dog in Sussex's Ashdown Forest. Hair samples and spoor casts were positively identified at London's Natural History Museum as having derived from a spotted hyaena *Crocuta crocuta*. The animal was later shot but eluded capture and disappeared (*Fortean Times*, December 1976). And on 8 February 1983, Ian Linn, the Senior Lecturer in Biological Sciences at Exeter University, announced that after having first examined the skull from the body of a supposed mystery "big cat" exhumed earlier from a Berkshire garden, it was "definitely that of a badger" (*Express & Echo*, 7 and 8 February 1983).

5: The (Not So) Missing Lynx?

Finally: in his annotated checklist of apparently unknown animals with which cryptozoology is concerned (*Cryptozoology*, 1986), zoologist Bernard Heuvelmans, the "Father of Cryptozoology" himself, stated regarding the possibility of an unknown large-sized cat form existing in the British Isles: "This, of course, is quite unacceptable to most zoologists." And certainly, as a zoologist myself, I personally do not consider such an option tenable.

Heuvelmans then went on to say "…but it should be remembered that lynxes survived until recent times in Britain, and may even have lingered into the present. After all, cats are most elusive creatures."

Yet just how recent is recent? In the 1989 edition of this book, I noted that the European lynx had died out in Britain around 10,000 years ago, using as my reference for this date a paper by R.D.S. Jenkinson that had appeared in a multi-contributor scientific tome entitled *In the Shadow of Extinction: A Quaternary Archaeology and Palaeoecology of the Lake, Fissures and Smaller Caves at Creswell Crags SSSI*, published in 1984 by the University of Sheffield's Department of Prehistory & Archaeology. Two decades later, however, the lynx's British extinction date was significantly revised.

Results published in 2005 by the *Journal of Quaternary Science* from radiocarbon dating of lynx bones recovered from two sites in the Craven area of northern England suggest that this species had survived in Britain until much more recently than previously supposed, dying out in medieval times around 1,300 years ago (i.e. c.700 AD). These findings, obtained by a research team headed by David A. Hetherington, Britain's leading lynx expert, also support the view that the mysterious game animal whose occurrence in the nearby Lake District is described in the early 7th-century Cumbric text *Pais Dinogad* (whose translation to date has been problematic), where it is named a llewyn, is a lynx. All of this in turn makes the prospect of this species' persistence into the present rather less remote than before.

Even so, if the lynx, unquestionably a highly elusive species, has indeed lingered on into modern times in Great Britain, one might surely expect a fair amount of tangible evidence to have been obtained and preserved down through the intervening 13 centuries between 700 AD and today—most especially during the Victorian era, when hunting, trophy collecting, and preparation of taxidermy specimens were such popular British pursuits. Worth remembering is that during that same period, those pursuits virtually exterminated the much smaller, less noticeable Scottish wildcat. Yet such evidence is conspicuous only by its absence—unless the three dead lynxes from Inverness and/or Cobbett's lucifee were examples?

Of course, if plans currently under consideration to reintroduce the lynx into Scotland are one day officially put into action, much of this speculation instantly becomes academic. Then again, in the opinion of a fair few British mystery cat investigators, Scottish reintroduction plans are themselves academic because the lynx is already there...

To date, there is no indication that Britain's long-running mystery cat saga is drawing to a close. On the contrary, new characters continue to make their debut—other notable examples have included the Durham Beast, Cornish puma, the panther-like Black Beast of Moray, the Bodmin Beast, and the Fen tiger. Quite a few physical specimens representing several different non-native species have also been procured, some living, others dead, as noted previously. In most cases, their origins remain unknown. Moreover, when the first edition of this book was published in 1989, the two most recent British mystery cat cases of note also featured physical specimens, which belonged to the same non-native species.

In the first of these two cases, a long-limbed mystery felid was killed by a car on 26 July 1988 at Hayling Island, in Hampshire, southern England, and was correctly identified by Marwell Zoo Director John Knowles as a jungle (aka swamp) cat *Felis chaus*. Although outwardly similar in some ways to the caracal *Caracal caracal*, the jungle cat is readily differentiated by several noticeable features, including shorter ear tufts, more rounded ears, wildcat-like stripes on head, vaguely striped limbs, and white upper lip (all readily discerned in a good news report photograph of the Hampshire felid that appeared in various media accounts). As the jungle cat is native to the Middle East and

The Ludlow jungle cat (Karl Shuker)

Asia, and uncommon in British zoos, the Hampshire specimen was most likely an escapee/released pet.

Six months after the Hayling Island specimen, a second dead jungle cat was found—this time by farmer Norman Evans, close to his home near Ludlow, in Shropshire, central England. It was very emaciated; a back injury (perhaps due to a collision with a car) may have prevented it from hunting. Both jungle cats were subsequently preserved as taxidermy specimens.

The existence within Great Britain of a variety of out-of-place felids—which for the most part unquestionably constitute escapees/releases from captivity in the form of black panthers, pumas, and lynxes—can surely no longer be doubted. It is time for officials to abandon their futile attempts to dispute the indisputable, and instead to use every means at their command to demonstrate their existence conclusively, so that this significant subject can finally receive the serious scientific attention and interest that it so richly deserves.

In *Proper Studies*, Aldous Huxley stated: "Facts do not cease to exist because they are ignored."

Neither do mystery cats.

Chapter 2
Ireland and Continental Europe: Daemon Cats and Scimitar Cats

The Naming of Cats is a difficult matter;
It isn't just one of your holiday games;
— T.S. Eliot, "The Naming of Cats" in *Old Possum's Book of Practical Cats*

One would surely imagine that a mystery cat ceases to be a mystery once at least one specimen is finally obtained and submitted to science for detailed examination and formal naming. Remarkably, however, this is not always the case, as will be demonstrated in this chapter by both the Transcaucasian daemon cat and the Corsican wildcat (not to mention, infamously, the onza in Chapter 6). Equally, it may seem a little strange to discover the Irish wildcat ensconced within a chapter dealing primarily with mystery cats from mainland, continental Europe. The reason for my inclusion of this felid here rather than in the previous chapter is to maintain intact a notable link between it and various mystery cats that are found in continental Europe.

Like the British Isles, modern-day Europe's official contingent of cat species is a pallid reflection of its Pleistocene plethora. Yet judging from the variety of unofficial felids being reported here in more recent times, Europe may not be quite so devoid of cat life after all.

Transcaucasian Daemon Cat and Desmarest's Obscure Cat

During the early 1980s, I was surprised to uncover a 1904 scientific paper published in the *Proceedings of the Zoological Society of London* concerning a black cat that appeared very reminiscent of Scotland's then-unexplained feline enigma the Kellas cat but which inhabited Transcaucasia. Also known as the South Caucasus, this is a geographical region near the southern Caucasus Mountains on the border of eastern Europe and western Asia, roughly corresponding to the area occupied by modern-day Georgia, Armenia, and Azerbaijan.

In this paper, the cat had been formally described and named *Felis daemon* by the eminent Russian zoologist Konstantin A. Satunin based on his description of two mounted specimens, three skins, and three skulls, all housed in what is now the Russian Academy of Sciences (headquarters in Moscow, archive and library in St Petersburg). Other scientists, however, did not share his view that this felid form warranted separate taxonomic status. In 1917, fellow Russian zoologist Nestor Smirnov referred to it as *F. silvestris caucasicus* aber. *daemon* (i.e. treating it as a morph or aberrant form of the Caucasian wildcat), whereas another Russian zoologist, Sergey Ognev, relegated it even further—to the level of a mere feral domestic cat (even though he did concede that a melanistic morph may exist within the Caucasian wildcat population). In 1951, within his major taxonomic review of Small Cats, *Catalogue of the Genus Felis*, British felid taxonomist Reginald I. Pocock also classed *F. daemon* as a feral domestic cat, and it has remained thus ever since. Yet is this a truly accurate assessment of its status? And how (if at all) is *F. daemon* related to the Kellas cat?

Feral domestic cat or not, the Transcaucasian daemon cat certainly displays some marked morphological similarities to Scotland's Kellas cat, as the following description of this contentious mainland European felid demonstrates. Based upon the museum specimens examined by Satunin, its length from nose to tail base ranged from 22.5 to 30 inches; its tail length from 13.5 to 15 inches. Fur color varied from black with a slight reddish tinge to reddish-brown, slightly paler on underparts, inner surface of extremities, and distal under-surface of tail. Viewed at certain angles, black transverse stripes were visible upon the flanks of the body's foreparts, more conspicuous on faded skins. Very long white hairs were scattered scantily all over the body. Vibrissae and eyebrows were brown, claws white. With regard to cranial features, Satunin noted that *F. daemon* differed from the wildcat both in the possession of a somewhat narrower frontal region and in the extension of the upper jaw bones further back than the nasal bones (which is the reverse condition to that exhibited by the wildcat).

Clearly, the dark pelage flecked with long white hairs exhibited by Transcaucasia's daemon cat compares with that of the Kellas cat, and especially with certain specimens of the latter felid form that share its cryptic striping and even its brown eyebrows and vibrissae. But how closely do their relative body proportions compare?

With the exception of one extra-large skin (but which may well have been stretched during preparation), *F. daemon* specimens examined by Satunin do compare favorably in head-and-body length with the Scottish wildcat and Kellas cat specimens measured in detail. Conversely, the tail lengths recorded from the *F. daemon* specimens are rather longer than those documented for Scottish wildcat and Kellas cat. Worth mentioning here is that an increased tail length is one condition proposed by wildcat expert Mike Tomkies and others as evidence for wildcat x domestic cat hybridization within the Scottish Highlands.

Possibly the Transcaucasian daemon cat is a simple (i.e. first generation) melanistic hybrid, as opposed to an introgressive hybrid resulting from several generations of interbreeding and backcrossing (like the Kellas cat has apparently done). It would be interesting to see what genetic analyses conducted upon DNA samples extracted from the preserved *F. daemon* skins would uncover.

Also of potential relevance here is *Felis obscura*—aptly-named inasmuch as this mysterious South African cat form has long since faded into zoological obscurity, with even its name ultimately given by palaeontologist Q.B. Hendey to an entirely different felid, a fossil species from South Africa's Miocene and Pliocene epochs (*Annals of the South African Museum*, January 1974). However, the original *F. obscura*, as formally dubbed by French zoologist Anselm G. Desmarest in his renowned tome *Mammalogie ou Description des Espèces des Mammifères* (1820), was a living felid, almost uniformly black or exceedingly dark brown, with faint striping on its limbs, tail, flanks, and cheeks, as depicted in a color plate from 1834, reproduced in black and white here.

This dusky felid was first brought to popular attention by the eminent French zoologist, Baron Georges Cuvier, who briefly documented it in Vol. 8 of *Dictionnaire des Sciences Naturelles* (1817), and referred to it as "Chat noir du Cap" ("black cat of the Cape"). This *F. obscura* (as opposed to the wholly different fossil one) is nowadays synonymized with the African wildcat *F. lybica*, so if museum specimens of it exist, they presumably constitute melanistic African wildcats.

Plate from 1834 illustrating *Felis obscura* (public domain)

Mediterranean Wildcats—A History of Mystery

The taxonomy of the wildcat *Felis silvestris* has been at the centre of much chaos, confusion, and controversy over the years. To begin with, it was once referred to scientifically as *F. catus* (the name by which the domestic cat is known today), but in the early part of the 20th century Reginald Pocock re-named it *F. silvestris*, by which it has been known ever since. Furthermore, at one stage several different species and numerous subspecies of wildcat were recognized.

In 2017, however, the Cat Classification Task Force amalgamated all of these into only two species and five subspecies. Namely, the European wildcat *F. silvestris* (of which as many as 22 subspecies were once recognized at one time or another, but nowadays once again reduced to just two); and the African wildcat *F. lybica* (ancestor of the domestic cat), nowadays consisting of only three subspecies (but once as many as 25), including the Asiatic wildcat *F. l. ornata* (formerly deemed a separate species in its own right).

Although, as would be expected, most wildcats in Europe have always been deemed unequivocally to belong to the European species, it was long considered that not all of them may do so. Those existing upon certain Mediterranean islands were believed by some authorities to belong to the African species instead. The felids in question are those wildcats inhabiting the Balearic Isles, Crete, Sicily, Sardinia, and Corsica. In view of this zoogeographically significant situation, it is therefore very surprising to discover that until recently hardly any detailed studies dealing with these unexpected residents of Europe had ever been carried out, making them amongst the most neglected and mysterious of all officially-recognized felids for many years.

The Majorcan wildcat (formally described from Santa Margarita in 1930, and originally named *F. s. jordansi*) has been hunted relentlessly by humans due to its liking for domestic chickens. According to James Parrack in *The Naturalist in Majorca* (1973), no recent record of trapped wildcats existed, and sight records are always rather ambiguous due to the plentiful supply of feral domestics on this island. However, Parrack noted that a few descriptions of clearly-observed, unusually large cats were on record, and he himself had seen one felid of characteristic wildcat build and tail near Puerto de Pollensa. Having said that, eminent mammal taxonomist Colin Groves subsequently measured the cranial volume of this cat form's holotype (type specimen), and found it to be that of the domestic cat *F. catus*. So it would appear that the Majorcan wildcat is not taxonomically distinct after all and is no longer believed even to be native here but to have been brought here by humans instead.

No less controversial is the Cretan wildcat *F. s. agrius*, which is variously believed to be a valid taxon, a feral domestic cat, or a hybrid of feral domestic cat and either African wildcat or European wildcat (both wildcat species exist on Crete). Irrespective of its identity, however, its numbers have also fallen, although not through hunting this time but instead through genetic dilution, due to interbreeding with feral domestic cats. If measures are not taken to preserve the pure-bred form (first described in 1905) in the very near future, it is inevitable that this wildcat form will soon be replaced totally by hybrids—always assuming, of course, that it is truly distinct taxonomically.

The wildcats of Sicily and Sardinia were traditionally classified together as a distinct subspecies, dubbed *F. s. sarda*. Since 2006, conversely, the Sicilian wildcat has been formally subsumed within the nominate subspecies of *F. silvestris* rather than retaining separate subspecific status. Moreover, by virtue of zooarchaeological discoveries first made public in 1992 by Jean-Denis Vigne in a *Mammal Review* paper, the Sardinian wildcat is now considered to have descended from domestic cats introduced here from the Near East around or during the 1st century AD.

Sardinian wildcat (Gurtuju/ Wikipedia)

Corsican Lynx and Corsican Wildcat

More mysterious for quite some time than these documented island forms was their Corsican counterpart, formerly assigned distinct subspecifc status as *F. s. reyi* because this French island's wildcat was long known only from a handful of specimens, all obtained within a few months.

December 1929 saw the publication of a paper by French zoologist Louis Lavauden in which he noted that although several previous articles had referred to the supposed existence on Corsica of a lynx, none had apparently mentioned the presence here of a wildcat form. While undertaking some research concerning this alleged lynx, however, Lavauden had received the skull and skin of a female wildcat, sent to him by a Mr. Rey-Jouvin (a teacher at the Lycée de Bastia), who had obtained it in February 1929 at the Forêt d'Aunes, south of Bastia.

The Corsican lynx was never discovered because Lavauden recognized that the wildcat before him constituted a hitherto unknown form, and concentrated his attention thereafter on this instead, resulting in his scientific paper. In honor of its discoverer, he named the new wildcat *Felis reyi* (demoted to *F. s. reyi* in 1951 by Reginald I. Pocock in his previously cited felid taxonomic review).

Lavauden noted that it differed by way of its very short tail, lack of russet shading on the backs of its ears, and very dark pelage from the Sardinian wildcat; and from the continental European wildcats by way of its dark coat once again, its slender body, and a black mark on the sole of each hind foot that extended between the toes. This latter mark seemed especially significant because it is a morphological characteristic of African wildcats, and Lavauden suggested that the Corsican form was indeed most closely related to these, despite its European locality.

Some time after having sent the first skin and skull to Lavauden, Rey-Jouvin obtained two further Corsican wildcat skins, which he passed on to the Muséum de Grenoble at Isère. According to Rey-Jouvin and also a Mr. Rotges (warden of the lakes and forests at Ajaccio at that time), the Corsican wildcat was not rare and could be encountered not only in the mountains and forests but also in the hilly scrub and the thickets of the open country. In his paper, Lavauden suggested that the reasons for this felid form not having been reported earlier were that it must be very shy of humans and that its pelage was in any case not of sufficient worth to warrant any active hunting. In 1986, moreover, two specimens were procured, and the Corsican wildcat's official rediscovery was duly recorded two years later via a *Mammalia* paper authored by J. Arrighi and M. Salotti.

In 1992, however, the zooarchaeological study by Vigne noted earlier that revealed the Sardinian wildcat to be merely a descendant of introduced domestic cats from the Near East drew the same conclusion relative to the Corsican wildcat, so it was subsequently reclassified accordingly. As for the Corsican lynx, conversely, it remains just as mysterious today as it was back in Lavauden's time.

Then again, so too it would seem is the Corsican wildcat because a most bewildering announcement was made in media reports published worldwide during June 2019. The reports stated that following studies of DNA samples taken from 12 recently examined specimens, this controversial felid (dubbed the cat-fox in these reports due to its bushy tail) may actually constitute a new species. But even more baffling was the claim that although Corsican shepherds had long known about such creatures, the first Corsican wildcat specimen was not procured until as recently as 2008! Needless to say, this bizarre statement entirely ignores all of the documented findings and writings of Lavauden and Rey-Jouvin concerning this cat form that date back as far as 1929 (not to mention the two specimens from 1986 recorded by Arrighi and Salotti). All very mystifying.

Ile du Levant Wildcat

No less mysterious is the wildcat form inhabiting the Ile du Levant, one of the Iles d'Hyères, located off the coast of the Var, France. Although there is no question that it exists, no specimen has apparently been submitted for scientific examination, hence it continues to evade formal recognition. Nevertheless, one fact is well-established concerning this elusive felid—it attains a considerable size. In 1932, veteran cryptozoologist Bernard Heuvelmans recorded one specimen that seemed to the local inhabitants who observed it to be so large in size that they named it the lynx of Paille. Moreover, during World War II, several specimens were actually snared in rabbit traps, and some of these allegedly weighed over 22 pounds. If only they had been sent to a museum! Heuvelmans himself spied one specimen several times in 1958, observed while it savagely attacked some feral domestic cats.

As no specimen has apparently been formally examined so far, the precise taxonomic status of the Ile du Levant wildcat is unknown. In addition, as this island lies much closer to the European mainland than do the other wildcat-inhabited Mediterranean islands noted here, it is particularly difficult to speculate whether its mystifying felid form is most closely allied to the African wildcat or to the European. Regardless of which of these two species lays the greater claim to its taxonomic allegiance, however, the Ile du Levant wildcat was believed by Heuvelmans almost certainly to constitute a distinct subspecies for which he even proposed *F. s. levantina* as a suitable name. Yet in light of the Cat Classification Task Force's recent respective mass lumping of both wildcat species' subspecies into just a very small number of distinct ones, it is highly unlikely that Heuvelmans's proposed name and subspecific delineation for the Ile du Levant wildcat would be accepted today as warranted.

Irish Wildcat

Unlike Great Britain, relatively few reports emerge from Ireland concerning mystery cats of the very large puma-like or black panther-like varieties. Yet this island is far from bereft of feline mystery on account of the Irish wildcat. It was traditionally thought that wildcats had never existed in the Emerald Isle, but in more recent times fossil evidence has emerged to confirm that such cats did indeed exist here up until approximately 3,000 years ago. Moreover, there is a sizable albeit highly controversial archive of reports on file claiming that bona fide Irish wildcats have actually been sighted right up to and including the present day. More remarkable still is that these reports of alleged Irish wildcats have suggested a closer relationship for these felids to the African wildcat than to the European.

Some of the material in support of Irish wildcats stretches back centuries, interwoven with ancient Celtic mythology. For example, an archaic poem believed to date from the 9th century (translated by Eugene O'Curry and published by Sir William Wilde) tells of the Irish hero Fin mac Cumhaill being held captive by the king of Erinn, Cormac mac Art, who pledged to free him only if a male and female of every species of wild animal inhabiting Ireland were brought to him at the ancient city of Tara. The poem subsequently lists many different animal forms, including a pair of cats brought from the cave of Cruachain.

African wildcat—once thought to be the identity of supposed Irish wildcats (copyright free)

While on the subject of Irish mythology: in a *Field* article of 6 December 1941, Irish writer Patrick Chalmers argued that despite the wildcat supposedly being unknown in both Ireland and the Western Islands, the warrior King Cairbar of Connacht was surely called "Cinn Chait" on account of the pelt of wildcat that he bore on his casque.

This leads to another aspect of the Irish wildcat mystery. Chalmers's comments drew a response by letter from A. MacDermott, who maintained that this cat form has never existed in Ireland, and that in his own boyhood the name "wildcat" was actually applied not to any felid but instead to the pine marten *Martes martes*, an arboreal relative of the weasel.

Compare this with information obtained a century earlier by William Andrews. As briefly noted by John R. Kinahan at the very end of a *Proceedings of the Natural History Society of Dublin* communication on 9 December 1853, Andrews had discovered that the inhabitants of the Fiadhghlenana or wild, remote glens of Kerry's western reaches knew of both the pine marten and an apparently genuine wildcat form.

They even had separate names for the two creatures, calling the marten "cat crann" ("tree cat" or "cat of the woods") and the wildcat "fiadhachd" ("hunting cat"), thereby destroying MacDermott's notion that the Irish wildcat was nothing more than the result of an etymological ambiguity.

The usage of "hunting cat" in Ireland relative to supposed wildcats was also noted within the mammalian tome of the Reverend J.G. Wood's three-volume *Illustrated Natural History* (1859-63), together with an anecdotal account taken from *Notes on the Irish Mammalia* by the well-known Irish naturalist William Thompson. After having noted on several occasions grouse feathers strewn near a water-break in his Irish beat, as well as a number of grouse corpses beheaded but otherwise undamaged, the gamekeeper responsible for that area set a trap and caught two specimens of what appeared to be bona fide wildcats, one adult and one juvenile.

Pine marten (public domain)

Thompson had taken a particular interest in reports of alleged wildcat sightings in Ireland, notably in the mountains of Erris in the county of Mayo. He had himself seen a very large cat, weighing 10 pounds 9 ounces, which had been shot in the wild at Shane's Castle park, County Antrim. Apparently this specimen resembled the European wildcat in every way except for its tail (which was not bushy at its tip like the European wildcat's) and its fur (which was of a finer texture). Consequently, when the *Larne Journal* reported in February 1839 that the wildcat occurred in Tullamore Park and also used to be found along the shores of Ballintrae, Thompson naturally sought out further details. He questioned Lord Roden's gamekeeper, who informed him that he had never seen wildcats in Ireland.

All through his researches, Thompson encountered similar conflicts of opinion on this subject, and even his own ultimate conclusion is somewhat paradoxical. Even though he never became entirely convinced (at least not in print) of the wildcat's occurrence here, after comparing the Shane's Castle specimen with two Scottish wildcats Thompson nonetheless offered the opinion that it was probably a wildcat x domestic cat hybrid. Needless to say, however, in order for a wildcat hybrid to occur in Ireland, there must be pure-bred wildcats there in the first place!

In his own *Illustrated Natural History*, Wood mentioned that William H. Maxwell's book *Wild Sports of the West* (1838) contained several accounts concerning a fierce, wild-living felid form in Ireland that was depopulating the rabbit warrens. Apparently, one of these cats was killed after a severe battle, and was, according to Wood:

> ...of a dirty-grey colour, double the size of the common house Cat, and its teeth and claws more than proportionately larger. This specimen was a female, which had been traced to a burrow under the rock, and caught in a rabbit-net. With her powerful teeth and claws she tore her way through the net, but was gallantly seized by the lad who set the toils. Upon him she turned her energies, and bit and scratched in a most savage style until she was despatched by a blow from a spade.

Although certainly fierce, feral domestics typically do not attain sizes larger than their tame counterparts (although as noted elsewhere in this book, in recent years evidence has begun to accrue that extra-large melanistic ferals may be responsible in Britain and various overseas regions for certain sightings of unidentified medium-sized pantheresque beasts). Similarly, in his own wildcat write-up, Wood did not attempt to ally ferals with the much larger and mysterious cat form typified by the beast in the incident documented by Maxwell. In any case, both in Wood's time and in the present day, feral domestics are very familiar animals in Ireland, not likely to be mistaken for anything else.

In or around 1883, while shooting rabbits near County Galway's Annaghdown, F.C. Wallace sighted an animal that in his opinion seemed to be a magnificent specimen of a genuine wildcat. As no physical evidence was obtained, however, no formal identification could be made.

If the Irish wildcat controversy were ever to be resolved, it was evident that a specimen would have to be procured and submitted for official scientific examination. Such an event appeared, at least initially, to have finally taken place in 1885. For on 28 January of that year, English naturalist William B. Tegetmeier exhibited the skin of an alleged wildcat from Donegal at a meeting of the Zoological Society of London and subsequently permitted English zoologist Edward Hamilton to examine it thoroughly. In the Society's *Proceedings* for 3 March 1885, Hamilton's report on this specimen was published. In it he unhesitatingly classified the skin as merely a feral domestic cat's and included excerpts from letters by earlier researchers interested in the Irish wildcat saga, all supporting his own belief that the latter felid did not exist.

Irish wildcat R.I.P.? Not quite, because a most unexpected discovery was made just a few years later that added a completely new dimension to the mystery. In the report of the Irish Cave Committee sent to the British Association meeting in 1904, zoologist Robert F. Scharff, specializing in Irish fauna, announced that he had discovered among a collection of felid sub-fossil remains obtained from the Edendale and Newhall Caves near County Clare's Ennis, a number that constituted two distinct series, one small in size, the other larger, and that he considered the larger to be of wildcat identity. Moreover, in a short report published by the *Irish Naturalist* during April 1905, Scharff dramatically reopened the case for the modern Irish wildcat by stating that the position and nature of the bones found suggested that the felid was not long extinct in Ireland, and that it was even possible that a few specimens still survived in the western regions' more remote mountainous areas.

Scharff then went on to comment that, until then, it had always been assumed that if a wildcat did actually exist in Ireland it would naturally belong to the Scottish form. However, as a startling climax in his report, Scharff disclosed that the County Clare cave remains were comparable not with the Scottish but with the African wildcat, and that its tail was not bushy at its tip but pointed—just like that of the sizable cat observed by William Thompson.

This complete turnabout in the tale of the Irish wildcat resulted in a series of letters on the subject by other interested parties appearing in the *Irish Naturalist* during subsequent months. Some received Scharff's findings favorably and contributed further news regarding the Irish wildcat; others were more critical and remained skeptical.

For example, R. Welch related an old fisherman's account originally given to Irish entomologist William F. de Vismes Kane concerning the wildcat's supposed existence in some numbers on the banks of Lackagh. Conversely, Robert Warren poured cold water on this report, arguing that if such cats were indeed so common within this wild and little-traversed region, then some representatives should still be there today. Yet as noted later by Scharff, in view of the similarity between feral domestic cats and African wildcats (both having tapering tails), perhaps they are.

Warren also attested that the finding of sub-fossil bones of a wildcat in Ireland did not prove that the wildcat was a native of Ireland. This was a quite paradoxical statement to say the least, which the editors of the *Irish Naturalist* were swift to point out in a footnote at the end of his letter of June 1905.

The following year, Scharff published his findings as a scientific paper in the *Proceedings of the Royal Irish Academy, Series B*, describing fully the unearthed remains. Meticulous comparisons made by Scharff between these and specimens of Scottish and African wildcats were also included, and which demonstrated conclusively in Scharff's opinion that the Clare cave remains were indeed most closely related to the latter wildcat.

Similarly, Major Gerald E.H. Barrett-Hamilton, a prominent British-Irish natural historian, had planned to include historical evidence favoring this felid's existence in a forthcoming book on British mammals. Unfortunately, however, his untimely death in his early 40s prevented this data from being published.

Nevertheless, one would have expected Scharff's researches to have been of sufficient importance in themselves to have initiated a new surge of interest and investigation regarding the Irish wildcat enigma. Instead, the possible existence of modern-day wildcats living in Ireland is nowadays totally dismissed—but why?

In an *Irish Naturalists' Journal* paper of July 1965, Belfast-born wildlife authority Arthur W. Stelfox re-examined Scharff's findings and offered a very different explanation for them. First of all, he considered that Scharff was too willing to accept anecdotal evidence of Irish wildcats unconditionally, and gave the Tegetmeier specimen as an example, which Scharff had used in support of their existence (even though it had been denounced as a feral domestic by Hamilton). And as far as the cave remains were concerned, Stelfox was convinced that a much simpler explanation than Scharff's was available for them. Namely, that instead of the smaller ones being domestic cats and the larger ones being wildcats, both sets were of domestic identity—the smaller being females and the larger being males (especially as the two sets were found at the same geological level and in the same mineralized condition).

Stelfox also noted that although one would expect remains of fossil wildcats to be associated with those of other wild fossil species of mammal in Ireland if it did indeed harbor wildcats at one time, no such find had been discovered. Instead, all cat remains known from Ireland had occurred only at levels where the bones of domesticated mammals had been found, and Stelfox reported that he had not uncovered evidence of any cat remains in Ireland dating back further than the Bronze Age.

However, this is no longer true, as revealed in an extensive *Quaternary Science Reviews* paper published in 2014, whose co-authors included Queen's University Belfast biologist W. Ian Montgomery and the University of Manchester's veteran British mammals expert Derek W. Yalden. They disclosed that fossil European wildcat remains dating variously from 9,000 to 3,000 years old had indeed been discovered in Ireland.

Moreover, sightings of large felids not readily explained away as feral domestics have continued to emerge from Ireland. In 1968, for example, while seeking lake monsters in western Ireland, Captain Lionel Leslie and his team were taken aback when a very sizable felid suddenly appeared on the opposite side of Connemara's Lough Nahooin (i.e. only about 100 yards away) from where they were standing. According to team member F.W. Holiday, who subsequently documented this encounter in his book *The Goblin Universe* (1986), Captain Leslie stated afterwards that he had never seen anything like it before. Dublin zoologists later contacted by Leslie concerning this sighting were equally bemused—a suggestion that it might simply have been a fox was flatly rejected by the team.

More recently, in their excellent book *The Mystery Animals of Ireland* (2010), authors Gary Cunningham and Ronan Coghlan noted that in May 2003 a very elderly man from Connemara named Francis Burke affirmed that wildcats had been reported here. He claimed that they were bigger than domestic cats and occurred mostly in wooded areas.

They also recorded a sighting dating from as recently as two decades ago that indicates wildcat-like felids (possibly even hybrids of original pure-bred wildcats and domestics) may still exist in Ireland:

> In February 2002, Sandra Garvey saw an animal while driving at night at Knockfune (Co Tipperary) which shocked her so much she nearly drove off the road. She described it as larger than your average moggy with a very striking tail. It transpired that Mrs Garvey's sighting was not an isolated one, with eyewitnesses coming forth, including park ranger Jimmy Greene who spotted such an animal with its two kittens while patrolling the Slieve Bloom Mountains in Co Offaly.

Appended to this report in their book was a detailed drawing by Gary Cunningham based upon Sandra Garvey's description of the cat that she had spied that evening, and the result is a burly tabby-striped felid with a sharply-pointed tail that looks very like a bona fide African wildcat. Even today, therefore, it would seem that with leprechaun-like elusiveness, the Irish wildcat continues to evade explanation.

Cave Lions and Scimitar Cats

Unlike its wildcats and lynxes, Europe's larger Pleistocene felids have long since vanished. Even so, various theories and discoveries suggest that some such extinctions may have occurred far more recently than previously supposed.

From fossil remains, paleontologists know that the European cave lion *Panthera (leo) spelaea* was still alive at the time of the Upper Palaeolithic artists (about 40,000 years ago), becoming extinct around 27,000 years later. Although not commonly portrayed in their cave paintings, some pictures of this felid are known, including the four painted and engraved frontal views in the Cave de Trois Frères, France, and also a splendid frieze of lions (again engraved) from the Cave de la Vache at Ariège. One of the lions shown in this latter depiction possesses a well-defined tail tuft, thereby eliminating any possibility that it represents a tiger; at one time, some paleontologists considered that the cave lion may really have been a cave tiger.

Cave lion, as depicted in a modern-day reconstruction (Semhur/ Wikipedia)

The frieze at the Cave de la Vache is approximately 11,000 years old. However, it has been postulated that not only the lions still existing in Greece in 480 BC (some of which famously attacked the baggage train of Xerxes during his march through Macedonia) but also some of the Assyrian lions of the ancient monuments may actually have been cave lions rather than any modern-day subspecies.

Equally, there is a most captivating piece of evidence to support the claim that another of Europe's larger Pleistocene felids may also have survived several millennia longer than many paleontologists once thought. During the Upper Palaeolithic, our ancestor, Cro-Magnon Man, had attained a level of culture termed the Aurignacian (flourished 37,000-33,000 years ago), from which date some of the earliest of humankind's artistic creations, including not only cave paintings but also sculptures. In 1896, an Aurignacian stone statuette measuring about 6.5 inches long was discovered in a Pyrenean grotto at Isturitz. Although slightly damaged (the limbs' lower portions were missing), it was still recognizable as a sculpture of some form of felid. A photograph was taken of this intriguing item—very fortunately, as it turned out, because shortly afterwards the statuette itself was lost and has never been recorded since.

The creature that it portrayed had a very large heavy head, rather short body, long powerful limbs, and short tail. The most noticeable feature, however, was its lower jaw, which was very powerful, heavy, and exceedingly deep

For a long time after the statuette's discovery, paleontologists simply assumed it to be a representation of the cave lion, but Czechoslovakian scientist Vratislav Mazak was not at all satisfied with this classification. The felid's curiously-shaped lower jaw, as well as its short tail (which appeared to have been deliberately created as such by the Paleolithic sculpturer concerned, rather than being the result of damage to an originally longer sculptured tail), did not agree with a cave lion identity. The more that Mazak studied the photo of the statuette, the more he came to realize that instead of a cave lion, it most closely resembled another, very different Pleistocene felid. Namely, the scimitar cat *Homotherium latidens*. This was a species of saber-tooth (machairodontid) that existed until the late Pleistocene in Great Britain but died out rather earlier on the European mainland (as did its larger relative *H. sainzelli*—although some researchers deem this latter species to be conspecific with *H. latidens*).

Scimitar cat—a modern-day reconstruction (Karl Shuker)

The scimitar cat was a most peculiar animal. Based upon paleontological evidence, in life it would have borne a superficial resemblance to a giant lynx, but it was instantly distinguished from this felid and even from mainstream saber-tooths by its teeth.

The upper canines of scimitar cats were not enlarged to such a notable extent as those of other saber-tooths. Instead, they were very curved, extremely flattened, and razor-sharp, as they were used not for stabbing but for slashing and slicing. In order to house these most unusual upper teeth, the lower jaw was deep in shape and curved down on each side to serve as a pair of sheaths for them. This latter modification was peculiar to the machairodontids; none of the true felids (e.g. lion, leopard, tiger, lynx, etc.) display this because they have no need of such—their upper canines are much smaller

Thus, the appearance of the scimitar cat's lower jaw as verified from fossil evidence very closely matches that depicted by the Isturitz statuette and, in combination with the other morphological similarities of the latter with this felid, supports Mazak's exciting theory. Moreover, if this is indeed correct, it provides evidence for believing that on mainland Europe the scimitar cat did not die out about 200,000 years ago (as hitherto believed), but instead lingered for at least a further 165,000 years.

Mazak's paper, published in 1970, attracted the interest of French scientist Michel Rousseau, who published his own account and views on the Isturitz statuette the following year. In this, he came out in favor of Mazak's belief but felt that the subject warranted a more detailed investigation—which he himself supplied at the end of 1971, by way of a formal scientific paper published in *Mammalia*. In this, Rousseau not only re-explored in depth the degree of morphological correspondence apparent between the statuette and the scimitar cat, but he also documented details of various other Paleolithic works of art that could possibly represent this felid species. One of these consists of an engraving on a fragment of stone, again from Isturitz, which depicts the shoulders and head of a felid with a chin even deeper than the jaw of the statuette's cat. An engraving on limestone at Badegoule portrays three partly superimposed profiles of animal heads; they share the eye and ears but the muzzles are separate, of which two have deep chins. And an engraving at the Cave de Trois Frères displays the profile of what appears to be the head of a lion, but possessing a fairly deep lower jaw, although less prominent than the previous examples.

The problem to be faced when attempting to categorize these as scimitar cat representations, however, transcends mere morphology. For they all belong to time periods even more recent than that of the Isturitz statuette, and during which (at least as far as Rousseau was concerned) the survival of the scimitar cat would therefore have been impossible. Indeed, for a long time there was no stratigraphical evidence even to support this felid's existence during the period when the Isturitz statuette was made, let alone during even more recent times.

However, that changed dramatically in March 2000, when the partial lower jaw of a scimitar cat was trawled by the fishing vessel UK33 from an area southeast of the Brown Bank in the North Sea, an area known for yielding mammalian fossil remains dating from the Pleistocene and Holocene epochs. When this jaw was subjected to radiocarbon analysis, it was found to date back to a mere 28,000 years BP (*Journal of Paleontology*, 2003), i.e. slightly more recent than the age of the Isturitz statuette, thereby adding weight to the possibility that this enigmatic figurine truly was a representation of *Homotherium*.

And here the tale, unfinished, currently remains. Were the above works of Paleolithic art nothing more than inaccurate or idealized depictions of cave lions, or do they constitute genuine proof that the extraordinary scimitar cat was a contemporary of our ancestors for a far longer period of time than hitherto believed?

An intriguing footnote on this subject of saber-tooth iconography was supplied by Charles Berlitz in his book *Atlantis* (1984), noting that on an ancient piece of Scythian goldwork (thus dating from approximately 12,000 years ago) a struggle is portrayed between hunters and a beast that markedly resembles a saber-tooth. Yet the most recent species of mainstream saber-tooth in Eurasia, *Megantereon inexpectatus*, died out at least 500,000 years ago (and the related *M. megantereon* even earlier), according to current belief. Evidently the artistic creations of ancient humans are worthy of close attention not only from the archaeologist and paleontologist but also from the cryptozoologist!

Pumas, Lions, Panthers, and other Out-of-Place Oddities

Just like Great Britain, continental Europe continues to file reports on a wide range of mystery felids. One of the best sources of material dealing with such creatures, including some notable medieval examples too, is Véronique Campion-Vincent's book *Des Fauves Dans Nos Campagnes: Légendes, Rumeurs et Apparitions* (1992). Another useful source, albeit one that examines the subject more from a folkloric than a cryptozoological standpoint, is *Les Félins-Mystère: Sur Les Traces d'Un Mythe Moderne* (1984) by Jean-Louis Brodu and Michel Meurger.

Also worth a read is Karl-Hans Taake's book *The Gévaudan Tragedy: The Disastrous Campaign of a Deported 'Beast'* (2015), in which he proposed that rather than a bloodthirsty wolf, an escapee hyaena, or a human serial killer (or even all three), as most commonly suggested by investigators of this gruesome historical episode, the ravaging Gévaudan

Beast that reputedly attacked more than 200 people in south-central France during the 1760s was actually a lion on the loose:

> The description of size, appearance, behaviour, strength—it all fits together: the comparison of size with a one-year-old bovine animal and a donkey; flat head; reddish fur; a dark line along the spine occasionally occurring in lions; spots on the sides of the body that appear especially in young lions; a tail which appears to be strangely thin (since shorthaired); a tassel on the tail; enormous strength that allowed the animal to carry off adult humans and to split human skulls as well as to jump nine metres [30 ft]; the use of a rough tongue to scrape tissue from the skull so that it appeared as if it were polished; roaring calls described as terrible barking or as dull sounds like from a dog that tries to bark; a paw print of 16 centimetres [5.5 in] length; using claws during an attack; throttling victims, that is: killing by interrupting the air flow; a preference for the open country.

Having studied this cryptozoological case in great detail over the years, however, I am by no means convinced by his hypothesis, as in my view the evidence points much closer towards the hyaena identity (*ShukerNature*, 22 August 2015).

In modern times, the mystery cats of continental Europe are strongly reminiscent of exotic species such as pumas, lions, and black panthers. So closely, in fact, do these European enigmas correspond with their British counterparts in terms of both appearance and behavior, and so comparably do the explanations offered in Chapter 1 for the origins and identities of the British cats apply to their European equivalents too, that there seems little point in devoting too much space here to continental examples, as this would simply involve repetition.

Gévaudan Beast in 18th-century depiction (public domain)

Even so, certain cases are worthy of brief mention. For example, 1927 saw reports of the discovery of a non-fossilized (i.e. modern-day) leopard skull within a cave in Ariège, France. Its origin was unknown. Two decades later, an outbreak of mystery cat accounts and stock farm killings hit the headlines in central France.

Not only was a "lion" killing sheep, but wolf-like, lioness-like, and even leopard-like beasts were also being reported during the later 1940s, all from this same region. All of these were frequently lumped together to yield a protean-like mystery creature dubbed the beast of Cezallier. In 1951, a genuine wolf was killed near to Grandrieu, Upper Loire, and the Cezallier Beast episode came to an end. Nevertheless, it is rather difficult to believe that a wolf could be responsible for the mystery cat sightings too. More likely is that it served as a convenient "official" solution to them.

A similar French episode occurred in 1978, involving a pair of panther-like beasts sighted in the Vosges area; whereas the equally elusive Beast of Valescure (also nicknamed the Beast of Esterel and classed as a lioness, wolf, puma, and various other carnivores) occupied newspaper headlines during February 1983 but was never caught, and eventually disappeared. At much the same time, German zoologists were examining the tracks of a puma-like animal sighted in Hanover; sightings surfaced again briefly in 1985, but the mystery was never resolved (1982 had seen a similar outbreak of German puma reports, but centered instead around Hamburg).

Italy's contribution to mainland Europe's cryptic cats includes a pair of pumas reported in 1983 at Bari; whereas Switzerland's Graubunden was the scene in 1974 for the observation by two hunters of a supposed tiger! The history and sociological implications behind various other modern-day continental cases, such as the Pornic panther and especially the Noth Beast, are dealt with succinctly by Jean-Louis Brodu and Michel Meurger in *Les Félins-Mystère* (1984).

Perhaps the single most memorable and extensively-publicized European mystery cat episode to occur since the publication of this book's original edition in 1989 took place 30 years after that, in northern France. Moreover, it certainly deserves inclusion here because it featured something infinitely more tangible than anecdotal evidence.

In mid-September 2019, residents in Armentières, near Lille, were amazed to see a juvenile black panther (melanistic leopard) walking along a second-floor ledge near the roof of a tall building and stepping in and out of an open window for almost an hour. Nor was this a misidentified black domestic cat (as has transpired so often in the past with supposed panther encounters in Europe and other locations where such cats should just not be), because it was eventually captured alive by the local fire brigade and conclusively identified as a black panther after a veterinary surgeon had successfully subdued it with a tranquilizer dart.

Weighing approximately 45 pounds, female, and estimated to be 4-5 months old, this young panther was the size of a small Labrador but was very tame, with clipped claws, and was in good health. Clearly an exotic escapee/released pet, it was transferred to an undisclosed location while police sought its owner, but the saga had a final twist still in store. Just a week later, media reports revealed that the panther had been stolen from where it was being held—a zoo in Maubeuge, whose staff coming into work on that fateful morning were shocked to find the gate of its pen broken and the panther gone. As far as I'm aware, it has not been rediscovered, so Armentières's feline anomaly remains precisely that—anomalous.

Not all recent accounts of out-of-place continental felids involve non-European species; some refer to species reappearing in localities from which they supposedly died out many decades before. In Jura, for example, where the wildcat has officially been extinct for many years, cats continue to appear which, when scientifically examined, prove not to be feral domestics but instead genuine wildcats. Peter Lüps of the Berne Natural History Museum has documented several examples; it is likely that they are migrants from bordering France. Similarly, the last reliable report of a lynx in the Italian Alps dated from 1915—until 1983, that is, when a new record of a lynx was made from Belzano in the eastern Alps. And in Switzerland, where the last formally documented lynx was officially taken in 1872 (a few unconfirmed later reports also exist), continuing reports of roe deer and small stock kills by mysterious predators since World War II suggest that the lynx may be re-establishing itself here too, always assuming, of course, that it ever died out to begin with.

Russian Mystery Beast

During the 1980s, Britain's most publicized mystery felids were the Exmoor Beast and the Kellas cat. In the former Soviet Union, comparable creatures were active many years earlier. We met *Felis daemon*, a Transcaucasian counterpart of sorts to the Kellas cat, at the beginning of this chapter. Now let us bring it to a close with an examination of an eastern European equivalent of the Exmoor Beast—the unaccountable carnivore of Orel Province.

The saga reached Britain quite inauspiciously, via a brief note contained in *The Field* for 12 August 1893:

> For some days past a panther, which escaped from the grounds of a landed proprietor who had a fancy for keeping wild animals, has been at large in the province of Orel, Russia, to the terror of the villagers. The panther has already killed six persons; and although repeated battues have been organised by various sportsmen of the district, they have not succeeded in slaying the beast.

A very unpleasant but nonetheless unremarkable incident.

In the months to come, however, it was to transform into a thoroughly bewildering affair. Whole pages were soon devoted to this subject, yet few real developments occurred. Sightings of the creature and attacks upon people and animals continued, but repeated searches by soldiers following instructions from the Province's Governor, together with interested sportsmen from near and far (plus contingents of the local populace), all failed totally in their attempts to shoot or snare it. By the end of the year, however, the sightings had petered out, and nothing further was heard of Orel's mystery assailant. Apparently the creature ate two poisoned sheep and disappeared, its tracks leading into the forest beyond the River Vetebet.

A black panther (melanistic leopard) absconding from captivity may explain Russia's Orel Beast (Karl Shuker)

Thus it was never caught, which is a great tragedy because by far the most extraordinary aspect of this whole incident is the thorough confusion that reigned concerning the creature's supposed morphology. Despite many sightings having been reported, no-one seemed at all sure of the its identity—a fact underlined by the shooting of animals as diverse as dogs, hens(!), and even a striped pig by soldiers who began to see savage carnivores everywhere.

For example, the panther identity soon gave way to a lynx, a wolf, and (helped along by a sighting made by the commander of a massive 1000-man beat) a tiger. As it happens, tigers are not unknown in Russia; they are sometimes driven by starvation during harsh winters to prowling villages and towns (especially in the Russian Far East—very notably, in January 1987 two Amur (Siberian) tigers were shot dead on snow-covered streets in Vladivostok). And just over a century ago, tigers did exist not all that far south of the Orel region. Nevertheless, what makes a tiger identity particularly bizarre in the case of the Orel beast is that none of the eyewitnesses on record could actually recall seeing stripes on the animal. In view of this, it is hardly surprising that not everyone believed the tiger theory.

The Moscow sportsman Prince Shirinsky-Shikhmatov was invited to seek the creature, which he and his companions duly did, but without success. Nevertheless, he did observe spoor allegedly left behind by it; and based upon this, together with the manner of its attacks and the types of wounds that it inflicted upon its victims, he judged the animal to be a black panther. This was an opinion shared by R.G. Burton, who prepared a detailed history of the Orel Beast that appeared in *The Field* on 9 December 1893.

Conversely, another sportsman who had also been pursuing this selfsame mystery beast closely considered it to be nothing more than a large dog. Combining these schools of thought were the opinions of various other hunters, who felt that there were two totally different beasts involved: one large and one small. Worth noting is that in this area at that time dogs were often kept together with panthers in menageries. Perhaps this gave rise to a further belief—namely, that a dog and a panther had escaped together.

Certainly, little can be ascertained from the descriptions documented of the Orel Beast. In a *Field* report of 19 August 1893, it was described as being ". . . the height of a wolf, of a yellow color, with a blunt muzzle. His tracks are round, like a wolf's, about 3½ inches in diameter, but the claws imprint pointed tracks." In Burton's later account, he added that its ears were round and pricked, and its tail was long, smooth, and hanging, whereas in overall size the beast appeared not to have been very large or remarkably powerful, no more so than a dog. It is quite likely that the original creature was an escaped felid but later became confused with sightings of large dogs (perhaps even wolves), and ultimately disappeared out of the district, with its presence replaced thereafter by canids. Unfortunately, as no scientific investigation was put into operation, there is little hope that this baffling case will ever be solved.

In 1939, Sir Winston Churchill described the action of Russia in relation to World War II as: "A riddle wrapped in a mystery inside an enigma." The same could also be said of its mystery beast from Orel Province!

Chapter 3
Asia: Multicolored Tigers and Mint-Leaf Leopards

Through the Jungle very softly flits a shadow and a sigh—
He is Fear, 0 Little Hunter, he is Fear!
— Rudyard Kipling, "The Song of the Little Hunter"

In terms of felid diversity, the continent of Asia is unrivaled, laying claim to a total of 16 species of Small Cat, 2 species of clouded leopard, and no less than 4 species of Big Cat (tiger, leopard, lion, and snow leopard). In addition, a tiny number of Asian cheetahs exist in the Middle East (including a small population rediscovered in Iran's Yeylaq mountain region, bordering Iraq, in 1988). Moreover, in November 1990 a specimen was sighted by Similipal Project Tiger's field director, S.P. Nagar, in India's Bulunda forest block (*Oryx*, April 1991); the cheetah had previously been deemed extinct in the wild in India.

However, as will now be revealed, some of these species may exhibit color morphs whose existence or genetic identity has yet to be recognized by science. Moreover, there could even be some totally novel, still-undiscovered Asian cat species too, eluding scientific detection and description.

Technicolor Tigers

Many of Asia's felids are striking, handsomely-marked creatures, but none more so than the tiger *Panthera tigris*, surely the living embodiment of Asia itself—exotic, elusive, majestic, mysterious. Eight modern-day subspecies have traditionally been recognized, ranging in shade and size from the pale Amur (Siberian) tiger *P. t. altaica* of average total length 11 ft to the dark orange Bali tiger *P. t. balica* not generally exceeding 7 feet. Having said that, in 2017 the Cat Classification Task Force of the IUCN Cat Specialist Group lumped most of these together to yield just two distinct subspecies—the continental tiger *P. t. tigris* and the Sunda Islands tiger *P. t. sondaica*.

In addition, there is much evidence to suggest that the "burning bright" black-and-orange pelage that illuminated the imagination of William Blake is far from being the only color combination that can be exhibited by the tiger.

White, Snow, Golden, Brown, Red, and Unstriped Tigers

The most familiar of all aberrant tigers are the famous white tigers of the old Indian State of Rewa (now part of Madhya Pradesh), well known from various priceless specimens displayed at several of the world's major zoos. These tigers are partial albinos and were originally thought to be homozygous for a recessive allele of the Full Color gene dubbed the chinchilla mutant allele, but this allele appears not to exist, although such tigers are still commonly referred to as chinchilla (or chinchilla-phenotype) white tigers.

However, on 23 May 2013, via a paper published (initially online) in the scientific journal *Current Biology*, a team of Chinese scientists that included Shu-Jin Luo of Beijing University revealed the long-awaited genetic basis of such tigers. Mapping a family of 16 tigers living in Chimelong Safari Park, which included both white and normal tiger specimens, the team discovered that the white coat coloration is caused by a single amino acid change, A477V, in a particular transporter protein known as SLC45A2 (which mediates pigment production), in turn induced by a recessive gene allele. This change inhibits the synthesis of red and yellow pigment (phaeomelanin), but not black (eumelanin), thereby explaining why white tigers still possess dark stripes.

Rewa-type white tiger (public domain)

In white tigers, their coat's background color is creamy white, and overlaid by ash-grey stripes; in addition, their eyes are blue. White tigers of this type, although most widely known from Rewa, have been documented from various other Indian States too, and even beyond India (e.g. China, the Korean peninsula), although sadly they have long been believed to be extinct in the wild state, killed off by trophy hunters. In July 2017, however, *The Hindu* and other media reported that a single white tiger had recently been observed and photographed by photographer Nilanjan Ray from Bengaluru in an undisclosed location within India's Nilgiri Biosphere Reserve, alongside a normal tiger. This suggests that the genetic mutation causing the above-mentioned amino acid change may have either reappeared spontaneously in the wild tiger population or even survived in it but had remained hidden until two wild tigers each carrying the recessive gene allele mated, yielding this new white tiger specimen.

Having been well documented and well studied, these "true" white tigers (i.e. A477V–engendered with chinchilla phenotype) are no longer anomalous. Conversely, the possible existence of complete albino tigers (i.e. homozygous for the Full Color gene's recessive complete albino mutant allele) is a subject still shrouded in mystery and controversy. In his definitive work *Le Règne Animal* [*The Animal Kingdom*] (1817; 1829-1830; 1836-1849), the eminent French naturalist Baron Georges Cuvier described a white tigress whose stripes were visible only when viewed from certain angles (i.e. shadow stripes), implying that this was a bona fide complete albino individual. A comparable specimen was on display at the Exeter Change Menagerie in London during the 1820s, and was described in the Reverend J.G. Wood's encyclopedic *Illustrated Natural History* (1959-63): "The colour of this animal was a creamy white, with the ordinary tigerine stripes so faintly marked that they were only visible in certain lights." Wood's tome also included an engraving of this specimen (reproduced here).

Complete albino tiger exhibited at Exeter Change Menagerie, 1820s (public domain)

Nevertheless, some years later the famous British zoologist Richard Lydekker still doubted whether such tigers really existed, a doubt that persisted in the minds of some other authorities too—until 1922. During that year, two pure white sub-adult tigers were shot in the former north-east Indian State of Cooch Behar. And as confirmation of their genetic status, they possessed the characteristic trademark of all complete albinos—pink eyes. For in such mutants (irrespective of species), even the pigment of the eye's iris is absent. Regrettably, the eye color of the Exeter Change specimen was never documented, but judging from its pelage morphology this tiger was probably a complete albino.

Having said that, there is another tiger morph that looks very like a complete albino tiger except for its eyes, which are not pink but blue, like those of "true" (A477V–engendered) white tigers. This tiger morph is called the snow tiger and totally or virtually lacks any degree of striping. Perhaps the most famous examples are the magnificent individuals owned by former American stage magicians Siegfried and Roy (d. May 8, 2020). They own a number of "true" white tigers too, as well as white lions, which can be seen on display in their private wildlife park, "The Secret Garden of Siegfried & Roy," at the Mirage Hotel on The Strip in Las Vegas. I stayed at the Mirage during a holiday there in 2004, and thus was able to see these exotic Big Cats personally. Other notable snow tigers exhibited in captivity include various specimens at Cincinnati Zoo (the first zoo, back in the 1980s, to have such animals); two cubs born at Liberec Zoo in the Czech Republic during the 1990s; and Artico, born in May 2004 to normal tiger parents at a wildlife refuge in Alicante, Spain. Snow tigers may conceivably arise via a specific amino acid change, but different from A477V that engenders "true" white tigers.

In addition to the "true" white tiger, the complete albino, and the snow tiger morphs described above, somewhat darker versions of white tiger also exist, in which the coat's background color is a very pale counterpart of that of normal tigers and the stripes are dark brown rather than black. These very unusual, pallid tigers are variously referred to loosely as red tigers, brown tigers, or leucistic tigers (although sometimes the true white tigers are also termed leucistic, a confusing state of affairs).

Snow tiger (Hodari Nundu)

A brown tiger with stripes only a little darker than its coat's background shading was recorded in 1929 by British zoologist Reginald I. Pocock in the *Journal of the Bombay Natural History Society*. According to felid geneticist Roy Robinson, this tiger's unusual coloration may have been due to expression of the Extension of black gene's non-extension mutant allele.

Interestingly, as I documented in my book *Cats of Magic, Mythology, and Mystery* (2012), a tiger color morph known variously as the golden tabby, ginger, or strawberry tiger also exists, and has come to public prominence due to several specimens having been exhibited in captivity recently. These very large animals are characterized by a mellow golden-hued pelage, patterned with faint, darker stripes, and complemented by snowy-white underparts. Such tigers are all either homozygous or heterozygous for the dominant wild-type allele of the Full Color gene, but some investigators have claimed that they are also likely to be homozygous for a gene dubbed the Wide Band gene, arguing that this would explain their paler, golden fur dorsally, their reduced striping, plus their snow-white underparts and lower flanks. In reality, however, there is no firm evidence that a Wide Band gene actually exists, even in domestic cats, let alone wild felid species, and I have yet to see any published scientific research verifying its presence in tigers. This genetic notion is explored with regard to golden tabby domestic cats as follows in [Roy] Robinson's *Cat Genetics for Breeders and Veterinarians* (1999):

> Non-silver agouti cats (ii) [i.e. homozygous for the recessive inhibitor-preventing non-silver allele of the Inhibitor of Colour gene, and either homozygous or heterozygous for the dominant agouti allele of the Agouti gene] bred from heterozygous chinchilla cats (Ii) are not identical to other tabbies...but are much brighter in color. They have been given the name of golden tabby, Chinchilla golden or shaded golden. In goldens, the agouti band of the tabby pattern is widened. Examination of hair's from goldens reveals that these are nearly all yellow with a dark tip and a slight gray undercolor at the base.

Although specimens of golden tabby tigers were formerly recorded in the wild in India, with records dating back to the onset of the 20th century, they suffered the same tragic fate as wild white tigers—shot by hunters as unusual trophies, the last just outside Mysore Pradesh during the early 1930s. Today, only around 30 individuals exist in zoos around the world, the first one being born to normal Bengal tiger parents in 1983 at Josip Marcan's Adriatic Animal Attractions in Deland, Florida. All of today's golden tabby tigers are descended from Bhim, a male white tiger sired by Tony, who was a part-Siberian white tiger.

As with white tigers, golden tabby specimens are unquestionably very eye-catching. Nevertheless, they have a price to pay for their strange beauty. As I learned from Big Cat specialist Graham Law, they suffer from weak pelvic girdles, thus rendering them less able to climb than their normal-colored brethren.

In a letter published by the London newspaper *The Times* on 16 October 1936, W.H. Carter commented that in one of the official gazetteers of Bengal, a local tiger form was mentioned that did not possess stripes at all but was instead uniformly brown. Carter suggested that this would serve as camouflage in the open sandy tracts of Sunderbans that it inhabits. Further data on tigers with reduced/absent striping was documented by Willy Schroeter in a *Sitzungsberichte der Gesellschaft Naturforschender Freunde zu Berlin* paper (1973).

In 1989, the *Indian Forester* published a short article written by S.R. Sagar (field director of Orissa's Similipal Tiger Reserve) and Lala A.K. Singh (the reserve's research officer), noting four different sightings of stripeless tigers. They took place between 1961 and 1988, and two of them (one in 1977, one in 1979, both in northern Similipal) were by the same person—a very experienced tracker named Shri Shailesh Ho, from the Kolha community. The most recent sighting occurred at 6:40 a.m. on 27 July 1988, when the guard at Brundaban (North Similipal) observed a stripeless tiger walking away from a salt lick, leaving an imperfect pug mark noticed by the guard when he inspected the area afterwards.

Blue Tigers

Even more exotic and ethereal than white, red, or even unstriped tigers is the concept of a blue tiger. Yet it would seem that such a felid morph exists not only in reverie but also in reality, at least as far as Harry R. Caldwell was concerned—because he almost shot one! Moreover, as both a Methodist missionary and a very experienced big-game hunter who bagged dozens of tigers during his stay in China, Caldwell was nothing if not a reliable eyewitness.

In September 1910, while in the Futsing region of Fujian (formerly Fukien) Province, southeastern China, he was watching a goat when an attendant directed his attention to something else:

> I glanced at the object, which appeared to be a man dressed in the conventional light-blue garment and crouching...I simply whispered to the cook, "Man," and again turned my attention to watching the goat. Again the cook tugged at my elbow, saying, "Tiger, surely a tiger," and I once more looked...Now focusing upon what I had altogether overlooked in my previous hurried glances, I saw the huge head of the tiger above the blue which had appeared to me to be the clothes of a man. What I had been looking at was the chest and belly of the beast.

Caldwell described this uncanny tiger as follows:

> The markings of the animal were marvellously beautiful. The ground colour seemed to be a deep shade of maltese, changing into almost deep blue on the under parts. The stripes were well defined, and so far as I was able to make out similar to those of a tiger of the regular type.

Just as Caldwell was about to pull the trigger on this near-fairytale felid (which he had aptly nicknamed Bluebeard), he perceived that it was clearly very interested in something in the ravine directly below the spot where it was sitting. Caldwell, however, was unable to spy the subject of the tiger's attention, so he leaned forward to obtain a better view and spotted two boys gathering bundles of vegetation in the ravine. Caldwell realized that if he shot the tiger from his present position, he may endanger the children, so he decided to move to a different position in order to alter the direction of his shot. This he did, but when he arrived there he discovered to his horror that the tiger had silently departed, leaving Caldwell with only a memory of his encounter with what must surely be the most marvelous of all mystery cats.

In fact, that was the second time that Caldwell had spied a blue tiger in Futsing. His first encounter had occurred in late spring of the same year, when one such cat had been attracted by a tethered goat that Caldwell (hiding in some bushes) had placed in a clearing as a lure. However, it did not approach close enough for him to be certain that if he shot it he would kill it outright, and an injured tiger would be even more aggressive and dangerous to the local people than a fit tiger. So he let it depart unscathed.

Caldwell noted in his book *Blue Tiger* (1925) that other sightings of such creatures had been reported in this region of China—later reiterated by Bernard Heuvelmans in his 1986 cryptozoological checklist.

In his own book, *Our Friends the Tigers* (1954), Caldwell's son, John C. Caldwell, recalled seeing on several occasions the beautiful maltese hairs of Futsing's evanescent blue tigers along the mountain trails when accompanying his father during his many searches for them. However, they never succeeded in bagging one of these "blue devils," the name given to such creatures by the locals on account of their alleged man-eating tendency. Speaking of names: in his comprehensive book *Mystery Creatures of China* (2018), David C. Xu stated that the blue tiger's native name is the *lanhu*, which translates literally as "blue tiger." He also claimed that in 1922 a hunter actually killed a bluish-greyish South China tiger within Fujian Province, but he made no mention of what happened to this potentially highly significant specimen.

In my 2012 cat book, I reported two more recent blue tiger reports, including a claimed sighting not in China but within the Korean Peninsula, made in 1952 by US Army Lieutenant-Colonel James McKee, who was stationed there during the Korean War.

Today, some zoologists doubt that such extraordinary animals could even exist, and skeptics have suggested that perhaps Bluebeard was merely a normal tiger that had rolled in bluish mud or had become stained while bathing in a pool polluted with some form of blue dye. Yet if the latter were true, surely other animals bathing or drinking there would have also become stained, but no reports of any such creatures have ever been reported. Also, such mud or dye would have obscured the tiger's stripes, yet according to Caldwell they were clearly visible on the blue tiger(s) that he observed.

Science has yet to examine one of Futsing's blue tigers, but taking into account Caldwell's precise description, this morph's genetic basis may be deducible. The domestic Maltese cat owes its smoky-blue pelage, termed "blue dilution," to a combination of two mutant alleles—the Agouti gene's recessive non-agouti mutant allele (responsible for melanism in leopards and various other wild cat species) and the Dense Pigmentation gene's recessive dilute (aka Maltese) mutant allele. It is this combination when present homozygously (i.e. with each of the two mutant alleles being represented by two copies as opposed to only one or none at all) that is responsible for the characteristic smoky blue shade of coat color in four domestic cat breeds—the Chartreux, British Blue, Russian Blue, and Korat. A tiger possessing this same combination homozygously is likely to sport a background pelage color corresponding closely to Caldwell's description of the Futsing blue tiger (this genetic scenario may also explain various confirmed pelts of blue lynxes and bobcats on record—see Chapter 5).

Its black stripes, however, pose something of a problem. As stated in Robinson's *Genetics for Cat Breeders and Veterinarians* (1999) regarding the coloration of the tabby markings in blue tabby domestics (which possess the recessive dilute mutant allele homozygously), the tabby markings are a slate blue. However, this does not constitute a direct comparison with the blue tiger because the blue tabby domestic possesses the dominant agouti allele for the Agouti gene, not the recessive non-agouti allele. The answer to the blue tiger's black stripes may either be that the stripes are not actually black but appeared so to Caldwell due to lighting conditions at the time of his sighting, or, if they are genuinely black, are of polygenic creation, i.e. they are due to the action of modifying genes functioning independently of the combined effect of the non-agouti and dilute alleles.

Finally, as documented in my 2012 cat book, what may have been a bona fide blue tiger was actually born in captivity! In the January

1998 issue of *Mainly About Animals*, a fact-filled natural history magazine edited by British zoologist Clinton Keeling, Dave Case of Bexleyheath in Kent, southern England, revealed that the May 1984 issue of *Zoonooz*, published by San Diego Zoo, contained the following tantalizing snippet written by Mary Van Nostrand: "Today, there are about 60 white tigers in zoos and private collections around the world, and one partially melanistic, smoky blue tiger born in Oklahoma City in 1964."

Dave Case lost no time in contacting the Oklahoma City Zoo to enquire about this remarkable tiger, but he learned from Don Whitton, the zoo's registrar, that far from being alive there, it had been a stillborn cub, born with two living normally-colored siblings on 16 October 1964. Oddly, the zoo's records listed it as "melanistic," and no photograph of it exists. So, had it survived, would this unique cub have been the first (and still the only) bona fide blue tiger ever exhibited in captivity, or would it have been a melanistic tiger, which, as will be seen, would have been just as extraordinary? Sadly, it seems unlikely that we shall ever know for sure.

Black Tigers

Melanism in felids is most frequently exhibited by tropical species, as exemplified by the leopard and jaguar. Consequently, one would expect the tiger to exhibit an all-black morph too. Yet, very surprisingly, not a single specimen of a melanistic tiger has ever been submitted for official scientific examination. Indeed, in recent books dealing with felids, mention of black tigers is either omitted or limited to a couple of lines in which alleged sightings of such cats are generally discounted as black panther misidentifications.

Supported as it is, however, by a very extensive history of observations and reports, the subject of black tigers is one that should not be dismissed so lightly, as now revealed here in what is the most extensive documentation of black tiger reports ever prepared.*

Probably the most significant black tiger record was provided by the noted 18th-century British artist James Forbes. In a *Country Life* article (17 September 1964), author/adventurer Charles Stonor noted that many watercolors by Forbes were housed within the museum of the seminary at Oscott College near Walsall in the West Midlands,

---

* Since its first publication in this present book's original 1989 edition, this documentation has been shamelessly copied and plagiarized by a number of online websites—as, indeed, has much else in that edition, sadly.

England. One of these watercolors, dating back to 1772, depicted in very great detail a black tiger; it was painted in south-west India during the period when Forbes was in the service of John Company (aka the British East India Company) there. A description of this felid featured in his painting also exists, contained in one of Forbes's own handwritten letters from that same period. He wrote:

> I have also the opportunity of adding the portrait of an extraordinary Tyger, shot a few months ago by the Nairs in this neighborhood, and presented to the Chief as a great curiosity. It was entirely black yet striped in the manner of the Royal-Tyger, with shades of a still darker hue, like the richest black, glossed with purple. My pencil is very deficient in displaying these mingled tints; nor do I know how to describe them better than by the difference you would observe in a black cloth variegated with shades of a rich velvet.

The current whereabouts of this intriguing painting are unknown. In October 1987, I learnt from Oscott College's then Rector, the Rt. Rev. Mgr Michael J. Kirkham, that in 1965 the entire Forbes collection was sold at Sotherby's, and no record appears to exist of the black tiger painting's new owner.

A follow-up letter concerning black tigers, written by Lieutenant-Colonel N.M. Hughes-Hallett, appeared in *Country Life* three weeks after Stonor's article and referred to a book entitled *Sophie in London*, 1786. This was the diary of German novelist Maria Sophie von La Roche, translated into English from the original German edition and published in 1933. In it, she recorded her impressions of a trip to the Tower of London's menagerie, mentioning lions, leopards, tigers, and wolves, and also noting: "The all-black Tiger, which Mr. Eastings brought with him from the East Indies is most handsome, but his tigery glance is horrible."

James Forbes's painting of a black tiger (public domain)

The next report that I have on file is a short item from London's *The Observer* newspaper for 27 January 1844. It records that a black tiger (originally intended as a present from the King of Java to Napoleon) was currently on display at Kendrick's collection of exotic animals opposite St James's Church, in Piccadilly, London.

Two years later came an account from the noted British naturalist C.T. Buckland. Early in 1846, a black tiger was reported from the Chittagong hills of what is now Bangladesh, which not only was raiding cattle but also had killed a local villager. In March, Buckland recorded in his diary that news had reached him of its dead body having been discovered, lying near to the edge of a road; it had been killed by a poisoned arrow. In the company of a number of other interested persons, Buckland set out to view it. Upon arrival, they inspected the carcass closely, and in a published letter of 23 March 1889 in *The Field* Buckland described it as follows:

> It was a full-sized tiger, and the skin was black, or very dark-brown, so that the stripes showed rather a darker black in the sunlight, just as the spots are visible on the skin of a black leopard...by the time that we arrived the carcase was swollen, the flies were buzzing about it, and decomposition had set in so that those of our party who knew best, decided that the skin could not be saved...Captain Swatman, who was in charge of the Government elephant kheddas, and Captain Hore (afterwards Lord Ruthven), of the 25th N.I., were well-known sportsmen, and had each of them killed many tigers. No doubt was expressed about the animal being a black tiger.

Moving into the 20th century, T.A. Hauxwell (then the Conservator of Forests at Maymyo in Burma, now Myanmar) recorded in 1914 the wounding of a supposed black tiger at a range of 15 yards by his son, in the Bhamo district. At first, his son took the all-black creature to be a pig, whereas his shikari insisted that it was a bear. Upon approaching very cautiously the area in which the growling wounded beast was moving, however, they observed enormous round pug-marks, each with a circumference measuring 20 inches, which in their judgement seemed closely similar to those of a tiger. Sadly, the body of this mysterious animal was never recovered.

On 25 October 1928, the editors of *The Field* noted that a dead black tiger had been reported in the Lushai Hills south of Assam in northeastern India. Frustratingly, however, this specimen's skin was too decomposed to be saved.

During the early 1930s, a reward was offered for the destruction of a savage creature inhabiting India's Imphal Manipur State. When eventually dispatched, the latter proved to be an adult black bear—but the Mikirs insisted that it was a black tiger! In the opinion of J.C. Higgins, however, this peculiar classification was largely influenced by the fact that the reward to be gained if the creature proved to be a tiger was more than twice that on offer for a bear!

Also during the 1930s, a series of letters concerning black tigers appeared in *The Times* (London). The earlier items dealt with cases noted above, but the later letters provided several additional examples, including the following:

British zoologist Captain Guy Dollman of London's Natural History Museum reported that a young black tiger had been shot in what was then British India's Central Provinces some years earlier; it had sported a dark brown background coat color overlain by black stripes. He also commented that a magnificent black tiger had been shot in 1915 by some natives in Assam east of Dibrugarh; of especial interest was the fact that this specimen was completely black, no hint of striping being visible even in sunlight.

Colonel S. Capper placed on record a very clear sighting of a black tiger made through a deer-stalker's telescope by himself and his hunting companion C.J. Maltby on 11 September 1895, on the Cardoman Hills of southern India. The tiger was initially lying on a rock at the jungle's edge, but eventually it rose and disappeared from sight into the jungle. When they reached this spot, Capper and Maltby discovered the clear pug-marks of a tiger; both men were well acquainted with tiger and leopard spoor.

The first few lines of a letter to *The Times* by R.G. Griffiths contained the news that science had been awaiting for so long—a black tiger had been captured alive! A colossal creature 12 foot long, it had been caught on 4 September 1936 in the vicinity of Dibrugarh in Assam. So far, so good. As one might have expected of a mystery felid, however, the truth—as Griffiths would revealed—was rather different. Further enquiries (made by Baini Prashad, Director of the Zoological Survey of India at Calcutta's Indian Museum) not only revealed that it was a lot less than 12 feet long but also discovered that it was not even a tiger at all. Instead, it was nothing more than an 8-foot-long black panther. This muddled melanistic was later sold and transferred to Calcutta, where Griffiths had recently seen it. Further details were given by Captain Dollman in a letter published by *The Times* on 4 February 1937.

Regrettably, it seems as if the misidentified black panther of Dibrugarh sounded the death knell for the bona fide black tigers. Until then, the existence of such cats had never been seriously doubted, but following this sorry episode the black tiger appeared to fall out of fashion.

In a letter published by *The Field* on 9 January 1937, A.A. Dunbar Brander of the British Indian Forest Service recorded that a couple of years earlier, the sportsman and Calcutta barrister J.A. Clough saw in the Deputy Commissioner's bungalow at Betul, Central Provinces, a tiger skin whose background color was chocolate brown. When questioned about this, the Deputy Commissioner mentioned that a second tiger of this type had been sighted in the area where his own specimen had been killed.

By and large, however, reports dried up, and the black tiger became yet another felid of fiction and fable rather than fact. To complete this account, I will come full circle, back to the 1964 *Country Life* article by Charles Stonor. In it, Stonor also included a personally collected report of particular interest. He recalled that while in the Mishmi Hills, at the foot of the Assam Himalayas, he learnt from a local headman that the Mishmi tribespeople know of an occasionally seen creature referred to as the bear tiger, which resembles a normal tiger in all respects but one—it is completely black. At Stonor's suggestion that it could simply be a panther, the headman reasserted vehemently that it was a black tiger, stating that not only was it of full tiger stature but also, when observed in sunlight, one could discern the familiar tiger pattern of stripes upon its otherwise inky coat.

During the late 1980s, I was informed by English correspondent Phil Bennett that in India during 1952, the people of one village complained that an exceedingly large black tiger was killing their cattle. According to these villagers, the cat in question was 20-24 feet long! Although the authorities searched for this extraordinary beast in vain, some unusually large pug-marks were found, which could not be explained. Further reports of this anomalous animal continued to emerge until 1967 (Bennett, pers. comm.).

With the exception of this last-noted colossus (whose length is clearly an immense exaggeration on the part of the locals; its size presumably increasing in direct proportion to their fear of it!), the alleged black tigers that figure in this collection of reports conform to a standard appearance. The only degree of variation on record concerns the pelage's background color—dark brown in some individuals, totally black in others—but such variation is commonly seen among melanistic individuals of other species too, including the leopard.

Is it possible, therefore, that genuine melanistic tigers really do occur? Let us first consider some other phenomena that could explain alleged black tiger sightings.

For example, might they be, as with the captured Dibrugarh specimen from 1936, nothing more than misidentified black panthers? This could certainly be the case with sightings involving only the briefest of observations, or made in conditions of poor lighting or deceptive shade, or made by observers with little/no experience of tigers. Conversely, most of the reports mentioned in this section have taken place in good light, with ample observation time (in some cases the specimen was dead, and therefore unlikely to run away!), and sometimes involving eyewitnesses with superb credentials, including experienced tiger-hunters Captain Swatman and Lord Ruthven; naturalist and Fellow of the Zoological Society of London C.T. Buckland; and painter and Fellow of the Royal Society James Forbes. The two reports incorporating these eyewitnesses are in themselves enough to verify that black tigers as well as black leopards exist. The experience of various eyewitnesses in distinguishing readily between tiger and leopard spoor also supports the validity of tiger identifications in relation to certain other black tiger reports.

In addition, the description of black tigers contained in the reports given here—a tiger-sized black (occasionally dark brown) cat whose stripes can be discerned in sunlight—is exactly what we would expect for a true melanistic (i.e. non-agouti) tiger.

Certainly, it is inconceivable that anyone could fail to distinguish (at least on a dead specimen) between the cryptic spots of a black panther and the long, continuous shadowy stripes of a melanistic tiger. This being the case, how, therefore, did the infamous Dibrugarh "black tiger" fiasco come about? Most probably not through any morphological muddle or misconception, but instead through an error of etymology.

As Assam-based Anglo-Indian naturalist E.P. Gee pointed out in his book *The Wild Life of India* (1964), "bagh," the word for "tiger" in various parts of India, is in fact commonly used for almost any cat species. In short, the locals will refer in their language to a striped tiger (the true tiger), a spotted tiger (leopard), and so on, right down to the smaller cat species. Consequently, when the Dibrugarh "tiger" was captured, it would simply have been referred to locally as a black tiger, even though it was a black leopard, and even though the locals knew it was a black leopard! The error came from the loose translation of "bagh" not being familiar to non-locals, who naturally took 'black tiger' to mean precisely that.

It is also worth noting that in the early 19th century, English people referred to both tiger and panther as tiger, with the epithet "royal" being added when speaking of the genuine striped tiger. Thus, the Piccadilly "black tiger" was probably a black panther; this may also have been the identity of the specimen noted in *Sophie in London*, 1786. Excluding etymological ambiguities, however, it is evident that few of the black tiger reports documented here could have been based upon panther sightings.

Another explanation sometimes put forward to deny the existence of genuine black tigers is that sightings of such cats were really of normal tigers whose coats had become darkened in various ways by external factors. One such manner involves deceptive shadows cast upon terrestrial beasts in dense jungles during early morning and evening, as mooted by R.G. Burton in a letter to *The Field* (25 October 1928). Needless to say, however, such shadows cannot account for the skins of dead black tigers reported here, which were examined carefully by competent observers.

A suggestion by A.A. Dunbar Brander was that black tigers are actually normal tigers that have rolled in the charcoal and ashes of forest fires, recalled in P. Turnbull-Kemp's book *The Leopard* (1967). If this were true, the cryptic stripes so characteristic of the black tigers reported would have been hidden by the charcoal. Moreover, observers of the dead specimens would surely have noted the presence of charcoal on such skins, but no mention of its presence has ever been made.

Dunbar Brander once watched a large tiger devouring a kill and, in so doing, becoming covered with its victim's blood, which rapidly turned from red to black. He commented that had he not seen this transformation taking place before his very own eyes, he would have been firmly convinced that he had stumbled upon a genuine black tiger. Once again, however, such an occurrence would hide the cryptic stripes clearly observed by many black tiger eyewitnesses.

This is also true relative to yet another of Dunbar Brander's suggestions. In various jungle pools used by tigers for bathing and by tanners for skin preservation, certain tree barks are introduced to poison fish that become marooned there when the pools shrink during hot weather. These barks also stain fur a deep brown shade, and could be responsible, according to Dunbar Brander, for reports of brown tigers, dead or alive. Yet again, however, this is an unsatisfactory explanation, for the same reasons already given.

Only one plausible possibility is left: that sightings of alleged black tigers have really involved melanistic tigers after all—and why not? As vouched for by felid geneticist Roy Robinson, the descriptions of cryptic stripes provide the most telling piece of eyewitness evidence in favor of such creatures' existence. What remains to be explained is why they should be so rare, in contrast to the relative abundance of the melanistic morph in leopards and other tropical felids.

The answer is probably connected to the fact that whereas such species have normally originated in the tropics, the tiger evolved in a much colder climate, only invading Asia's tropical zones in relatively recent times (geologically speaking). Black is the most efficient heat radiator, hence black coat coloration is advantageous to a felid inhabiting a tropical locality, because it would effectively assist it in releasing heat from its body, especially in shade, where dark colors operate most efficiently. Consequently, melanistic coat coloration would be selected for as a result of this inherent survival advantage, and thus the frequency of the non-agouti mutant allele would increase within populations of leopards, etc. Once the tiger also entered such locations, this same allele would then begin to increase through selection in tiger populations too. Yet because this allele would have been exposed to positive selection for far less time in tiger populations than in populations of leopard, etc (due to the tiger's much more recent settlement in the tropics), its frequency (and hence its expression in the form of black tiger individuals) would not be as great as in longstanding tropical felids—and therefore far fewer black tigers would be born.

Furthermore, due to the extremely severe depletion of the normal tiger in recent times (a proportion of which must be heterozygous for the mutant non-agouti allele in order to explain the origin of black tigers in the first place), the numbers of black tigers being born would again be reduced.

Incidentally, it is worth noting that most black tiger reports derive from localities in fairly close proximity to one another and to the north-east of the Bay of Bengal. Such restricted distribution (unusual for a simple mutant) presumably marks its origin (this compares well with that of the king cheetah, white lion, and blue tiger).

Having said that, on 19 March 2013, I received a fascinating email from a *ShukerNature* reader suggesting that black tigers may have been known over 2,000 years ago in China. Published for the very first time, here is the relevant section from that email, which was sent to me by Luke Cockle:

I was idly browsing your blog when I came across the post concerning black tigers. You commented that "the fact that its stripes could still be discerned against its fur's black background colouration is precisely what one would expect with a genuine melanistic tiger". It put me in mind of the following entry from the Erya (爾雅), an ancient Chinese encyclopaedia written c. 300 BC, in which one of the commentaries specifically mentions this fact. As this work has not, to my knowledge, been translated into English, and Classical Chinese is a somewhat inaccessible language, I thought the following excerpt might be of some small interest.

From the chapter 釋獸 "Explaining the beasts":

虪黑虎。郭璞注:晉永嘉四年建平秭歸縣檻得之狀如小虎而黑毛深者為班[=斑]。

Which I translate as:

"The shu is a black tiger. Guo Pu notes: In the fourth year of Emperor Huai of Jin [310], one was obtained in the border regions between Jianping and Zigui counties. It resembled a small tiger, but it was black; its fur was dark and luxuriant, but bore stripes/spots [the character can mean both, but in the context of a tiger, stripes seem more likely]."

Note that here is a specific (albeit very obscure) character meaning 'black tiger' (虪 shu), rather than simply the characters for 'black' and 'tiger' separately, by which the Erya defines it (黑虎). But looking in the Kangxi dictionary, I find yet another 'black tiger' character following it, this one so long-forgotten that it isn't even in Unicode.

http://www.kangxizidian.com/kangxi/1076.gif (18th from the right).

Of course, ancient Chinese literature is awash with mythological beasts, but I thought the particular observation of stripes in this instance lends credence to a real animal. And that there are at least two discrete words for a black tiger (the second one would be pronounced *teng*) suggests that they were well acquainted with melanistic tigers even further back than the Warring States period.

Similarly, in his book *Mystery Creatures of China* (2018), David C. Xu stated that references to melanistic tigers in China date back 2,400 years.

In summary: everything points to reports of black tigers being based upon melanistic specimens. Ironically, however, the black tiger nowadays could actually be a felid of fable after all because this fascinating mutant morph may have slipped quietly and tragically into extinction while science has been casually contemplating whether or not it ever existed in the first place!

Finally, on 17 February 1993, less than four years after the publication of this book's original 1989 edition, Warren D. Thomas, the then recently-retired director of the Los Angeles Zoo, wrote me a fascinating letter describing the birth at the Oklahoma City Zoo of what was apparently an incompletely melanistic tiger cub, i.e. exhibiting melanism on certain (but not all) portions of its body. Here is the relevant section from his letter:

> During the period of time that I was director of the Oklahoma City Zoo, during the early '70s, we had a pair of tigers of unknown origin (probably Bengal). During a period of about four years, the female had six litters...All of the cubs were normally colored, except litter number three...she became quite disturbed and killed all but one of the cubs [in litter number three]. We rescued the remaining cub, which was raised on a bottle and later sold.
>
> Now let me describe the four cubs in litter number three. One cub was completely normal in color and was one of the ones that were killed. The second cub had a normal ground color with some darkening on all four legs. However, the stripes were still visible. This too was killed. The third cub was the most remarkable of all. It had a normal ground color, but considerable darkening over the shoulders, down both front legs, over the pelvis, and encompassing both back legs. The darkening was essentially the same coloring as the stripes. Over the areas of darkening, the stripes were only partly visible. Unfortunately, this animal too was killed by the mother. I preserved the animal in formalin and to the best of my knowledge it is in the possession of the Los Angeles Zoo...The fourth cub was perfectly normal in color except for the feet. All four feet were darker than normal, though not to the intensity of cub number three or cub number two. This is the cub that survived, and by the time we sold it at 18 months old, the darkness on the feet had disappeared and it looked perfectly normal.

Thomas very kindly enclosed with his letter three color slides of the third cub, preserved in a formalin-filled jar. Even allowing for some formalin-induced color fading that had occurred to it over the years, the prominent extent of pelage darkening could still be readily perceived. What a tragedy that this truly unique tiger cub was killed—imagine what a spectacular animal it would have become if it had retained its remarkable coloration into adulthood. Having said that, there is no guarantee of course that it would have retained its coloration into adulthood, especially in view of how the darkening on the feet of the fourth, sole surviving cub eventually disappeared.

Indeed, in what must surely be the most extreme example of color change ever recorded from a single tiger individual, on 6 June 2010 a male white tiger cub named Sembian, one of three such cubs born in a litter to two white tiger parents at the Arignar Anna Zoological Park in Vandalur, Chennai, in India, subsequently began turning black. By the time he had reached maturity, however, Sembian had gradually turned white again! During his temporary black phase, Sembian was largely but not completely melanistic, making it difficult to decide whether his melanism was of the true type, i.e. exhibiting abnormally dark background coloration, or whether it was of a lesser-known reverse version, known variously as pseudo-melanism or reverse melanism, and which, as will now be seen, has definitely been confirmed to occur in the tiger.

Oklahoma City Zoo's black tiger cub, preserved (Warren D. Thomas)

Pseudo-Melanistic Tigers

There is one form of unusually dark-furred tiger whose reality can no longer be doubted because several skins from such tigers have been procured, and living specimens have lately been photographed and videoed in the wild. For various reasons, this tiger form is known as a pseudo-melanistic tiger (although in media reports they are often incorrectly referred to as black tigers, thereby inciting confusion with melanistic tigers).

Sometimes reported from Sunderbans are tigers whose coats exhibit an abnormally dark background color patterned with orange stripes, i.e. the reverse of the normal tiger pelage, which suggests that they may be pseudo-melanistic tigers.

In 1993, moreover, news—and photos—emerged of two such tigers killed that year. One was a young tigress killed in July 1993 by a tribal boy in self-defense at the village of Podagad within the Bhandan river valley, in the west of the Similipal Tiger Reserve. The pelt of the other one had been confiscated from a New Delhi hunter-smuggler and was acquired on 20 February 1993 by India's National Museum of Natural History on the orders of a Delhi court. Remarkably, as I learnt from Lala A.K. Singh, and as shown by both skins, it was as if the normal tiger coat color and patterning had been reversed—because instead of exhibiting an orange background color and black stripes, both of these tigers sported a black background color and orange stripes.

Looking at them more closely, however, it became evident that the black color was actually caused by an abnormal widening and coalescing of the normal black tigerine stripes, yielding a solid black mass of color. Furthermore, the orange stripes were not true tiger stripes at all but were actually gaps in the black mass of amalgamated striping through which the tigers' normal orange background coat color was still present and visible. This condition is characteristic of pseudo-melanism and is also dubbed "reverse melanism," because in true melanism (as commonly caused in felids by the expression of the mutant non-agouti allele of the Agouti gene) it is the pelt's background coloration that is abnormally dark, with any pelt patterning such as stripes or spots being unchanged.

Painting of pseudo-melanistic tiger (William M. Rebsamen)

Since 2007, several pseudo-melanistic tigers have been photographed and videoed by camera traps in India's Similipal Tiger Reserve. To quote a *DownToEarth* article of 15 April 2017 penned by Moushumi Basu:

> A curious genetic phenomenon, found nowhere else in the world, is occurring in the Similipal Tiger Reserve (STR) in Odisha. Normal tigers are giving birth to black or melanistic tigers and even normal cubs are being delivered by the black or melanistic tigresses. A census carried out by the Odisha forest department in STR in 2016 found six-seven melanistic tigers out of a total of 29, including cubs that have so far been recorded through camera traps. "We are witnessing a growing trend of black tigers in STR in the past few years. In a litter comprising three-four cubs, one or two are born dark," says Ajit Kumar Satpathy, deputy field director, STR.

As confirmed, however, by the excellent close-up color photograph included in Basu's article that depicts one of STR's so-called "black" or "melanistic' tigers," they are in reality pseudo-melanistic, not melanistic, tigers.

In *Mystery Creatures of China* (2018), David C. Xu reported that some time subsequent to 1951, Qin Fubang, a retired secretary of Miyun County (now Miyun District, northeastern Beijing), had bought the pelt of an unusual cat that had been killed by farmers of the neighboring Xinglong County, northeastern Hubei Province. It was described as a relatively small tiger, with black, white, and grey fur, which suggests that it may have been a pseudo-melanistic specimen (a melanistic tiger would not have possessed any white fur), but where its pelt is now was not recorded.

A Green Tiger in Vietnam?

On 27 June 2017, I received a fascinating email from correspondent James Nicholls of Perth, Australia. It included a link to a thoroughly extraordinary account on the website *Reddit*, which had been posted on Christmas Day 2016 by someone with the username AnathemaMaranatha and seemingly of American nationality (judging from their style of grammar and spelling, and various other *Reddit* posts by them). This consisted of their supposed first-hand eyewitness description of a truly unique mystery cat. It reads as follows:

> Okay. I saw a green tiger. I wasn't alone.
>
> We were out towards the Cambodian border in summer of 1969, an American light infantry company of about 100 or so guys. We were operating in flatlands, thick jungle, along a river. (Saigon River? Not sure.) Bright, sunny day.
>
> We were proceeding single file when point platoon came to a stop, there was some yelling (we were stealthy - yelling is bad) from the point, then point platoon radioed for the Command Post (CP - the company commander and his people) to come up to point.
>
> When we got there, we found the point team glaring at each other - some kind of tussle. Point and drag were standing in the machine gunner's line of fire glaring at him. The machine gunner had wanted to shoot. Point and drag stopped him. He didn't like that.
>
> The object of discussion was across a jungle opening maybe 15 meters away, just peeking at us over the elephant grass. It was a big

> tiger - biggest I've ever seen, Frank Frazetta-style big, but without the lady.
>
> Here's the insane part. The tiger was white where a tiger is white and black where a tiger is black, but all the orange parts were a pale green. We all saw it, maybe twenty grunts and me. The machine gunner was arguing that we have to shoot it, because otherwise no one would believe it. He had a point.
>
> But the rest of us were just awestruck. I mean, it might as well have been an archangel, wings halo and all. I felt an impulse to kneel. I don't think I was alone.
>
> The tiger stood there checking us out for maybe 15 minutes, not worried, not angry, just a curious cat. Then he turned and disappeared.
>
> Don't believe me? That's okay. I don't believe it myself. I mean WTF was that? Hallucinogenic elephant grass? Some trick of the light? The tiger walked through some kind of green pollen just before we saw it? No freakin' idea.
>
> There it is, OP. I don't believe it, and I saw it. Or hallucinated it. Me and all my blues. Make of it what you will. I'm done.

In fact, this person did make a few additional, minor comments in reply to various responses from other *Reddit* readers, of which the following one is well worth noting here:

> I apologize for not making clear that the tiger was scaring the shit out of all us. He did NOT look sick or malnourished. He looked like he could be right in the middle of all of us in no time flat. He thought so, too. Didn't seem the least bit scared of us.
>
> And I guess he wasn't hungry.

Not surprisingly, faced with an account from someone claiming to have encountered a green tiger, my initial reaction was to assume that it was just a spoof, a joke, not to be taken seriously. But then I decided to investigate the credentials of the person who had posted it, especially as their account did sound as if it had been written by someone familiar with military action in Vietnam, and I was very intrigued to discover that they had written a number of other, much more mainstream and very detailed accounts on *Reddit* concerning their alleged time and military service there during the Vietnam War that all seemed entirely authentic and had been well-received by Vietnam veterans who would surely spot and soon expose any imposter. Consequently, it seems both reasonable and parsimonious to assume that this person's

Vietnam-related testimony is indeed genuine.

But a green tiger? Really? I noticed that the green tiger account had attracted an interesting response (by someone with the unfortunate username eggshitter):

> It was a bright sunny day right? Is there any chance that there was some murky green pool that reflected the light on to the tiger? Maybe he had just been rolling around in the grass?

Other, later posters made similar comments. They reminded me of a suggestion that has been put forward in the past concerning the blue tigers of Fujian, China—namely, that perhaps their distinctive fur coloration was simply due to their having rolled in bluish-colored mud or having been stained with dye when bathing in a polluted pool. However, as I pointed out when responding to this suggestion previously, if that were true the entire tiger would look blue, whereas eyewitnesses have specifically mentioned seeing their black stripes and pale underparts, which of course would have been obscured if they had rolled in mud or been stained with dye. The same logic, therefore, can be applied to the green tiger had it merely been rolling around in grass, or perhaps bathing in an alga-choked jungle pool.

Conversely, an optical illusion induced by reflected light is certainly possible. Yet bearing in mind the substantial length of time of the observation (15 minutes) by several different people simultaneously rather than just a single observer, this might initially seem somewhat improbable too (always assuming, of course, that the report is indeed genuine)—but not impossible (see Chapter 4 regarding an alleged green leopard).

Harimau Jalor

Natives of the Malaysian State of Trengganu repeatedly informed Lieutenant-Colonel Arthur Locke of the harimau jalor—a larger-than-normal form of tiger immediately distinguished from all others by the fact that its stripes do not run downwards (from back to underparts) but instead stretch longitudinally (from head to tail). No-one had apparently seen a dead harimau jalor, but as it was so frequently mentioned by the natives, Locke felt reluctant to discount its existence.

Nevertheless, as he noted in his book *The Tigers of Trengganu* (1954), he considered that an optical illusion may be responsible, involving normal tigers viewed head-on (thereby concealing their stripes) and overlain by longitudinal shadows from branches overhead.

Bali Tiger and Javan Tiger

Three of the tiger's eight formerly recognized modern-day subspecies (prior to their recent afore-mentioned amalgamation into just two subspecies) are or were confined to Indonesian islands. Two of these, the Sumatran tiger *P. t. sumatrae* and the Javan tiger *P. t. sondaica*, are exceedingly rare (if not already now lost in the case of the Javan). Neither, however, is more so than the third, the Bali tiger *P. t. balica*, because most authorities believe that this is definitely extinct and has been for almost a century—or has it?

It was (or is?) indigenous to Bali, and during the early part of the 20th century it was believed to be relatively abundant but was decimated by the spates of uncontrolled killing by locals and Dutch colonials alike that occurred throughout the island during the intervening years between the two World Wars. Indeed, by the middle of the 1930s, hardly any tigers were left in West Bali, although others allegedly survived in its northwestern and southwestern regions. At the close of the decade, however, the Bali tiger seemed to have become extinct throughout the island; the last known specimen, an adult tigress, had been shot at Sumbar Kima in West Bali on 27 September 1937.

Even so, unsubstantiated rumors continued to circulate for many years that a few specimens still survived in a West Bali reserve, thereby suspending this erstwhile subspecies in an uncertain limbo between existence and extinction for over 30 years.

Then in 1972, two visitors to Indonesia reported that during conversations with Balinese Forestry Department officials, they had learned that at least one tiger had been seen very recently. Nevertheless, by 1975, zoological explorations of central and western Bali instigated in search of this individual and any others that may also exist were forced to concede defeat. Similarly, although animals apparently killed by flesh-eating creatures were discovered in northwestern Bali, there was no positive indication that tigers were responsible. Another dead end, but in 1980 more encouraging news emerged. The World Wide Fund For Nature (WWF, at that time known as the World Wildlife Fund) announced that claw marks had been observed on trees in Bali at a height that made them compatible with those of a tiger. They had been made 6-18 months earlier. Forty years have passed since then, however, with no confirmation of tiger survival on Bali, so the hope for its survival that had been fanned for a while by these earlier reports seems now to have been finally and permanently extinguished.

Worth noting, incidentally, is that its continuing existence is not the only mystery surrounding the Bali tiger. Some authorities have declared that the tiger was not native to Bali in the first place. Instead, they believe that tigers had been introduced here from some other Asian locality. Arguing against this, however, is the longstanding familiarity of the tiger to the Balinese people, plus the singular appearance of Bali tiger skins and skulls, which can be readily differentiated from those of other tiger forms.

As for the Javan tiger's putative survival, in Indonesia, Didik Raharyono, a biology graduate from Gadjah Mada University, has long been adamant that tigers still survive in Java, even though the Javan tiger was formally declared extinct in the 1980s. Yet sightings continue to this day among local villagers and poachers, who have been interviewed at length by Didik. They have even given him tiger teeth, skin, and whiskers, some of which may be from tigers killed only months earlier. And he has collected feces and spied footprints in caves and other localities that tigers allegedly still frequent.

Since beginning his search in 1997, Didik claims to have once seen a tiger himself, albeit briefly, during a 14-day survey in Meru Betiri National Park, and hair samples have apparently been identified as Javan tiger by the Indonesian Institute of Sciences. Even so, finding sponsors for Didik's quest is proving to be as great a problem as finding the tigers themselves.

And while on the subject of tigers in unexpected localities, please note that my book *Cats of Magic, Mythology, and Mystery* (2012) also documents unconfirmed but tantalizing reports of tigers in the Korean Peninsula, Borneo, Sri Lanka, and Turkey.

Iriomote Felids

Southeast of Japan lie the Ryukyu Islands, a long chain whose principal member is Okinawa, and whose southernmost member is a somewhat smaller island called Iriomote. Belying its apparent insignificance, however, this latter isle could be home to a completely unknown form of very sizable cat. Indeed, the name of Iriomote is already more than familiar to all felid researchers due to a noteworthy discovery of the cat kind made there in the 1960s.

During the winter of 1965, Japanese naturalist Yukio Togawa (sometimes spelled Tagawa) visited Iriomote to carry out some work, and while doing so he also followed up stories that he had heard concerning a strange cat form that existed here, known locally as the pingimaya. His efforts, however, were somewhat hampered by the regrettable tendency of the natives to eat any specimen of this mystery felid that they happened to trap!

Nevertheless, Togawa's perseverance eventually paid off, because with the help of Okinawan zoologist Tetsuo Takara, a professor of Ryukyu University, a skull and two skins were at last obtained and sent to the eminent Japanese zoologist Yoshinori Imaizumi of Tokyo's National Science Museum.

Following his detailed studies of these preserved specimens, together with further collected material and at least one living cat, in 1967 Imaizumi published a formal scientific description of the Iriomote cat, in which he claimed that it was indeed a new species (and genus), naming it *Mayailurus iriomotensis*. However, *Mayailurus* was later demoted from genus to sub-genus, and more recently still the Iriomote cat itself was demoted from a separate species in its own right to a well-marked island subspecies of the leopard cat *Prionailurus bengalensis*, so that it is nowadays known scientifically as *P. b. iriomotensis*.

Nevertheless, the scientific world was startled to learn that a totally new felid taxon had been discovered so recently and on such a small island, but even more startling news was to follow.

Clouded Leopard Controversies

Asia's clouded leopards are unquestionably among the most beautiful of all wild cats. Housed in the genus *Neofelis*, two species are currently recognized—the mainland clouded leopard *N. nebulosa*, and the Sunda clouded leopard *N. diardi*. The latter species is restricted to the islands of Borneo and Sumatra and until recently was deemed to be merely a subspecies of the mainland form.

Nebulosa translates from Greek as "clouded." Similarly in English they are referred to as clouded leopards, and in Chinese they are termed mint-leaf leopards (all three names are derived from the shape of the ornate blotches decorating their very handsome coat). These specialized species occupy their very own genus, *Neofelis*, due to their perplexing combination of Big Cat and Small Cat characteristics.

Collectively, the two species have a wide distribution across southeast Asia but are not known officially to have any representative on the Ryukyu island of Iriomote. Yet according to this island's inhabitants, the pingimaya is not the only felid form known to them here. Apparently it is also home to a much more sizable cat, which they refer to as the yamamaya or yamapikarya. From their descriptions, this still-unknown felid is as large as a sheepdog, yellow or orange in color, and with distinctive markings variously likened to the spots of a leopard or the stripes of a tiger. If real but not a bona fide tiger, might it be a clouded leopard with unusually elongate, vertical blotches, as suggested by Swiss zoologist C.A.W. Guggisberg?

Clouded leopard (Chris Brack)

Most reported yamamaya sightings occurred during the 1950s and 1960s, but there have been some more recent ones too. In September 2007, for instance, while sitting on a beach in Iriomote, Professor Aiyoshi from Japan's Shimane University had a startlingly close feline encounter, at a distance of only 8 feet away, when out of the thick forest stepped a 3-foot-long cat with a very lengthy tail and black spots that reminded him of a leopard. The cat stared at him for a moment before slinking back into the forest. In 2003, a cat of identical appearance had leapt down from the top of a large boulder, landing on the ground directly in front of a Mr. Shimabukuro while he was setting wild boar traps in this island's mountains, before vanishing into the underbrush. Apart from its very eye-catching markings, the clouded leopard's most noticeable feature is its extremely long tail, which accords again with the yamamaya.

In view of the Iriomote cat's own quite recent discovery, plus the fact that Iriomote is virtually unexplored and almost completely covered in dense mountainous rainforest whose fauna has hardly received any scientific attention, the existence here of an unrecorded tiger or clouded leopard form would certainly not be impossible.

Clouded leopards can be quite variable in terms of pelage patterning, but traditionally it was not generally supposed that either of the two species exhibited a melanistic morph. Yet around April 1946, the skin and lower jaw of a pure-black Sunda specimen were obtained in Borneo, and placed on record three years later by naturalist Tom Harrisson in the *Malayan Nature Journal.* Another black specimen was supposed to have existed at one time on this island's Mount Matang.

Still in Borneo, six reports of large black cats were recorded during a WWF faunal survey of Sabah in 1982, whereas in 1986 three such cats were sighted in Sarawak and a further account recorded from Sabah (*Oryx*, April 1987). The first photograph of a melanistic clouded leopard in Borneo was snapped in Brunei's Sg Ingei Protection Forest by a camera trap in July 2010.

Anomalous and Unexpected Leopards

Today, science recognizes just one species of true leopard, *Panthera pardus*. Yet until the 1920s, only those individuals with relatively short tails were actually classed as such; any with tails equalling or exceeding their head-and-body lengths were classified as a separate species, the panther. One authority went even further, splitting off those individuals of very small total length to yield a third leopard species, the pantherette. Zoologists now recognize that leopards come in all shapes and sizes, and the term "panther" is generally restricted in present-day usage to melanistic individuals, known in full as black panthers. However, even in modern times some very anomalous and unexpected leopard specimens have been reported, including the following selection of examples.

Doglas

Taking into account the controversial pygmy leopard from the then Somali Republic in east Africa, which at one time was classified as a separate subspecies, *P. p. nanopardus*, the leopard normally ranges from 3-to-5 feet in head-and-body length, plus a further 2-3 feet of tail. Nevertheless, much larger specimens have also been recorded occasionally, including one tiger-sized individual from India that measured in total length a mighty 9 feet 4 inches.

Speaking of tigers, as noted by F.C. Hicks in his book *Forty Years Among the Wild Animals of India* (1910), it is popularly believed by natives throughout India that the largest male leopards often mate with tigresses to yield mysterious hybrids known as doglas. Moreover, some sizable leopards have been recorded that bear very prominent stripes on their underparts, and which the natives allude to as support for this belief. Nevertheless, Hicks doubted very much whether individuals such as these were truly of interspecific origin. This opinion was based upon personal experience—because he once shot a felid that seemed very likely to have been a bona fide leopard x tiger hybrid, and this individual was very different morphologically from those leopards identified as doglas by the natives. In his book, Hicks described his specimen (later lost or stolen) as follows:

> ...the spotted head of a panther [old usage of term] of extraordinary size pushed its way through the grass, followed by the unmistakable striped shoulders and body of a tiger, though looking a bit dirty as if it had been rolling in ashes. I succeeded in dropping this extraordinary creature dead with a shot in the neck, and, on examining it, I found it to be a very old male hybrid, with both its teeth and claws much worn and broken; its head and tail were purely that of a panther, but with a body, shoulders and neck-ruff unmistakably that of a tiger, the black stripes being broad and long though somewhat blurred and breaking off here and there into a few blurred rosettes, the stripes of the tiger being the most predominant on the body. One of the peculiarities of this creature which I particularly noticed was that though it was male, it had the feet of a female and measured a little over 8 feet in length.

Could it have been both a hybrid and a hermaphrodite? Intersexual tigers and leopards have indeed been recorded. Notwithstanding this possibility, it possessed an undeniable combination of tiger and leopard characteristics. Hicks admitted that if such hybrids were themselves fertile and capable of breeding in turn with leopards, this could then explain the doglas of the Indian natives. However, most of the Big Cat hybrids on record have been found to be sterile, so this prospect is very unlikely.

A Melange of Morphs, and a Mystery From Malabar

In any event, when dealing with the leopard, the fact that some individuals have striped underparts should come as no surprise. After all, not only does this species vary in color and patterning from region to region, it also exhibits an unparalleled variety of morphs and freakishly-marked types over its equally wide distribution range. In Asia alone the leopard extends from Asia Minor to China and the Korean peninsula (but not Japan; *P. p. japonica*, the so-called Japanese leopard, was described from a single skin that was almost certainly derived not from Japan but instead from either China or Korea), and as far south as India, Sri Lanka, Malaysia, and Indonesia.

The leopard's most familiar color variety is its melanistic (non-agouti) morph, the black panther. This is very abundant in Asia and most frequently found in humid, densely-forested areas. Cream-colored leopards with pale markings and blue eyes (chinchilla-phenotype mutants) have been documented from the Indian States of Dumraon and Hazaribagh, and from southern China. In addition, an orange-colored (erythristic) specimen, in which the spots were virtually assimilated into the pelage's background color, was reported in 1930 by Reginald I. Pocock from India (probably a non-extension of black mutant). Even jaguarine leopards have been recorded, whose rosettes each contain one or more central spots like those of jaguars.

Few of these variants, however, can compare with the magnificent and mysterious skin purchased in December 1912 by Holdridge Ozro Collins from G.A. Chambers of Madras. Its predominant color was an elegant glossy black, and in a *Bulletin of the South Californian Academy of Science* paper from 1915 it was described by Collins as follows: "The wide black portion, which glistens like the sheen of silk velvet, extends from the top of the head to the extremity of the tail entirely free from any white or tawny hairs…"

He went on to say:

> In the tiger, the stripes are black, of an uniform character, upon a tawny background, and they run in parallel lines from the center of the back to the belly. In this skin, the stripes are almost golden yellow, without the uniformity and parallelism of the tiger characteristics, and they extend along the sides in labyrinthine graceful curls and circles, several inches below the wide shimmering black continuous course of the back. The extreme edges around the legs and belly are white and spotted like the skin of a leopard...The skin is larger than that of a Leopard but smaller than that of a full grown Tiger.

The cat had been killed in Malabar, southwestern India, earlier in 1912, and so unusual was its exceedingly handsome skin that Chambers had been totally unable to classify it, so that he wondered whether it could actually represent some hitherto unknown species of felid. To obtain an answer, Chambers sent it to Madras's Government Museum for official identification. He subsequently received a letter from J.R. Henderson of the Museum, who stated that although the species was certainly leopard, it constituted a variety that he had never before seen. Collins also sought scientific advice concerning its status, and learned from Gerrit S. Miller Jr, Curator of the Smithsonian Institution's Division of Mammals, that it was indeed a black leopard, but not of the normal melanistic type.

In fact, this remarkable skin was that of a pseudo-melanistic leopard, a rare mutant form known only from a handful of leopard specimens. In a normal melanistic leopard (i.e. black panther), its coat's background color is abnormally dark, but its coat's rosettes are unchanged (so that they can often still be spied in shadow-like form against its dark pelage). Conversely, in a pseudo-melanistic leopard, its coat's background color is normal (orange-yellow) but is largely obliterated by abnormal fusion (nigrism) and multiplication (abundism) of the rosettes. In extreme cases of pseudo-melanism, as demonstrated by Collins's specimen, this fusion and multiplication of the rosettes can be so extensive that virtually the entire upper body is covered in a mass of black color, with only occasional gaps through which the normal background color is visible (appearing as orange streaks or spots). Faced with such a bizarre skin, it is little wonder that its owners had wondered whether it constituted a major zoological discovery.

Nelliampatti and Bali Leopards

Nevertheless, Asia may indeed still harbor undescribed leopard types. For example, as noted by S.H. Prater in *The Book of Indian Animals* (3rd edition, 1971), it is possible that a small dark race of leopard from Cochin's Nelliampatti Hills awaits formal scientific recognition. Attention to this form was first directed by R.C. Morris.

When the WWF announced that evidence for the Bali tiger's survival had been obtained in the late 1970s, this was not the only surprise that it unveiled concerning felids. News also emerged regarding the fact that in 1979 a leopard had been heard calling at night in northern Bali, and a leopard's fresh pug-mark had been discovered in a dried-up river bed by zoologists working within the area in question. This created great zoological excitement because science had not previously been aware that the leopard actually existed on Bali. Moreover, it later transpired that two officers of the Indonesian Nature Protection and Wildlife Conservation Service had reported seeing a cat that seemed to be a black panther a full four years earlier on Bali, in the Prapat Agung area (*Fortean Times*, spring 1978). Suddenly, instead of being devoid of big oats, Bali now appeared to have two different species—a tiger subspecies resurrected from extinction, and a race of leopard raised from obscurity! To date, however, no further news has emerged regarding Bali's putative leopard, so whether it truly existed—or exists—remains unresolved.

Anomalous Asian Cheetahs

Today, *Acinonyx jubatus venaticus*, the cheetah's once-abundant Asian subspecies, is known to survive only in Iran and possibly also India and Pakistan, with probably less than a hundred individuals alive in total. (Prior to an earlier-noted November 1990 sighting of a cheetah in India, the last known Indian specimens, three in number, had all been shot in 1947.) In earlier centuries, conversely, it was abundant in the Middle East and southwest Asia, but most notably in India, where it was commonly used as a hunting beast during the 16th-19th centuries by the Mughal (Mogul) emperors. Indeed, Akbar the Great alone reputedly owned a thousand cheetahs—but the specimen under consideration here, and which challenges even the blue tigers of Fujian as the most eye-catching and exotic mystery cat morph ever reported from the Asian continent, was owned not by Akbar but instead by a different yet no less celebrated Mughal emperor.

The Blue-Spotted Cheetah of Jahangir

In 1608 AD, the specimen in question was brought by Raja Bir Singh Deo to the Emperor Jahangir, a keen naturalist, at Agra in India. Apparently, the fur of this marvelous creature was white, and its spots were blue!

Some researchers have sought to identify it not as a cheetah at all but rather as a snow leopard *Uncia uncia*. According to a detailed account authored by the modern-day Indian naturalist Divyabhanusinh and published in August 1987 by the *Journal of the Bombay Natural History Society*, however, this is a highly implausible identity. Bearing in mind that Raja Bir Singh Deo was from Orcha in central India, it is most unlikely that he had ever encountered a snow leopard. Conversely, central India was apparently replete with cheetahs in those days, and it is confirmed that cheetahs were formerly caught from Mughal hunting grounds northwest of Orcha and elsewhere in central India.

Equally implausible is an alternative explanation proposing that Jahangir had confused an albino leopard for a cheetah. It is surely inconceivable that as ardent a hunter as Jahangir could have made such a fundamental error in identification. Presumably, therefore, his white-furred, blue-spotted enigma was indeed a cheetah and may have been either a partial albino or a coat color mutant comparable genetically with white lions and tigers.

King Cheetahs in Asia?

In Chapter 4 of this book, dealing with African mystery cats, I document the history and discuss the genetic basis of a spectacular striped morph of the cheetah known as the king cheetah, which is unequivocally known in the wild state only from a restricted region within southern Africa. Officially, therefore, no specimens of this eye-catching variety are known to have arisen in wild cheetah populations anywhere across Asia—but some years ago I discovered not one but two possible pieces of evidence for the putative erstwhile existence of Asian king cheetahs.

I initially documented my thought-provoking findings in a *Fortean Times* article (February 2011), which I subsequently expanded and converted into a formal paper submitted to the *Journal of Cryptozoology*, currently the world's only peer-reviewed scientific journal devoted to mystery animals. My paper was duly assessed by two independent referees, accepted by them, and published by the *Journal* in 2013. (See Appendix 1)

Additional Asian Mystery Cats

During the three decades that have passed since this book's first edition was published in 1989, I have learnt of several additional Asian mystery cats, which I have variously documented in my later book *Cats of Magic, Mythology, and Mystery* (2012) and on my *ShukerNature* blog. Five of the most notable examples are as follows.

Pogeyan

While visiting the Western Ghats in India at the very end of the 20th century to film its varied fauna, wildlife photographer and environmentalist Sandesh Khadur sighted a large, still-unidentified cat known locally as the pogeyan—a name that translates as "the cat that comes and goes with the mist," which succinctly highlights its elusiveness. Khadur observed it in broad daylight in the high-altitude grasslands around the lofty Himalayan peak of Anamudi and described the animal as large and uniformly dark-grey in color, with big roundish ears and a long tail. This description does not correspond with any known species of cat. He set a camera trap in the hope of photographing the animal but was not successful. All that he could do, therefore, was sketch what he saw, and the resulting images call to mind a lioness or a long-legged leopard.

No Asian lions are known to occur in the Western Ghats, but leopards are, and a few freakishly-pale specimens (leucistic and/or chinchilla-phenotype) have been recorded from various parts of Asia (as well as Africa, see Chapter 4). Another possibility that occurs to me is a dark morph of Temminck's golden cat *Catopuma temminckii*, a morphologically-variable and highly-elusive species that is already known to exist in northeastern India, and which therefore may also do so, scientifically incognito, in the Western Ghats. Perhaps one of these identities explains the pogeyan, unless of course it is a scientifically undescribed species?

A White Saber-Tooth in China?

According to a fascinating online article published in Chinese on the tieba.baidu.com website, which was kindly summarized in English for me by Canadian cryptozoologist Sebastian Wang after alerting me to its existence, this vast country may be home to a truly remarkable mystery cat.

As recently as May 1994, in the Shennongjia region of China's northwestern Hubei Province, the article's author allegedly spied a giant cat measuring 12-16 feet long (and therefore much larger than any tiger, the biggest modern-day cat species known to exist), with white fur bearing vertical yellow stripes. As documented earlier in this chapter, an extreme white-furred morph of the tiger, known as the snow tiger, does possess very pale, yellowish stripes (indeed, some specimens bear no or virtually no markings at all), but it does not measure 12-16 feet long. Moreover, both the snow tiger morph and the more familiar "true" white tiger morph with brown or dark grey/black stripes have only been recorded from the Bengal tiger, not from the South China version.

In any case, the Shennongjia mystery cat also reputedly sported a pair of huge canine teeth up to nine inches long, and therefore reminiscent of those possessed by the prehistoric saber-tooths or machairodontids.

The sighting's precise location was on the tallest peak of the eastern Shennongjia region, at an altitude of just over 9,000 feet. The article's author later learned that a few such cats had previously been killed by local hunters. If only a pelt or skull had been preserved for scientific examination—but perhaps some hunter does possess such objects, as trophies displayed proudly in his home. If so, he may own specimens of immense cryptozoological significance.

Interestingly, in *Mystery Creatures of China* (2018), David C. Xu documents the alleged existence of what may be one and the same Chinese saber-tooth lookalike cryptid, known as the guoshanhuang. Reported from northwestern Hubei Province just like the Shennongjia's mystery cat, as well as from northern Hunan Province, and eastern Chongquin, it too is said to sport a pair of large downward-curving canine teeth and to be yellow-striped, but according to Xu its stripes are horizontal, not vertical. Several modern-day sightings were documented in his book, and back in the 1960s a yellow specimen was allegedly killed by local hunters at Huanhugang, Huping Mountains, northern Hunan Province, with its complete skin measuring 9.8-23 square feet when stretched across two tables. Inevitably, however, no record of the skin's current whereabouts apparently exists, let alone the skin itself.

Shennongjia white saber-tooth, based upon eyewitness descriptions (Karl Shuker)

Thailand's Seah Malang Poo

In January 1994, I was very interested to learn from independent television film-maker Martin Belderson of Four Winds Productions that Khao Sok National Park in Thailand supposedly harbors a strange cat known as the seah malang poo (*seah* is Thai for "tiger"). It is apparently of stocky build, with brown and black stripes, and it lives in the area's karst limestone mountains. One was reputedly shot during the 1930s, and its skin sent to Thailand's national museum in Bangkok.

Pursuing this potential lead, I was informed on 22 May 1995 by Sophy Day of the Tiger Trust in Suffolk, England, that while visiting the Khao Sok National Park with her husband, Trust director Michael Day, they and a ranger friend, Pee Aroon, journeyed to the precise locality where the elusive seah malang poo had last been sighted, but they failed to see anything there. However, they did learn from locals that a pair of these cats were supposedly shot and sent to Bangkok. Unfortunately, my attempts to discover whether any specimens were indeed present at Thailand's national museum were unsuccessful; nor have I been able to obtain any further details regarding this mystery felid itself.

Sumatra's Cigau

According to native testimony in western Sumatra, the wilderness region east of Mount Kerinci and south towards the market town of Bangko is frequented by a very sizable feline mystery beast known as the cigau. Yellow or tan in color, with a short tail and a ruff encircling its neck, the cigau is said to be slightly smaller, but more heavily built, than the Sumatran tiger, and is a source of terror to the local people, because it reportedly attacks without any provocation, displaying a highly aggressive, fearless dislike of humans.

In summer 2003, cryptozoologist Richard Freeman from Britain's Centre for Fortean Zoology, Chris Clarke, and Jon Hare spent three weeks in Sumatra's Mount Kerinci region seeking an elusive mystery man-beast known as the orang pendek ("short man"). While there, they also obtained some intriguing feline hair samples that they hoped would constitute the first physical evidence for the reality of the cigau. When later checked by Danish zoologist Lars Thomas against all but one of the known cat species from the region in which the team had been exploring, no positive match was made.

The only known felid left to check with these intriguing hairs was Temminck's (=Asian) golden cat *Catopuma temminckii*. Sadly, when this last, crucial comparison was made, a match did result, which means that if the Sumatran hair samples were indeed from a cigau, it is Temminck's golden cat. Even so, if the local descriptions of the cigau are accurate, it is unquestionably a much larger, shorter-tailed version of Temminck's golden cat than any currently known to science.

Sundanese Horned Cat

Stranger still, however, must surely be the unidentified cat that allegedly inhabits the islands of Alor and Solor in Indonesia's Lesser Sundas group, southeast of Sumatra and Java. According to native reports gathered by British explorer Debbie Martyr during a visit to southeast Asia in 1992, this very odd animal, the size of a domestic cat, has pronounced knob-like protuberances upon its eyebrows that resemble short, stubby horns. A cat with horns would be a wonder indeed, even in a land as famous for its mysteries and illusions as the inscrutable Orient.

Hong Kong Mystery Cats

During the last two weeks of October 1976, more than 20 dogs (some sizable) were bitten to death in Pik Uk and Junk Bay, sited in the Hang Hau area of Sai Kung (within Hong Kong's New Territories). A police search was instigated but drew a blank. And a gleaming-eyed, long-tailed beast sighted in this same vicinity by villagers on 30 October was thought by them to be a large leopard, not a dog. Yet its fur was described as blackish-grey in color (not even a black panther's fur matches this). During late November/early December, a villager alleged that within the past 10 days he had twice seen a creature referred to by him as a tiger "about 3 ft high, 4 ft long, and of a dark color" (surely too small and the wrong color, especially for a species that does not officially exist in Hong Kong anyway!). Once again a police search was made, and this time some pug marks were discovered, after which the subject was referred to the Agriculture and Fish Department and was not heard of again.

Even if the above two incidents did involve the same animal (which is far from certain), then it was surely neither panther nor tiger. Yet until (if ever) further information becomes available, little more can be said.

Interestingly, tigers are by no means unknown in Hong Kong. Indeed, when British naturalist Geoffrey A.C. Herklots, based in Hong Kong for almost two decades, was documenting several such cases in his book *The Hong Kong Countryside* (1951), he made the following very pertinent comment:

> Nearly every winter one or more tigers visit the New Territories; often the visitor is a tigress with or without cubs. The visit rarely lasts more than two or three days. A tiger thinks nothing of a 40 mile walk and in a couple of nights could walk from the wild country behind Bias Bay to Tai Mo Shan or the Kowloon hills. Because their visits are usually of such short duration and because most people exaggerate, little credence is given to tiger rumours. Most that I have investigated have been founded on fact.

So too, albeit on decidedly unexpected facts, was the following strange tale of a tiger that barked!

Set once again in Hong Kong's New Territories, and as described within an article published in May 1966 by the British magazine *Animals*, it all began in July 1965 when a senior girl of the Diocesan Girls' School, on a picnic with school friends upon the slopes of the mountain Taimoshan, saw a tiger stalking through the undergrowth nearby. She was naturally alarmed but also very surprised (because tigers are not supposed to exist here). A search party was organized at once but succeeded only in spotting some flattened grass where the creature may have rested. Inevitably, further sightings were soon reported, and so for the next three months hunting parties trekked far and wide seeking this evanescent animal—by now referred to as the Shing Mun tiger, after the valley that it was allegedly raiding—but always in vain.

During the last week of October, it was sighted at the village of Shatin, near the Chinese border (and the site of a fully authenticated tiger encounter 50 years earlier), but the eyewitness in question, local builder Chan Pui, caused confusion and consternation by stating flatly that the animal was not a tiger at all—it was a wolf instead! This identification found little favor, especially with the original schoolgirl eyewitness, who was adamant that she had seen a tiger, and a naturalist announced that wolves did not exist in this region anyway. In reply, Chan Pui retorted that he would vindicate his statement by capturing the creature—and to everyone's surprise, he proclaimed on 1 November that he had done just that! Moreover, he described the fabled Shing Mun tiger as being a fierce wolf with eyes that "burn like lamps," and he even allowed interested (albeit apprehensive) newspaper reporters to photograph this savage creature.

Yet the final metamorphosis in this tale of tiger-into-wolf was still to come. For upon his inspection of the ambiguous animal in question, zoologist I.W.B. Thornton announced that it was nothing more fabulous or fearsome than an Alsatian-chow crossbreed! And far from being a rapacious predator, it was only too happy to lick the hand of another of its visitors, a representative from Hong Kong's Society for the Prevention of Cruelty to Animals.

Clearly it is most unlikely that this docile dog was really the same animal as the creature responsible for the schoolgirl's sighting and the Shing Mun raids. Almost certainly the genuine culprit, whether tiger or something else, had simply moved on, with the capture of the dog being nothing more than coincidence. Also, as seen in various earlier cases documented in this book, once a mystery beast has been sighted several times there is a tendency for the episode to be brought to an "official" end by the capture or shooting of something—regardless of whether it actually bears any resemblance to the original creature reported. Such "substitutions" (whether accidental or deliberate) are very frequent components of a cryptozoological case and can cause a great deal of problems for any investigator attempting to distinguish the relevant from the inconsequential.

When Oscar Wilde wrote "the truth is rarely pure and never simple," I doubt very much that he was contemplating any cryptozoological conundrum. Yet even if he had been doing so, he could hardly have encapsulated its vital essence within a more accurate and appropriate epigram!

Chapter 4
Africa: Cheetahs with Stripes and Lions with Spots

> *Ex Africa semper aliquod novi. [There is always something new out of Africa.]*
> — Pliny the Elder, *Natural History, VIII*

As far as cryptozoological cats are concerned, this quotation is singularly apt. For in addition to possessing several species of Small Cat and two species of Big Cat (lion and leopard) plus the taxonomically-aloof cheetah, there is substantial evidence to suggest that the Dark Continent is also sheltering a number of other types of felid— felids that still await formal scientific recognition. There should be no reason for considering such a possibility unlikely. Almost all of the last century or so's tally of major zoological discoveries from Africa—e.g. pygmy hippopotamus, okapi, mountain gorilla, giant forest hog, Congo peacock—have occurred as a result of explorers and naturalists following up native reports of creatures not resembling any species known to science at that time. Yet, bewilderingly, by continuing today to dismiss as superstition and folklore this primary source of information concerning the possible existence of undiscovered African animals, science indicates that it has still not learned its lesson.

Striped, Woolly, Speckled, Melanistic, and other Unexpected Cheetahs

Based upon the findings of various laboratory-based studies (reported later in this present chapter), genetic diversity in the modern-day cheetah *Acinonyx jubatus* is officially deemed to be markedly depauperate—but how accurate is this assessment? After all, as now revealed, some truly extraordinary and highly mysterious, controversial variations upon the typical spotted cheetah theme have been recorded down through the decades, whose genetic identities for the most part remain as perplexing and unresolved today as they were when these aberrant but fascinating forms were first brought to scientific attention.

King Cheetah

Reports of a strange and exceedingly shy forest creature, described as being half-leopard and half-hyaena by the native Zimbabwean people and termed by them the nsui-fisi (leopard-hyaena), were simply rejected by scientists as fantasy during the early part of the 20th century. Such notions, however, would soon be revised.

In 1926, *The Field* published a short letter and accompanying photograph sent by Major A.C. Cooper of Salisbury (now Harare) that concerned a most unusual skin, obtained from a peculiar felid trapped near Salisbury and presented to its Queen Victoria Memorial Library and Museum by a Donald Frazer. It should be noted here that despite Major Cooper's original statement that this cat had been trapped in the Umvukwes Range, the Museum's records reveal that it had actually originated from Macheke, about 62 miles southeast of Salisbury. Apparently it was one of several specimens of this cat form sighted here. Cooper suggested that it was a hybrid of leopard and cheetah, and described it as follows:

> ...a very stockily built leopard, with powerful limbs and a comparatively short thick tail. As against this there are the non-retractile claws of the cheetah, also the ruff round the neck, which is totally absent in the leopard. The background of the skin is the full yellow of the leopard, not the sandy yellow of the cheetah. The markings are like nothing on earth (note the longitudinal stripes down the back and shoulders).

Its markings were indeed exceptional. A series of longitudinal black stripes extended from its neck along the entire length of its back to the upper part of its tail, whereas its flanks and upper limbs were ornately decorated with thick black blotches and curved stripes of irregular shape and size, plus notable rings around much of its tail's length. Cooper wondered if any similar skins had been recorded.

His letter came to the attention of felid expert Reginald I. Pocock, who judged from the photograph (which did not reveal the ruff and non-retractile claws) that the cat was most probably an aberrant leopard. When the skin was subsequently loaned to London's Natural History Museum for his direct examination, however, Pocock realized immediately that it was not a leopard at all but was instead a most remarkably-patterned cheetah with unusually long silky fur and a prominent mane. Taking note of the latter feature and its regal appearance as a whole, Pocock formally named it in 1927 in the *Proceedings of the Zoological Society of London* as *Acinonyx rex*, the king cheetah, classifying this striped form as a distinct species in its own right, totally separate from the normal spotted cheetah.

As a result of Major Cooper's zealous researches, further king cheetah skins were discovered and publicized, and one was bought for £150 by Lord Walter Rothschild (at that time the world's foremost private collector of animal specimens, which were housed in his own natural history museum at Tring in Hertfordshore, England). This latter skin was particularly interesting in that its markings were less ornate than those of other king cheetahs, so that it bore more of a resemblance to the typical spotted form than did any of the earlier "kings," as duly noted by Pocock (*The Field*, 15 April 1928).

Indeed, this somewhat intermediate specimen inspired Abel Chapman in the very next issue of *The Field* to voice the opinion that the king and spotted cheetahs belonged to the same species after all. Over the next few years, an increasing number of zoologists subscribed to this view, until in 1939 Pocock himself finally conceded that the king cheetah was indeed most likely to be merely a freak variety of the normal cheetah. In just over a decade the king had been crowned and dethroned. It was now referred to simply as *Acinonyx jubatus* var. *rex,* and scientific interest waned.

Nevertheless, skins and sightings of king cheetahs continued to emerge, and almost invariably from Zimbabwe. In a *Field* communication of 24 May 1962, Sir Archibald James listed eight skins (of which three had been lost or destroyed), and by 1980 the total had risen to 13. By 1987, 38 specimens were known, recorded south of the Zambezi in a triangular portion of Africa enclosing eastern and southern Zimbabwe, northern South Africa, and eastern Botswana, with the habitat housing the kings consisting of thorn forest and woodland, usually of high elevation—very different from the typical savanna habitat of normal cheetahs.

King cheetahs (David Pepper-Edwards)

This feature was not lost upon the researchers responsible for the modern-day upsurge of scientific interest in the king cheetah—biologist Lena Godsall Bottriell and her husband, Paul Bottriell. Until the late 1970s, the only known photograph of a living king cheetah was one that had been taken in 1974 within the Kruger National Park and published by G. de Graaf that same year in *Custos*. Five years later, in 1979, and with the aid of a hot-air balloon, the Bottriells conducted an intensive aerial investigation of king cheetah country in its triangle of distribution.

Moreover, they succeeded in tracking down and photographing a number of skins as well as subsequently filming live specimens in captivity (more on which follows), and they also discovered a most surprising fact concerning this striped variety of cheetah.

Hair samples that the Bottriells had submitted to the Institute of Medical Research in Johannesburg, South Africa, revealed that the cuticular scale pattern of king cheetah guard hairs compared much more closely with that of leopards than with that of normal spotted cheetahs. Yet if the king cheetah were nothing more than a freak coat-pattern morph of the normal cheetah, one would have expected its hair structure to correspond (or at least to be most similar) to that of the latter felid. The striped cheetah mystery deepened.

It was no longer simply an abnormally-marked mutant—it also differed from the spotted form in the structure and texture of its fur, in that it inhabited forests rather than savannas, and as the Bottriells also discovered, in that it tended to be more active during the night than did the typically diurnal spotted cheetah

Worth mentioning here is that according to East African legend noted in *Animal Kitabu* (1967), leopards and cheetahs will sometimes mate, yielding crossbreed progeny. Arguing against a hybrid identity for the king cheetah, however, the Bottriells remarked that leopards actively attack cheetahs; and even if they did mate, the two species are so different genetically that the production of offspring would surely be impossible.

In May 1981, a litter of cubs was born to normal spotted cheetah parents at the de Wildt Cheetah Breeding and Research Centre of Pretoria's National Zoological Gardens. One of these cubs proved to be a king cheetah, thereby proving categorically that this variety was neither hybrid nor separate species. A few days later, a king cheetah was also born in a litter produced by the sister of the first litter's mother (*Wildlife*, February 1982). Now, by carrying out monitored breeding studies, the genetics of king cheetah production could be studied.

By 1984, a recessive mutant allele was suspected to be responsible, and in 1986 a formal scientific paper on the subject appeared in the *Journal of Zoology*, containing the findings of king cheetah researchers R.J. van Aarde and Ann van Dyck. They proposed that this felid's remarkable coat pattern was indeed due to a recessive mutant allele, one quite probably homologous (equivalent) to that responsible for the blotched tabby pattern in the domestic cat. In short, the entire king pattern was due to a single genetic difference.

Moreover, in a paper published by the journal *Science* in September 2012, a team of American genetics researchers from several different institutes revealed that they had identified the specific gene responsible for the king cheetah's striped coat pattern and the blotched coat pattern in domestic tabby cats. Both are caused by a recessive mutation in a gene dubbed Taqpep by the researchers.

Lena and Paul Bottriell, however, were not satisfied with the simple expression of a single mutant allele as a complete explanation for the king cheetah's very exotic morphology. As they pointed out, the change from spotted to king in the cheetah involves several markedly different components (body stripes, body blotches, more pronounced mane, distinctive hair, striped upper tail, and rings around the remainder of the tail), hardly compatible with either a single color change (as in albinos, or melanistic individuals) or the single markings change (as from stripes to blotches in domestic tabby cats) that normally results from a single allele change. Moreover, whereas blotched tabbies are born everywhere, king cheetahs have a very precise and limited distribution within the total range known for the cheetah.

In combination with the king cheetah's environmental and behavioral idiosyncrasies already noted, all of these anomalies led the Bottriells towards a dramatic but fascinating conclusion. Namely, that the king cheetah is demonstrating evolution before our very eyes—the development within a specific region of a completely new strain of cheetah, adapted not for daytime sprinting across open grasslands as its hunting mode but instead for nocturnal stalking through dense forests where its more heavily-patterned coat would provide it with excellent camouflage. In short, natural selection has acted upon the original "freak" king individuals to bring about a habitat extension for the cheetah, which (given sufficient time and continuous isolation between spotted and king individuals) could result in so great a genetic and morphological divergence between the two forms that they will ultimately cease to belong to the same species.

Thus the king cheetah would seem to be more than a mere freak. Instead, and as labelled in the title of the Bottriells' presentation at the 1987 International Society of Cryptozoology's annual members meeting, it may well be a new race in the making. The Bottriells' hypothesis, presented fully in Lena Bottriell's stimulating book *King Cheetah* (1987), is eminently plausible, fascinating, and fits the facts very succinctly—the nsui-fisi has been vindicated after all!

Mpisimbi

What makes the king cheetah so memorable in addition to its incredibly beautiful coat is its extremely limited distribution. Many freak mutations of coat color or patterning in mammals are spontaneous, i.e. they can arise abruptly in any population of a given species, regardless of geographical location. Yet whereas the typical spotted cheetah occurs in southern, eastern, central, northern, and western Africa, king cheetahs have never been reported conclusively outside southern Africa—or have they?

There are two possible and potentially extremely significant exceptions to this widely-assumed rule. (Perhaps three if we consider some remarkable evidence that I have uncovered for the erstwhile presence of at least one king cheetah specimen in the wild not anywhere in Africa, but instead in Asia, and which in 2013 I formally documented in the *Journal of Cryptozoology*—whose logo, very aptly, is a king cheetah—reprinted here in Appendix 1.)

The first possible exception to the king cheetah's strict zoogeographical limitation to southern Africa is a king cheetah skin that in 1988 turned up in the West African country of Burkina Faso (formerly Upper Volta). It supposedly came from a specimen that had been shot by a poacher in the northern end of the Singou Total Fauna Reserve. However, some researchers wonder whether this mystifying skin is one that in reality originated in southern Africa but later travelled northwest via itinerant poachers or other skin traders. Alternatively, however, could a "king" strain of cheetah have spontaneously arisen in West Africa?

As for the second putative exception to the rule of the king cheetah being exclusively southern African in distribution, this is one that had never been publicly revealed until I did so via my *ShukerNature* blog in 2014. It features an extremely obscure East/Central African mystery beast known as the mpisimbi

In 1927, *Chambers's Journal* published a fascinating article on East and Central African mystery beasts entitled "On the Trail of the Brontosaurus and Co." It was written by "Fulahn," the pen-name of Captain William Hichens, a man whose name should already be familiar to mystery cat aficionados. For he was none other than the Native Magistrate at Lindi, Tanganyika (now Tanzania) during the 1920s and 1930s who investigated a succession of particularly gruesome murders there attributed by the local people to a giant brindled mystery cat known as the mngwa or nunda, a subject I will take up later in this chapter.

Most of the cryptids documented by Hichens in his *Chambers's Journal* article are relatively famous ones, with one notable exception. Contained in his account are a couple of tantalizing lines that have fascinated and frustrated me in equal measures for many years: "But such are the mystery animals. There are others—the mpisimbi, the leopard-hyaena, which eats sugar-cane, and which I have hunted many a weary night without success..."

Despite numerous searches, I have never been able to uncover any additional information concerning this enigmatic creature. So what exactly is the mpisimbi?

The quoted lines offer no morphological description whatsoever of the mpisimbi. Conversely, its name's English translation—"leopard-hyaena"—is very intriguing because it corresponds precisely with the English translation of the king cheetah's native South African name, "nsui-fisi." Moreover, Hichens's unusual claim that the mpisimbi eats sugar-cane adds further to a putative link between the mpisimbi and the king cheetah because in a second article, published under his own name a year later in *Wide World Magazine*, Hichens stated: "The Nsuifisi, or striped cheetah...was also reputed to be a raider of grain and sugar-cane."

Of course, as Hichens went on to discuss, because cheetahs are carnivores it seems improbable that they would raid grain-plots. And even though hyaenas are notorious scavengers with an extremely catholic diet, they are not known to attack standing crops, but they will certainly devour cooked grain, vegetables, and even boiled flour.

Such considerations and qualifications, however, are not significant with regard to the cryptozoological mystery under review here. What is significant is that both the mpisimbi and the nsui-fisi were claimed, rightly or wrongly, by the native tribes in their respective, separate areas of Africa to consume the very same unexpected foodstuff—sugar-cane.

Is it conceivable, therefore, that the mpisimbi and the nsui-fisi are indeed one and the same creature—namely, the king cheetah? If so, it suggests that at some time in the distant past, striped cheetahs did exist in East and/or Central Africa—although, with no modern-day reports of such beasts on file, even if they once did exist there they seemingly no longer do. Put another way: whatever it may have been, tragically the mpisimbi is now apparently extinct.

Of course, skeptics may well claim that this is all supposition, but the presence of those brief lines regarding the mpisimbi in Hichens's article means that the possibility of mpisimbi and king cheetah synonymity, however remote it may seem, cannot be discounted.

Moreover, who can say whether, in the future, a king cheetah or two will not spontaneously arise in the East African population of the normal spotted cheetah? That is, after all, what spontaneous mutations do!

At present, however, the mystery of the mpisimbi's zoological identity remains yet another enigma in the eventful history of Africa's extraordinary striped cheetah—the once (and future?) king.

Woolly Cheetah

Whereas the king cheetah has enjoyed a well-earned scientific renaissance in modern times, for over a century a second equally perplexing cheetah form seemed to have vanished without trace into the mists of scientific obscurity—until I restored it to public attention in the 1989 edition of this book and in various later writings of mine. This second controversial form is known as the woolly cheetah.

On 19 June 1877, Philip L. Sclater, longstanding Secretary of the Zoological Society of London, recorded in its *Proceedings* (i.e. the *PZSL*) the acquisition by London Zoo of a most unusual cat—male and apparently not fully grown—which he described as follows:

> It presents generally the appearance of a cheetah (*Felis jubatus*) [its old scientific name], but is thicker in the body, and has shorter and stouter limbs, and a much thicker tail. When adult it will probably be considerably larger than the Cheetah, and is larger even now than our three specimens of that animal. The fur is much more woolly and dense than in the Cheetah, as is particularly noticeable on the ears, mane and tail. The whole of the body is of a pale isabelline colour, rather paler on the belly and lower parts, but covered all over, including the belly, with roundish dark fulvous blotches. There are no traces of the black spots which are so conspicuous in all of the varieties of the Cheetah which I have seen, nor of the characteristic black line between the mouth and eye.

His report was accompanied by a full-color chromolithograph that revealed just how unusual but beautiful the cat was.

Evidently this brown-blotched felid appeared very different from the usual form, to the extent that Sclater stated that it was impossible to associate it with this. Instead, he proposed for it the temporary name of *Felis lanea*, the woolly cheetah. It had been obtained from Beaufort

West, South Africa, and as Sclater himself remarked: "It is difficult to understand how such a distinct animal can have so long escaped the observations of naturalists."

One other matter is also difficult to understand, and remains a source of confusion concerning this mystery cat. Sclater referred to its markings as blotches, but in the chromolithograph that accompanied this report the creature was depicted with numerous tiny spots.

London Zoo's woolly cheetah, depicted in a 1877 chromolithograph (public domain)

A year later, on 18 June 1878, Sclater noted in the Society's *Proceedings* that he had received a letter from a Mr. E.L. Layard, informing him that a second woolly cheetah was currently preserved in the South African Museum. As with the first one, it had been procured from Beaufort West and had been killed by Arthur V. Jackson who, like Layard himself, assumed that it was an erythristic (abnormally red) variant of the normal cheetah. At the end of this report, in answer to an enquiry by Layard, Sclater recorded that the claws of the London Zoo specimen were non-retractile.

Sharing Sclater's own bewilderment as to how so large and unusual an animal could have evaded scientific detection until then, many zoologists had grave reservations concerning his optimism that the woolly cheetah constituted a totally separate species. In 1881, English biologist St George J. Mivart commented that the noted American zoologist Daniel G. Elliot regarded this felid simply as a variety of the known cheetah species. (Curiously, Mivart added a stripe to one side—but not both sides—of the woolly cheetah's muzzle when describing this felid form in his book *The Cat* (1881), a feature not mentioned by Sclater and in any event highly abnormal, thereby confusing the issue even further.) By then, the London Zoo's specimen had died, and Elliot's opinion received support from the discovery by the eminent mammalogist Oldfield Thomas of London's Natural History Museum that this cat's skull did not differ from those of other cheetahs.

On 4 November 1884, Sclater recorded in the *PZSL* a woolly cheetah skin sent to him by the Reverend G. Fisk, again obtained from

Beaufort West. Yet in comparison with the Zoo specimen, this example was more distinctly spotted, less densely furred, and rather smaller in size. Reverend Fisk believed that these differences were due to the specimen being a female, an explanation accepted by Sclater, who felt that this new skin consolidated his opinion concerning the woolly cheetah's separate status. The rest of the scientific world, conversely, remained unconvinced; and since then, the woolly cheetah has been regarded as merely an unusual non-taxonomic variant of the typical cheetah species.

True, the woolly cheetah may indeed be nothing more surprising than an atypical color morph—perhaps an erythristic version, as opined by Jackson and Layard, or even a partial albino, as suggested independently by Lena Bottriell and Roy Robinson. At the same time, Sclater's more radical views can also be appreciated because this cat form differs from the typical cheetah not only in color and markings but also in fur density and even in relative limb length. Excluding the involvement of pleiotropic genes (genes that influence two or more ostensibly unrelated phenotypic characteristics), simple color variants do not generally exhibit such pronounced differences as these from normal individuals of the same species. Its shorter limbs suggest a non-cursorial life; could it possibly have been a forest form?

It is worth noting that a "lion-like forest cheetah" known as the kitanga was described in the 20th century's early years to Major G. St J. Orde-Brown by the Embu natives of south-eastern Kenya (as recorded by Kenneth C. Gandar Dower in his book *The Spotted Lion*, 1937, chronicling Dower's own searches for another of Africa's mystery cats, the elusive marozi, discussed later in my book). Furthermore, according to correspondent Owen Burnham who lived there for many years, a comparable felid has occasionally been reported from the little-explored forests of Senegal, West Africa, which is apparently very different from the cheetah's typical spotted form that, albeit very rare there nowadays, is known to exist in this country (Burnham, pers. comm.).

As already discussed with the king cheetah, the possibility of a cheetah form becoming modified for life in this type of habitat is by no means implausible. On the contrary, even the normal spotted form is not an exclusive denizen of the savanahs. This was well demonstrated in March 1983, when Lise Campbell spied a single cheetah at a height of 2.5 miles in the vicinity of the Sirimon Track in the moorland zone of Mount Kenya. She had a second sighting later that day of what may have been the same animal, even higher, amidst the tufted high-altitude grass, and documented her observations in an *East African Natural History Society Bulletin* communication (May-June 1983).

According to an *Arnoldia Zimbabwe* paper of 1980 authored by mammalogists Daphne Hills and Reay Smithers, the woolly cheetah no longer occurs in Beaufort West. Presumably, therefore, this odd form is extinct, and the chance to investigate further its precise nature similarly lost—or is it? London's Natural History Museum owns the skin of the London Zoo specimen, so now, with the ever-advancing techniques of DNA-based genetic analyses readily available to researchers, perhaps it may be possible to carry out genetic tests upon small tissue samples taken from this specimen and finally reveal the mystifying woolly cheetah's true identity.

Cheetalines, Black Cheetahs, and Desert Cheetahs

In a *Field* letter of 19 March 1921, Reginald I. Pocock brought to public attention the skin of a very unusual cheetah. Instead of being patterned with the normal cheetah's black spots, it was flecked with a dusting of only the finest, tiniest dots (rather like those of the servaline, a small-spotted pattern morph of the serval *Leptailurus serval*), and even these were absent on the neck (where its mane was remarkably sparse) and a large area behind the shoulders. This extraordinary cheetah had been shot in what is now Tanzania by Lieutenant-Colonel W.T. Gregg of the King's African Rifles. In view of its comparability to the serval's servaline phase, I proposed in my book *Cats of Magic, Mythology, and Mystery* (2012) that this remarkable speckled variety of cheetah should be officially termed a cheetaline.

Kenyan cheetaline (Guy Combes)

Until quite recently, the Tanzanian cheetaline appeared to be unique. On 5 September 2011, however, I received an extremely exciting email from wildlife artist Guy Combes who informed me of the current existence in the Kenyan wilds of a living cheetah exhibiting the very same finely-speckled coat markings! Not only that, Guy kindly enclosed with his email one of many color photographs that he had taken of this remarkable animal, a subadult male, which readily confirmed its morphological status as a second, bona fide cheetaline.

Guy has included a detailed account of this extraordinarily beautiful cat on his website (at: http://guycombes.wordpress.com/2011/06/15/the-phantom/), which also contains a spectacular painting of it that he has meticulously prepared.

Guy first learned of this modern-day cheetaline's existence in December 2010, and realized that such a rare creature urgently required not only monitoring and surveillance, but also some carefully selected publicity in order to secure its protection. So he decided to produce a painting of it in order to assist him in achieving these goals. Although he subsequently received a phone call confirming that the cheetaline had been sighted, various airplane searches of the area in question by him failed to track down his elusive quarry. Consequently, Guy then set forth by Land Rover, and this time he was successful—not only encountering the cheetaline and its mother (a normal cheetah), but also observing it closely for around 30 minutes and snapping plenty of photos to use as source material when preparing his painting, which he has formally—and fittingly—entitled "The Phantom."

Judging from photographic evidence and Guy's painting, the Kenyan cheetaline may well be a leucistic individual because its eyes, nose, and even the characteristic cheetah tear-drop mark down each side of its jaw are of normal coloration (which would not be the case were it an albino). Let us hope that the location containing this cheetaline will now receive full and continued governmental protection, thereby safeguarding one of the most wonderful mystery cats ever documented from Africa.

Also on file are a few records of all-black (melanistic) cheetahs. These include a specimen spied in the company of a normal cheetah by Lesley D.E.F. Vesey-Fitzgerald in Zambia during the first half of the 20th century, and another one sighted in Kenya's Trans-Nzoia District by H.F. Stoneham in 1925.

Even more elusive is the Qattara cheetah of Egypt, reports of which date back to the late 1960s and include a specimen taken by a Bedouin shepherd in 1967. None has been observed by scientists so far, however, although cheetah-like tracks have been photographed in the Qattara region on two separate occasions during the 1990s. It is said to be paler in color and with a thicker coat than the typical cheetah from eastern and southern Africa, and is probably similar to the little-known desert-adapted cheetahs *Acinonyx jubatus hecki* with speckled, sandy pelage living in Niger's Tenere Desert, West Africa, which were photographed for the first time as recently as the early 1990s, and again during summer 2010.

In short, the king and woolly cheetahs are by no means the only unexpected cheetah forms on record. In view of the wide range of colors and markings exhibited by certain other felids—notably the leopard—one could be forgiven for seeing little reason to be surprised at a similar trend in the cheetah. However, as discovered by Stephen O'Brien and co-workers via certain protein examinations (isoenzyme analyses), African cheetahs display a conspicuously low degree of genetic variation (lower than that for lions, tigers, and leopards), i.e. the cheetah populations are very inbred, especially in South Africa (*Science*, 9 April 1987).

Consequently, unless genetic drift is exerting a major influence here, one would hardly expect the appearance of genetic variants as markedly different from the normal type as are the king and woolly cheetahs (both inhabiting South Africa too!), which makes the reconciliation of these as mere freaks all the more difficult to accept. Even though we know that the king cheetah is not a separate species, the fact that this form and the woolly cheetah actually derive from South Africa strongly suggests that they may be more delineated genetically from the normal cheetah than many taxonomists currently believe. Certainly, in view of the latter's depauperate genetic variation, it is very difficult to explain away these forms as simple mutants. Evidently, cheetah genetics hold quite a few more surprises in store for science.

Spotted Lions

As with the woolly cheetah, at least one form of spotted lion is known to science via preserved skins, hence its actual existence is not in doubt. What is in question, however, is its precise identity.

Reports of such creatures had filtered into Europe from East Africa as long ago as 1903, via Colonel Richard Meinertzhagen, soon to be immortalized in zoological circles as the discoverer of the Congo's giant forest hog. Moreover, according to an Irish adventurer named C.J. McGuinness, writing in his book *Nomad* (1934), one of the hunters employed by the famous animal collector Carl Hagenbeck had himself sighted a spotted lion. Nevertheless, the scientific world took little notice of these testimonies. Then in 1931, Kenyan game warden Captain R.E. Dent observed four lions crossing a track ahead of him at a height of just under 11,000 feet near to the headwaters of Mount Kenya's Garcita River. He noticed that they seemed to be darker and smaller than normal lions, but simply assumed that this impression was due to a trick of the light. Yet some months later, his native attendant boys informed him that they had trapped a strange cat on the Aberdares mountains in west-central Kenya that seemed intermediate in appearance between lion and leopard. Unfortunately, they had not preserved its skin.

This situation was remedied a little later by Michael Trent, when he shot two lions that had raided his Aberdares farm, sited above Tomson's Falls. He was very surprised when examining the bodies to discover that although they were both adults (one male, one female), each possessed a very heavily spotted coat.

Normal lions display varying degrees of spotting when cubs, but this is generally lost by adulthood, and in any event it never approaches the intensity and extent of spotting sported by these two, covering neck, back, flanks, and legs. Equally perplexing was the presence in the male of only the sparsest of manes.

Very fortunately, Trent preserved these remarkable skins, which in themselves demonstrated unequivocally that adult lions with conspicuous spotting did indeed exist. At the same time, however, as he did not retain their skulls or skeletons, their precise taxonomic identity could not be conclusively established.

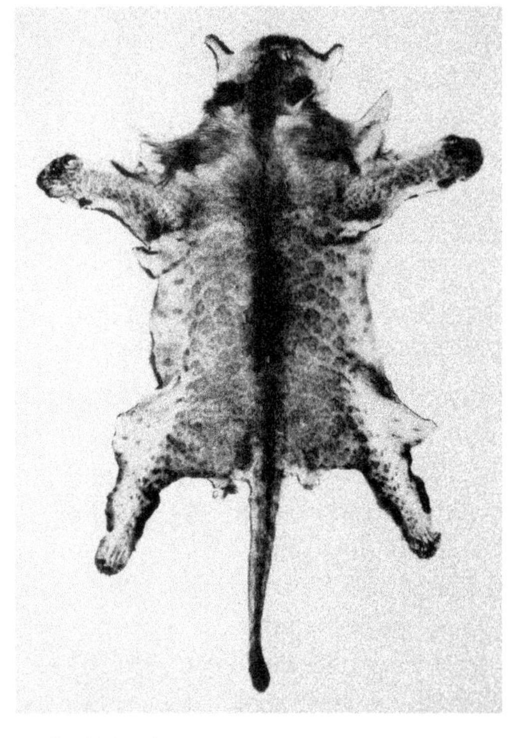

Michael Trent's male marozi skin (public domain)

Nevertheless, the spotted lion saga was steadily gaining momentum because Trent's pair of patterned pelts attracted the interest of wealthy young adventurer Kenneth C. Gandar Dower, who in turn brought this felid to popular Western attention by preparing an illustrated article on the subject, published by *The Field* on 6 July 1935. In addition, he was so attracted to the possibility of a new species of lion awaiting formal scientific discovery that he organized an expedition to the Aberdares in search of this mottled mystery cat, which, it transpired, was well known to the region's natives, who referred to it as the marozi.

Gandar Dower was accompanied by renowned hunter Raymond Hook, but despite their perseverance the quest did not succeed in its ultimate objective of obtaining a specimen. Nevertheless, some unusual tracks were sighted, which seemed too small in size for a lion's but too large for a leopard's and appearing to be thinner in shape than those of a young lion. Undaunted, Gandar Dower made further recourses to the Aberdares, and on one occasion, due to a misunderstanding, missed the opportunity of observing some spotted lions by a mere 24 hours.

Despite his very understandable frustration and disappointment at such events, in 1937 Gandar Dower published a very interesting and well-written book entitled *The Spotted Lion*, in which he described his search and included an appendix written by Reginald I. Pocock concerning Trent's male specimen. Pocock had examined this skin personally, and like other observers he had been struck by its exceptional degree of spotting, which he went so far as to describe as being jaguarine in quality. Furthermore, he confessed that not even the Natural History Museum in London had anything to compare to this felid.

Even so, Pocock preferred to reserve final judgement concerning its taxonomic status until he could examine skeletal material from such lions. A skull had been passed to him that was supposedly from one of Trent's specimens, but this has never been officially confirmed. It seemed to be that of a young lioness, although Pocock did not totally rule out the likelihood that it had come from a form of diminutive lion. Quite plainly, the scientific world was somewhat bemused by these paradoxical pelts.

The interest that Gandar Dower's exploits and explorations had stirred among the hunting and wildlife fraternities, as well as Kenyan inhabitants, resulted in a welter of new information and accounts of such animals coming to light. Gandar Dower himself, for example, was informed that in East Africa's Virunga mountains a man had encountered a lion that in his opinion had resembled a jaguar (remember Pocock's use of the word "jaguarine" in relation to Trent's skins). He also received word from a lady curious to know why he was so excited about "the ordinary common or garden Spotted Lion of the Aberdares which everybody had known about for years"!

Similarly, *The Field* published much correspondence on this subject during the 1930s and 1940s. On 1 June 1935, for instance, it reproduced a photograph and accompanying letter by Andrew Fowle concerning a normal-sized lion aged two years but still possessing distinct spotting. Nevertheless, this individual did not compare with the extensive patterning exhibited by Trent's cats. On 11 December 1937, it published a letter by B.V. Richardson, who did not share Gandar Dower's optimism that Trent's specimens represented something more than a mere freak form of normal lion.

For although Richardson's own safaris to the Aberdares did not traverse higher than the mountains' foothills and plateau, he had made contact with most settlers and a fair proportion of the natives who had at some time travelled through areas covered by Gandar Dower's searches; yet he had never heard any of these people speak of spotted lions. Richardson also remarked that native attendants were liable to exaggerate in an attempt to please their Western employers, fostering their hopes of locating something that may well not even exist. Could this be the origin of the marozi?

Apparently not, because on 9 October 1948, G. Hamilton-Snowball informed *The Field* that many years before Gandar Dower he too had learned of the marozi, from Kikuyu attendants in the Aberdares, who considered it to be a creature very different from the typical lion, being smaller, spotted, hunting in pairs, and living only at very high altitudes. In addition, during the spring of 1923, while crossing the Aberdares on foot across the Kinangop plateau from Liakipia to N'jabini, at a height of 11,500 ft and at about 4:00 p.m., he had observed a pair of strange cats approximately 200 yards away. The light was poor, and he thought at first that they were "two very tawny and washed out looking leopards," but when he turned to his bearer for his rifle he overheard an excited murmuring from his native attendants and the repetition of an unfamiliar word—"marozi." But before he could bag either of these curious cats, they had turned away, bounding to safety within the nearby forest belt. When questioned, his attendants firmly denied that lions ascend to such mountainous heights, but affirmed: "Marozis live here."

Painting depicting a pair of marozis (William M. Rebsamen)

On 13 November 1948, a further spotted lion letter appeared in *The Field*, written by J.R.T. Pollard, who was an acquaintance of Raymond Hook. Pollard recalled that Hook believed the spotted lion as such to be largely mythical, although he did suggest that the existence of a small race of lions, possibly descended from the plains form and driven by European settlements to take refuge in the mountainous forests of Kenya, would not be impossible. As far as a fully-fledged spotted race was concerned, however, Hook felt that at present there was insufficient evidence to support this possibility.

In contrast, Pollard himself was sympathetic to the mooted existence of an unknown felid in Mount Kenya's upland rainforests. He recollected that Powys Cobb of Elementeita (who possessed a wide knowledge of Africa's larger carnivorous mammals) had given chase to an unusual cat trespassing on his farm, near to the edge of the Mau Forest. Cobb observed that the felid was intermediate in size between a lion and leopard, and left behind spoor resembling a small lion's; thereafter he remained convinced that an unknown cat did indeed survive in this little-explored region

Conversely, G. Flett suggested that Cobb had been deceived by shadow effects into believing that he had seen a spotted felid (*The Field*, 15 January 1949). Flett based his opinion upon personal experience because on two separate occasions in Kenya this particular optical illusion had temporarily persuaded him that the lions that he had been watching at close range were genuinely spotted.

On 30 September 1950, *The Field* published a full-page review of the spotted lion saga, written by Major W. Robert Foran. Although greatly enjoying Gandar Dower's book, Foran remained highly skeptical of its subject's mooted identity as a full-blown spotted race or subspecies of lion. Instead he favored the idea that some individuals of the modestly-sized Somali lion *Panthera leo somalica* (which has little in the way of a mane) had somehow wandered into Kenya and up into the Aberdares and included amongst themselves some freak members with intense spotting. He concluded his review by stating that he would await further developments in the attempt to solve this riddle conclusively, but sadly no such development has occurred.

Reports of marozis were recounted by various persons operating in the Aberdares during the Mau Mau Uprising of 1920-1963 (which included the Aberdares invasion by the Kikuyu-dominated Mau Mau forces in 1952). Speaking of which, on 26 June 1999, I received an email concerning a marozi skin that had never been publicly documented. The email was from Jorge A. de Lima Jr, who informed me that his father—an African wildlife hunter and movie maker for almost 28 years, whose name appears in a number of books, including William Negley's *Archer in Africa* (1989)—had once owned the skin of a bona fide Kenyan spotted lion. Tragically, however, it was stolen by Mau Mau terrorists.

Also of note, yet not previously included in any cryptozoological document as far as I am aware, is a short but intriguing passage concerning the spotted lion that was included by the famous Kenyan-born authoress and journalist Elspeth Huxley in her book *On the Edge of the Rift: Memories of Kenya* (1962). While trekking one time in a party across a shoulder of Kingangop, one of the three peaks constituting the Aberdares, using a cart and ponies for transport rather than a motor vehicle, their guide Njombo complained that a car should have been used. When told that there hadn't been enough money (shillings) for a car, Njombo disparagingly replied:

> "What good are shillings if we are killed by elephants or buffaloes, or by the fierce marozi that lives with spirits in this bad bush?"
>
> [Huxley then asked him what a marozi was, because she had not heard that word before.] A kind of lion, he said, but spotted like a hyaena; it lived only in the mountains and attacked people who slept there.

Sadly, popular interest in the spotted lion has faded in more recent times, and nowadays it is chiefly remembered (if remembered at all) as a whimsical wildlife novelty of the World War II period. Yet is this really all that it was—the product of romantic dreams of adventure on the part of Gandar Dower, triggered by a pair of freak skins, perpetuated by optical trickery? All of these may well be involved, but the simple facts of this cryptozoological case, coupled with relevant information drawn from parallel zoological situations, suggest that something much more substantial was also present. Consider the following possibilities.

(1) Trick of the Light?

Although optical illusions may certainly be responsible for some spotted lion reports, the skins of Trent demonstrate unequivocally that they clearly cannot be offered as an explanation for all of them.

(2) Native Invention?

The suggestion that native attendants say anything to please their Western employers may be true on occasions, but it must be remembered that when Hamilton-Snowball sighted those two strange cats in the 1920s, his native attendants murmured "marozi" amongst themselves—not directly to him. Their reaction appeared to have been spontaneous, triggered by their surprise at seeing the cats, rather than as a means of pleasing Hamilton-Snowball. The vast number of native reports that have led to major new animal species being discovered should also be borne in mind here.

(3) Freak Individuals?

Andrew Fowle's spotted lion photograph demonstrated that spotting can be retained well into adulthood by normal lions, but the extent of spotting displayed by this specimen was much less than that of Trent's cats. Furthermore, its normal body size conflicts with their stunted appearance, and the meagre mane of the male member of Trent's pair also has to be explained. An ordinary lion cub may sometimes retain its juvenile pelage into maturity, but when it also retains its juvenile size and fails to develop one of its species' most characteristic sexual features, it is surely time to look for a better explanation.

(4) Leopons?

A possible solution that on first sight has rather more potential than the previous three is that the spotted lion may constitute a natural hybrid of lion and leopard. Such hybrids, known popularly as leopons, have been bred on several occasions in zoos, especially in Japan's Hanshin Park Zoo and Koshien Zoo, as well as in Italy more recently. These very striking animals bear more than a passing resemblance to the descriptions of marozis, possessing heavily-spotted bodies, poorly-developed manes in males, and attaining an overall size intermediate between leopard and lion.

One of Japan's male leopons (Warren D. Thomas)

According to North African legend, lion x leopard interbreeding does occur in the wild. However, their offspring, although exhibiting a leopard-like pelage, are supposed to be fully-maned and as powerful as the lion—a far cry from the marozi. In any event, naturalist Henry Scherren pointed out that wild-born leopons would probably not be produced because an encounter between a wild lion and leopard would most likely result in the leopard's death (*The Field*, 25 April 1908). This is very feasible, with numerous reports existing of interspecific felid aggression. But the biggest obstacle of all to be surmounted when attempting to reconcile spotted lions with leopons is the fact that such hybrids are almost certainly sterile. Very few Big Cat hybrids have ever been obtained that have themselves succeeded in producing viable offspring.

A notable exception was the mystery felid supposedly originating from the Congo and of a hitherto unknown species that was briefly exhibited at the London Zoo in 1908. Subsequent investigations revealed, conversely, that it had been bred in captivity in the USA, and was actually a complex three-way crossbreed—the offspring of a lion and a female jagupard (jaguar x leopard hybrid), i.e. a lijagupard!

Uneeka, London Zoo's lijagupard, Illustrated London News, 9 May 1908 (public domain)

The full history of this truly remarkable felid, whose owner aptly named it Uneeka ('Unique'), is presented online in my *ShukerNature* blog and also in my book *ShukerNature Book 2* (2020), together with the only known photograph of Uneeka in the living state, but which had remained wholly unknown to zoologists until I encountered it in a very old magazine.

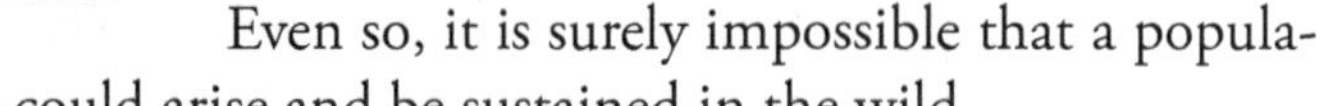

Even so, it is surely impossible that a population of leopons could arise and be sustained in the wild.

(5) Separate Taxonomic Form?

Could the spotted lion really be a distinct race, subspecies, or even full species of lion? The stark fact generally overlooked or ignored by marozi skeptics is that its appearance as described by many observers with no knowledge of zoology or ecological principles is exactly what a zoologist or ecologist would predict for a leonine type adapted for inhabiting dense forests. Namely, a smaller, less stocky form with dappled pelage, a more primitive hunting technique, and, in the male, little or no mane.

A bulky felid would not move as effectively through thick forests and would not be effectively hidden from its intended prey with an unpatterned pelage. The vast open spaces of the African savannas have facilitated the evolution in the normal lion of social hunting; it's not carried out by any other known felid form because such activity would be far less effective in densely wooded areas. Similarly, as it would obviously be very difficult for the two sexes of the lion to identify one another across huge expanses of grassland via the normal subtle signals and color differences exhibited by other felids, the development of highly visible and distinctive morphological features are required, hence the male lion's unique mane. With a lion inhabiting deep forests, conversely, this feature would be superfluous (in fact, a mane may even be deleterious—betraying the lion's presence to would-be prey).

From this, it can be seen that the marozi fits the bill very well as a forest lion, morphologically and behaviorally. Moreover, there is nothing unusual in the concept of a given mammalian species existing in different forms according to habitat. The African elephant exists in what were traditionally deemed to be two distinct subspecies (only recently upgraded to full species): one large and inhabiting plains and bush, the other smaller and inhabiting forests. In the Cape buffalo, its savanna subspecies is huge, black, and aggressive; its forest form, conversely, is small, reddish, and reclusive. Compared to these, a somewhat small and spotted forest lion is nothing extraordinary at all.

Finally, if the spotted lion truly constitutes a separate race rather than a few freak individuals, we would expect it to occur in parallel with the plains lion across Africa, at least in relatively undisturbed forests, rather than only in Kenya. And sure enough, felids closely resembling the Kenyan marozi have indeed been reported from various mountainous forests distributed widely across Africa, as now revealed.

Ikimizis and Wonder-Leopards

For example, reported in the Mufumbiro Volcanoes region of Rwanda is the ikimizi, a mystery cat that in 1921 attracted the attention of no less a personage than Prince William of Sweden. He later included the following description of it (based upon native reports) in his book *Among Pygmies and Gorillas* (1923):

> The ikimizi is said to be a cross between a lion and a leopard. No white man has ever seen it, and only a few black ones. It is said to be grey in colour, with darker spots and a beard under its chin.

Despite exploring the area where one such animal had been seen some time previously, and putting out bait in the hope of luring a specimen, Prince William did not spy an ikimizi, which dissuaded him from believing in its existence. Nevertheless, he did note that this felid even had its own native name, in spite of the fact that the natives' language had a very restricted vocabulary. This in itself implies that the ikimizi is indeed very different in appearance from other cats known to them (lion, leopard, serval, etc).

A similar animal, referred to as the bung bung, has been reported from the Cameroons; a comparable creature was spoken of by East Africa's Akamba tribe to the Swedish ethnographer Gerhard Lindblom; and an Ethiopian equivalent is known as the abasambo. Even the Embu natives' alleged forest cheetah (kitanga) may actually be a spotted lion.

In the Central African Republic's Ubangi region, French naturalist Émile Gromier heard tales of a strange and extremely fierce cat form called the bakanga, said to be intermediate between a lion and leopard. Although comparable with a maneless lion in overall appearance, its reddish-brown pelage was supposedly dappled like a leopard's. Its most curious feature, however, was that it did not roar, but instead barked rather like a dog! Although he was promised skins of the bakanga by white hunters, Gromier never received any, and his subsequent researches failed to uncover anything further regarding this very peculiar beast

In Uganda, the spotted lion is represented by the ntarargo. In actual fact, the term "ntarargo" (sometimes spelled "ntarago" or "enturargo") is a plural noun, of which "ruturargo" is the singular, at least according to Captain Charles R.S. Pitman (game warden and African wildlife authority), writing in his book *A Game Warden Among His Charges* (1931). Nevertheless, so many reports use "ntarargo" as a singular noun that it will be used in this sense here too for the sake of conformity. Also worth noting is that another version of this name exists—"kitalargo," which Captain Pitman concluded was a contraction of "kitalo-engo," or wonder-leopard. Regardless of its name, however, its appearance follows the same familiar pattern: "This creature is said to be a cross between a lion and a leopard [with] a long tail, a slightly spotted skin, and retractile claws."

This is the description given by the Kichwamba natives to E.A. Temple-Perkins, as documented in his own book, *Kingdom of the Elephant* (1955). Moreover, he once heard a most peculiar animal voice in this region that was harsh and guttural, most closely resembling the cough of a leopard but with an additional gurgling quality. Four different natives, questioned independently of one another by Temple-Perkins, all stated without hesitation that the creature he had heard was the ntarargo.

There seems little doubt that Africa is home to a distinct race of diminutive spotted lion, whose preferred habitat—remote mountain forests—is surely a major reason for its continued neglect by science. Certainly, its addition to the catalog of creatures formally recognized and classified by science is long overdue. Indeed, in an attempt to rectify this, Heuvelmans actually proposed an official scientific name for it—*Panthera leo maculatus*, thereby considering it to be of subspecific status. All that is now required is a complete specimen to determine definitely whether this felid is truly more than a freak and a fable. Perhaps someone with the exuberance and diligence of the late Kenneth Gandar Dower will one day complete his quest and bring back to the West a living spotted lion.

White, Black, Red, Golden, and Striped Lions

The spotted lion is certainly not the only unexpected variation upon the basic leonine theme to attract widespread attention. This was strikingly demonstrated in 1977, when zoologist Chris McBride revealed to stunned reporters across the world that for the past two years on a private South African game reserve called Timbavati he had been studying and protecting three blue-eyed, snowy white lions. In fact, sightings of white lions in the Kruger National Park and from the neighboring Transvaal had been reported for many years prior to this, but the Timbavati trio were the first to receive extensive media publicity and official recognition; several more have been formally documented since.

Tragically, as revealed in McBride's books *The White Lions of Timbavati* (1977) and *Operation White Lion* (1981), one of the original three Timbavati white lions, a female called Phuma, was subsequently shot by a poacher. As a result, the remaining two, Temba (male) and Tombi (female), together with their normal tawny-furred brother, Vela, were successfully captured and rehoused within South Africa's National Zoo in Pretoria for their own safety. Since then, many white lions have been bred from them and from others in captivity; there are currently around 300 individuals worldwide.

A white lion breeding program is presently underway at Inkwenkwezi Private Game Reserve in South Africa's Eastern Cape Province. In 2008, moreover, confirming that their coat coloration does not prevent them from hunting efficiently, an entire pride of captive-bred white lions was successfully introduced into the wild, in South Africa's Sanbona Wildlife Preserve. This was the culmination of a years-long White Lion Project that was sponsored by the Shamwari Dubai World Africa Conservation team. Whether these released lions are Timbavati-type white lions or, as discussed subsequently, so-called blonde lions with normal-colored eyes that may constitute leucistic specimens, however, has not been made clear in media reports that I have seen.

Timbavati-type white lions are recessive chinchilla-phenotype mutants, analogous to Rewa's famous white tigers, although (unlike the latter tigers) the precise genetic make-up responsible for their creation has yet to be determined.

A light-grey lion cub, also labelled (tentatively) as a chinchilla-phenotype mutant, had been born at Alabama's Birmingham Zoo in 1974 (as recorded during that same year by Roy Robinson in an issue of *Carnivore Genetics Newsletter*). However, this specimen was darker than those from Timbavati, so its genetic identity is probably different.

In contrast to the Timbavati-type lions, certain other white lions have normal eye coloration, as well as normal nose, mouth, and foot pad pigmentation, and merely appear extremely pale, even blonde, versions of normal lions. Some workers claim that these so-called blonde lions' abnormally pallid appearance is the result of leucism—a condition that has been reported from many other species.

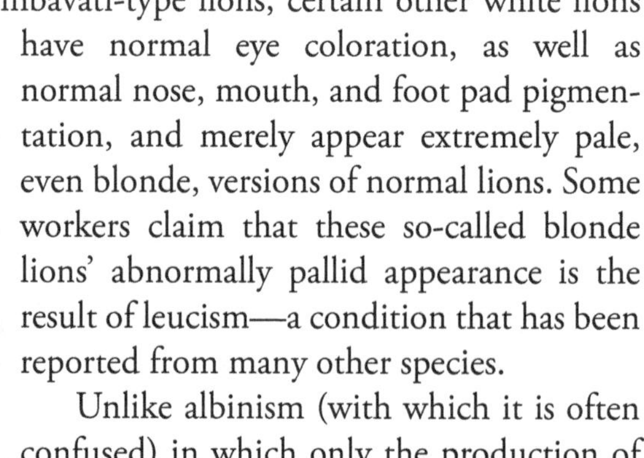

Leucistic white lion, with normal-colored eyes (Karl Shuker)

Unlike albinism (with which it is often confused) in which only the production of dark pigment (eumelanin) is affected, leucism is caused by the failure during an animal's embryonic development of some or all of the actual pigment cells themselves to differentiate or to migrate properly from the neural crest (where they originate) to the skin, fur, scales, or feathers. And because all types of pigment cell differentiate from the same multi-potent precursor cell type, this means that leucism can cause the reduction of all types of pigment, not just one, resulting in the affected animal exhibiting a characteristic faded or washed-out appearance.

Yet even though leucism is popularly claimed to be the correct explanation for the type of pale/blonde lion with normal eye color, pad color, etc., under discussion here, I am presently unaware of any published cytological studies confirming this, i.e. by showing that such lions lack pigment cells rather than merely lacking pigment. The same applies in relation to white tigers, as these are also claimed by certain workers to be leucistic. Consequently, if anyone can provide me with references to such studies, I would very much like to receive and read them.

Various non-leucistic explanations have also been proposed for these lions, but none has been verified to date. They include the action of a gene analogous if not homologous to the version inducing the champagne phenotype in horses; and the expression of the recessive dilute allele of the Dense Pigmentation gene in homozygous state modifying the action of the Red gene to yield the cream phenotype, as occurs in domestic cats.

I saw my first reputedly leucistic white lions in 2004 while staying at the Mirage Hotel during a holiday on The Strip in Las Vegas. As noted in Chapter 3, on the grounds of the Mirage is The Secret Garden of Siegfried & Roy—a wildlife park owned by those now-retired world-famous American stage magicians—where several of these very distinctive lions, as well as white tigers and even a snow tiger, are exhibited. Since then, moreover, I have seen specimens at a number of other locations around the world, where their pallid beauty never fails to fill me with awe.

Paralleling the situation with tigers and whereas white lions are now confirmed and studied, the likelihood of black lions existing is still undetermined. According to W.L. Speight in an *Empire Review* article from 1940, an experienced game warden once stated that he had spied a whole pride of pitch-black lions in the Kruger National Park. Half-a-century earlier, a very dark brown specimen had been killed by soldiers of the Luristan Regiment and was seen by archaeologist Sir Henry Layard at Ispahan in what is now Iran, as documented in his book *Early Adventures in Persia, Susiana and Babylon* (1887). An account of a black lioness observed at very close quarters was included in June Kay's book *Okavango* (1962). And celebrated lion conservationist George Adamson's autobiography *My Pride and Joy* (1986) noted that "an almost entirely black" lion was spied in Tanzania.

Several purportedly genuine photographs depicting black lions have been circulating online for some years, but via a series of investigative

ShukerNature blog articles (12 June, 1 October 2012; 15 February 2017) I have revealed that every one of them to date is a fake, created by Photoshop or other digital photo-manipulation programs from original photographs of normal lions (plus in one instance a white lion).

During 1975, at Glasgow (formerly Calder Park) Zoo in Scotland, a lion cub named Ranger was born with a black chest and a patch of black pigment on his right foreleg, plus a large, slightly paler dark patch on his left hind leg's rear upper portion (as revealed in a color photograph of Ranger snapped on 28 July 1984 by Peter Adamson of St Andrews, Scotland, and kindly shown to me by him on 28 March 2013). According to media news of the time, the Zoo's then-director, Richard J.P. O'Grady, planned to mate Ranger when old enough with his mother Kara in the hope of producing an all-black specimen. I later learned from O'Grady that such matings did occur on several occasions; but despite both lions being in excellent health, Kara never conceived. Ranger was also mated with another lioness, but again without success, suggesting that he may have been sterile. Both Kara and Ranger are now dead (Ranger having lived to the ripe old age of 22), thereby ending any further prospect of creating a black lion from matings between them. Nevertheless, the fact that Ranger possessed those sizable areas of black pigment (probably an example of mozaicism) implies that it is not beyond the realms of possibility that one day a totally black lion will be born.

In addition to white and black lions, reddish-brown individuals have also been reported in the wild, but whether their coloration is natural or merely the result of rolling in dust or even an illusion caused by sun-glare is difficult to determine based solely upon eyewitness accounts. Rather more exotic is the Senegalese chakpuar, a very large, mysterious West African cat said to resemble a lion but with red fur and a noticeably long neck.

In 2008, a young lion with golden fur was observed and photographed in Zambia's South Luangwa National Park by Egil Dröge. He was much paler than his normal-colored siblings, and even the pads of his feet lacked the usual black pigmentation. Soon dubbed Ginger, he attracted so much media attention that in 2016 his very own Facebook page was created for him, which currently has almost a thousand Likes.

Finally, zoologist C.A.W, Guggisberg noted that some lion cubs' spots are arranged in distinct vertical lines, and occasional examples have been recorded in which these spots have merged to yield definite stripes. Exceptionally, these are retained into adulthood; a good photograph of one such felid is included in Ivan Heran's book *Animal Coloration* (1976). Otherwise, however, the closest animals to striped lions known from actual specimens are various zoo examples of lion x tiger hybrids, known as ligers (if sired by a lion) or tigons (if sired by a tiger).

A Green Lion, or a Green Leopard—or Neither?

If a black lion or a long-necked red lion seem unlikely, how even more so a green lion? Remarkably, however, as very briefly mentioned in Guggisberg's book *Simba: The Life of the Lion* (1961), what was claimed to be a bona fide green lion was allegedly spied on one occasion by a prospector in the forests of western Uganda. Could it have been an individual covered in greenish slime from a stagnant, alga-choked pool in which it had recently bathed? Possibly, although Guggisberg took a rather more pragmatic view: "[It] no doubt emerged from a whisky bottle!" Yet whatever the answer may be, this is not the only green-furred Big Cat claimed for Africa.

In his annotated checklist of cryptozoological animals, published in 1986 in *Cryptozoology*, the scientific journal of the International Society of Cryptozoology (ISC), Bernard Heuvelmans included the following tantalizing sentence: "Anomalous felines, such as black, red, or white lions, green leopards, and striped cheetahs, reported from many African countries." (Heuvelmans, 1983b)

The reference that he cited there was his own, then-unpublished book manuscript *Les Félins Encore Inconnus d'Afrique* (*The Still-Unknown Cats of Africa*). In 2007, however, six years after his death, this important cryptozoological work was finally published, but although I perused it carefully, I was unable to find any mention of green leopards in it.

Consequently, I subsequently contacted French cryptozoologist Michel Raynal, who had been in close cryptozoological contact with Heuvelmans for many years, and asked his views on this subject. In his reply to me of 22 June 2011, Michel stated that Heuvelmans had been wrong, and confirmed that no mention of green leopards was present in his African mystery cats book. So the original source of this extraordinary claim remains unknown.

However, I strongly suspect that the answer may simply be that Heuvelmans had misremembered the short note concerning green lions in Guggisberg's *Simba* (with which he was familiar and had referred to in several of his own works), confusing lion with leopard, and thereby inadvertently inventing a mystery cat that had never existed even in legend, let alone in reality.

On 19 November 2012, after having blogged about the supposed green lion of Uganda on *ShukerNature*, I received an extensive, fascinating response from John Valentini Jr. Here is the summary of John's response that I added as a comment below my green lion blog article:

> One day, while visiting a local zoo, John photographed a lioness, of totally normal colouration, but when he received his negatives and prints back from the developers (i.e. back in the days before digital photography), he was very surprised to discover that in them the lioness was green! She had been walking through an expanse of grass with her body held low when he had photographed her, and at the precise angle that John was photographing her the green light reflecting from the grass had made her look green. (Some grass, noted John, can be around 18-26% reflective.) Having to concentrate keeping his camera focused upon her through only a small viewfinder and thick glass, however, John hadn't noticed this optical effect himself—not until the negatives and prints had subsequently revealed it. Consequently, John speculates that perhaps, if viewed at precisely the correct angle, a similar effect could occur with a lion observed in the wild in decent light conditions but with plenty of green foliage around it, and that this may explain the Ugandan prospector's claimed sighting of a green lion.

Needless to say, I am delighted to learn of John's extraordinary photographic experience, as it may indeed offer a very plausible, rational explanation for Uganda's alleged green lion—but one so remarkable that I would never even have thought of it, had John not brought it to my attention. Incidentally, it may also offer a plausible, rational explanation for the alleged green tiger of Vietnam (Chapter 3)—assuming the report's veracity.

Having said that, even if a green Big Cat truly existed, it is highly unlikely that we would ever know. After all, concealed amid leafy jungle foliage or verdant grasslands, such a creature would be an unrivaled master of camouflage!

A Bewilderment of Black Leopards

In spite of the expansive array of colors and patterns exhibited by the leopard in Asia, its African equivalents are even more dramatically diverse. On a morphic (as opposed to a subspecific) level, these range from red leopards, cream leopards, and striped leopards, to jaguar-rosetted leopards, buff leopards with orange rosettes, and even leopards without any distinct rosettes, to mention but a few.

Surprisingly, however, and in stark contrast to the extremely abundant black panther of Asia, very few records exist of African melanistic leopards. Considering that this latter continent contains numerous localities whose habitats and climate correspond closely with those in Asia that support black panthers, the reason for this anomaly is quite obscure. In fact, the only areas from which true (non-agouti) black leopards have traditionally been recorded with certainty are Ethiopia and Cameroon, plus the forests of Mount Kenya and the Aberdares (clearly the places to be when seeking unexpected African felids!). In mid-November 2013, camera traps set by Chris Kelly snapped several photos of a black leopard at Ethiopia's Bale Mountain Lodge (as revealed on Bale Mountain Lodge's official Facebook page). And in January 2019, photographic evidence confirming the presence of a black leopard in central Kenya's Laikipia County was published in the *African Journal of Ecology*.

Yet if we also take heed of the many unconfirmed reports of predominantly black, leopard-like cats from several other African regions, then it would seem that African panthers are more widespread—and varied—than science supposes.

Grahamstown Mystery Cats

A mysterious felid of quite remarkable appearance was killed during the 1880s at Grahamstown (since renamed Makhanda), a town in South Africa's Eastern Cape Province, and its skin was sent to German-born British zoologist Albert Günther. Its coat's background color was tawny, brightening to a rich orange gloss on the shoulders. Rosettes were virtually absent, being replaced mostly by numerous small separate spots, but these had coalesced dorsally to yield an unbroken expanse of black, stretching from its head right along to its tail base. In contrast to this specimen's richly-hued upperparts, however, its underparts were principally white with large black spots, as in typical leopards, and it also bore the facial markings characteristic of this species. Its total length was 6 feet 7 inches (including its 2.5-foot tail).

Günther had initially entertained the possibility that this singular cat constituted some bizarre hybrid. However, as he documented in the *Proceedings of the Zoological Society of London* on 3 March 1885, his detailed examination of its skin ultimately revealed certain very specific but taxonomically significant aspects, which, in combination with its already-noted leopard features, persuaded him that despite its exotic color scheme its owner had indeed been nothing more than a leopard after all—albeit of a very spectacular pseudo-melanistic variety (comparable with the Malabar specimen documented in Chapter 3).

A year later, Günther received a second, even darker specimen of this dusky form (sometimes referred to as *Panthera pardus* var. *melanotica*, the melanotic leopard), again from Grahamstown (*PZSL*, 6 April 1886). Other, less striking pseudo-melanistic examples have since been recorded, with a total of nine such individuals currently on record (although, tragically, some of these no longer exist).

First Grahamstown melanotic leopard documented by Albert Günther, 1885 chromolithograph (public domain)

They include two pelts and sightings of two living specimens as reported by Abraham in his letter to Günther, but only from South Africa's Eastern Cape Province and none at all since the 20th century's opening decade, as documented in 1987 by Jack Skead (a former director of the Kaffrarian Museum in King William's Town) within a major review entitled *Historical Mammal Incidence in the Eastern Cape*. Skead's work was brought to my attention via some references to it in a *CFZ Yearbook 1997* article on these exotic-looking leopards authored by Chris Moiser, who with fellow wildlife writer David Barnaby had viewed and photographed a mounted specimen at the Izoko South African Museum in Cape Town two years earlier. This specimen had been purchased from a professional taxidermist based in Grahamstown in November 1898, and had apparently been shot 15.5 miles south of that town. Although somewhat faded with age nowadays, appearing brown rather than its original black coloration (as depicted in a photograph of it featured in *The Mammals of South Africa, Vol 1*, 1900, authored by the museum's then-director, W.L. Sclater), it is still visually arresting.

Damasia—Dark Leopard or New Species?

Well worth considering is whether a comparable variety could be the explanation to a still-unidentified African felid known as the damasia, which dwells—need I say where?—in the Aberdares! The damasia was referred to by G. Hamilton-Snowball in his previously-mentioned letter of 9 October 1948 to *The Field* concerning his sighting of spotted lions on these mountains. In this letter, he also recalled that during the 1920s, he had shot a creature that he had taken to be a leopard, albeit a very large, dark specimen. Yet when his Kikuyu attendant boys saw it, they announced that it was not a chui (leopard) but a damasia, and that a damasia was as different from a leopard as a simba (lion) was from a marozi. Apparently the damasia is well-known to the Aberdares natives but is always mistaken by non-locals for a leopard.

Tropical Africa's native tribes frequently classify animals by way of very different criteria from those used by scientists. Often an individual animal that sports a different coat color, coat pattern, or body size from normal specimens of the same species, or an individual that is notably more aggressive than others of its own species, is given an entirely separate name by the natives and thought of by them as being a totally different form from the more typical members of its species. Consequently, it is certainly possible that despite the Kikuyus' firm denial, the damasia really is just a dark-colored (pseudo-melanistic?) leopard. Also, don't forget that genuine black leopards are on record from the Aberdares; it would be interesting to learn whether the natives class these as leopard or damasia. Alternatively, considering that the Aberdares' primeval forests already house one mystery cat in the form of the marozi, it is conceivable that they are hiding further zoological surprises too.

Kibambangwe and Uruturangwe—Cats or Composites?

The same may be true of Uganda's Bufumbira County, the mountain home of a quite ferocious beast that supposedly hunts both by day and by night and is known as the kibambangwe. Its name translates as "snatcher," a term assigned to the hyaenas in other parts of Africa where Bantu is spoken. But is this the true identity of the kibambangwe? As recorded by Captain Charles Pitman in his book *A Game Warden Among His Charges* (1931), during the 1920s one of these animals created terrible havoc in this region but was never captured. Suggested identities ranged from a notably dark-colored giant hyaena (in itself very deserving of cryptozoological attention, I would have thought!) to a melanistic leopard. Unfortunately, hardly any morphological details concerning the kibambangwe have been recorded, other than its possession of blackish markings and short ears, not much upon which to base a zoological identification.

Pitman also recalled a story regarding a pair of these creatures that once existed in lava caves sited loftily on the mountains, and which every so often descended to the plains to ravage the local inhabitants' livestock. Eventually, in desperation the locals banded together, and upon their next encounter with the kibambangwes they entered into a fierce battle with them, as a result of which both animals were finally destroyed.

Then there is the bloodthirsty uruturangwe, allegedly inhabiting the forests around Mount Muhavura and Mount Sabinio in Rwanda. It apparently resembles the leopard in size, but possesses a hyaena-like pelage. According to the natives, it can enter a hut by a very tiny hole (just like the leopard) and kills its human victim by suffocation—lying on him and taking his entire face and throat within its jaws. Pitman noted that this behavior again compares well with the leopard's mode of killing. Moreover, the uruturangwe is supposed to have a long tail and retractile claws, neither of which is a typical hyaena characteristic. Yet, paradoxically, when a skull said to be from an uruturangwe was brought forward for scientific examination, it was indeed found to be that of a hyaena—a spotted hyaena *Crocuta crocuta*, to be precise, albeit an exceedingly large one.

The most plausible explanation for this inconsistency of identity is one that I have already aired in relation to the British mystery cat situation, and exemplified by the mystifying Nandi bear of Africa (*Still In Search Of Prehistoric Survivors*, 2016). Namely, the uruturwangwe per se is probably a non-existent composite creature, rather than a real entity. Certainly, as if not complex enough already, the Nandi bear melange has even encroached upon various mystery cat reports on occasion, e.g. the skin of a supposed ntarargo was identified by Pocock as that of a young spotted hyaena (*Natural History*, 1930).

Comparably, it is not difficult to envisage sightings of abnormally-large hyaenas and killings made by savage leopard specimens in the same area being attributed by the latter area's alarmed native population to a huge and ferocious mystery beast, requiring the creation of a new name. The same applies to the equally vaguely-defined kibambangwe.

Ndalawo—Black-and-Grey Mystery Cat of Uganda

The ndalawo is a Ugandan mystery carnivore that in a December 1937 *Discovery* article authored by Captain William Hichens was described by him as "... a fierce man-killing carnivore, the size and shave of a leopard, but with a black-furred back shading to grey below." An ndalawo skin was actually procured but was sent out of the country before it could receive formal scientific attention. Consequently, its identity was never ascertained, and its whereabouts are now unknown. Captain Charles Pitman recorded in his book *A Game Warden Among His Charges* that it seemed to be a "partly melanistic leopard" (note the word "partly," indicating that it was not a normal black panther), practically devoid of spots, but displaying a few typical leopard markings on the extremities and round the lower jaw.

This more detailed description is reminiscent of that cited by Günther for the melanotic Grahamstown leopard. Certainly, pseudo-melanistic leopards have paler underparts, unlike the uniformly-dark black panthers. Based upon pelage considerations alone, it is not implausible that the ndalawo may indeed prove to be a pseudo-melanistic leopard (albeit a less showy version than those from South Africa).

However, there is more than just its pelage to consider; the ndalawo exhibits some rather unexpected traits for a mere leopard. For example, it allegedly hunts in threes or fours, and while hunting gives voice to a most peculiar laugh. These are indicative of a hyaena. Yet as Hichens pointed out, the ndalawo is very greatly feared as an exceedingly ferocious beast, whereas even the oldest woman in a native kraal is more than prepared to shoo away a hyaena that comes too close. If the ndalawo is a form of leopard, then it is a very unusual one; in fact, out of all of the black mystery cats of Africa discussed here, the ndalawo is surely the one most likely to constitute a hitherto unknown felid species.

Mngwa (Nunda)—Tanzania's "Strange One"

An even stranger felid—even its name, "mngwa," comes from the Kiswahili for "strange one"—may still survive within the temperate coastal forests of coastal mainland Tanzania (formerly Tanganyika). Also known as the nunda, this is a quite monstrous beast by all accounts, described by natives as being as big as a donkey but striped with grey like a domestic tabby cat.

The history of the mngwa extends far back into Swahili legend, appearing in many of its sayings and songs (including a very early warrior's song, dating back to c.1150 AD and attributed to the Swahili hero Liongo Fumo wa Ba-Uriy), and features prominently in the legend of the Sultan Majnun documented by Edward Steere in *Swahili Tales* (1870).

As a result of its frequent occurrence in folktale and myth, the mngwa was dismissed out of hand as purely imaginary by English inhabitants of what was then Tanganyika—until the raw reality of a horrific episode in the 1920s demanded a very different view to be taken.

In 1922, Captain Hichens was Native Magistrate at a small Tanganyikan coastal village called Lindi, the location of the following gruesome occurrence, which he documented five years later in a *Chambers's Journal* article under the pen-name "Fulahn":

> It was the custom for native traders to leave their belongings in the village market every night, ready for the morning's trade; and to prevent theft and also to stop stray natives sleeping in the market-place, an askari or native constable took it in turns with two others to guard the market on a four-hour watch. Going to relieve the midnight watch, an oncoming native constable one night found his comrade missing. After a search he discovered him, terribly mutilated, underneath a stall. The man ran to his European officer, who went with me at once to the market. We found it obvious that the askari had been attacked and killed by some animal–a lion, it seemed. In the victim's hand was clenched a matted mass of greyish hair, such as would come out of a lion's mane were it grasped and torn in a violent fight. But in many years no lion had been known to come into the town.

Hichens was still perplexed the next morning when he was visited by an old native governor of the district and two men brought with him. This latter pair seemed very frightened—and for good reason, as Hichens soon discovered. For he learned that on the night of the askari's death, these two men

> ...had slunk by the market-place lest the askari should see them and think them evil-doers; and as they crept by they were horrified to see a gigantic brindled cat, the great mysterious nunda which is feared in every village on the coast, leap from the shadows of the market and bear the policeman to the ground.

The native governor added that this creature had visited the village on several previous occasions and was neither lion nor leopard, being bigger than both. Until now, Hichens had disbelieved such accounts; the previous evening's events, conversely, were only too real. And so to be absolutely safe, he kept watch himself on the market the following evening, accompanied by two armed askaris—but nothing happened.

Mngwa, aka nunda (William M. Rebsamen)

Consequently, the next morning Hichens delivered a severe lecture to the askaris on the foolishness of superstition, little knowing just how soon he would be regretting his impatience. That same evening saw the slaughter of another askari, and when his hideously mangled body was examined, more of the same grey matted fur was discovered. For the next month, similar killings were reported up and down the coast at a number of small villages, and despite sending out search-parties, setting down traps, and laying out poison, the ferocious felid responsible was never obtained. Instead, it simply failed to appear one night, and did not return again—and the killings stopped.

Fur samples sent for analysis were identified very non-specifically as cat. If only the hair analysts had taken the trouble to record how similar the fur was to that of lion, and leopard—this would have been far more useful.

During the 1930s, another outbreak of nefarious killings began. Needless to say, this time Hichens was far less skeptical, especially when a horribly mauled man was carried on a stretcher to him at Mchinga, another small coastal village in Tanganyika, and informed him that his attacker had been a mngwa. The man was a renowned hunter, who had frequently tracked and killed lions and leopards. Hence it was most unlikely that he would mistake either of these familiar felids for anything else, or invent some far-fetched tale.

In any event, to testify even further how distinct a creature the mngwa is from lion and leopard in the eyes of the coastal Tanganyikan people, there is a famous native hunting song in which all three of these large-sized cats are mentioned within a single verse, clearly demonstrating that there is no confusion between them.

Hichens was not the only Westerner with experience of mngwa activity. As documented by wildlife author Frank Lane in his book *Nature Parade* (1955), Patrick Bowen actually saw the spoor of one such cat that had carried off a small child from another coastal village. He tracked the creature without success but did find some brindled hair on the stakes through which it had forced its way into the kraal to reach the child. The hair appeared to be quite unlike that of either lion or leopard—and the spoor? Bowen stated that it compared with that of a leopard as large as the largest lion. Any leopard of leonine proportions would truly be a creature to fear!

So is this the identity of the mngwa, an abnormally-large leopard? Its coat morphology argues against this theory. Diverse though the leopard's pelage can be, it is still very difficult to conceive how it could mutate into the brindled appearance described for the mngwa, as commented by Heuvelmans in his own consideration of this mystery cat.

A very different but extremely interesting identity on offer is one that was proposed by Heuvelmans in his cryptozoological checklist. Namely, that the mngwa could constitute a hitherto-undescribed giant form of the African golden cat. He gave no reasons for his opinion, but from a zoological standpoint it is not difficult to comprehend why he favored this identity.

Nowadays known scientifically as *Caracal aurata* and measuring 3.5-4.5 feet in total length (i.e. twice the size of a large domestic cat) with markedly long limbs and tail, the African golden cat is one of the most beautiful yet least-known of all felids. Its most notable feature is its pelage, which exhibits a quite outstanding range of variation—from the rich golden hue that has earned it its English and scientific names, to all manner of reds, browns, grays (even an occasional melanistic example), and sometimes adorned with varying degrees of dark spotting and streaks too.

African golden cat, spotted morph (Chris Brack)

The African golden cat's lifestyle and distribution have long been shrouded in mystery. For example, traditionally thought of as a predominantly West African felid, this secretive species gave science quite a jolt when Raymond Hook's investigations revealed that it also inhabited Kenya's Mau Forest (documented in *African Zoo Man*, a Hook biography from 1963, written by J.R.T. Pollard). Moreover, in the September-October 1979 issue of the *East African Natural History Society Bulletin*, I.W. Hardy published a record from the Aberdares. Worth considering is whether brief sightings of speckled individuals of this elusive felid may even be responsible for some marozi reports.

Although it approaches neither lion nor leopard in body size, the African golden cat inspires extraordinary fear and tribal superstition amongst the natives throughout its range—as observed, for example, by Jersey Zoo founder Gerald Durrell when collecting animals in Cameroon and subsequently recorded in his book *The Bafut Beagles* (1954). Similarly, as noted in his own book *Liberia* (1906), okapi discoverer Sir Harry Johnston was informed by Liberians that the golden cat was very bloodthirsty. They even referred to it as the leopard's brother.

How much greater, then, would be the terror of natives if faced with a golden cat that was comparable in size to a lion or leopard? Equal, perhaps, to the fear aroused by the mngwa? Additionally, of all African felid species, the bewilderingly variable pelage of the golden cat affords the greatest chance of giving rise to a coat-pattern variant possessing the unusual brindled appearance described for the mngwa. Also, although not officially chronicled from Tanzania so far, its relatively recent, highly unexpected discovery in Kenya and its known occurrence in neighboring Uganda suggest that the existence of this evanescent species in Tanzania too is by no means improbable.

There is one further mngwa feature, moreover, which, although seemingly insignificant in itself, gives the golden cat identity very great impetus in combination with all of the other correspondences mentioned here, but which I had not seen mentioned anywhere until I highlighted it in this book's original 1989 edition. In a *Wide World* article from December 1928, Captain Hichens mentioned in passing that when making its exit with its mutilated victim, on several occasions the mngwa had been heard purring. Yet whereas the Big Cats can purr between breaths, the cats with the correct throat structure and capability for normal continuous and loud purring are the Small Cats. As the golden cat is a Small Cat, it is perfectly capable of purring loudly and continuously, and young specimens in particular purr profusely when contented. Consequently, the purring prowess of the mngwa provides important evidence in favor of a Small Cat identity, taxonomically speaking.

Like the mngwa, the African golden cat is believed to be primarily nocturnal, and hunts crepuscularly (i.e. at dawn and dusk) with arboreal tendencies and the capacity to kill animals as large as medium-sized antelopes. Yet until more specific details regarding its lifestyle can be ascertained, it is both difficult and unwise to speculate too closely upon whether circumstances could exist whereby a giant-sized form would or could evolve. At present, the most that can be stated with safety is that such a felid would certainly constitute a persuasive mngwa identity. In any event, only an actual mngwa specimen can verify (or disprove) this theory, but if the latter mystery cat is found to be a giant golden cat, the fact that the world's largest felid form (which is what the mngwa may well be) is, taxonomically, a Small Cat rather than a Big Cat will undoubtedly represent one of the most startling zoological revelations for a very long time.

Another interesting mngwa identity is one that was very tentatively raised in 2001 by Spanish cryptozoologist Angel Morant Forés. In a communication of 30 July posted on the cryptozoology chat group cz@yahoogroups.com, he noted that maneless lions allegedly inhabiting Kenya's coastal regions and known locally as buffalo or river lions had been proposed in a 2001 *Swara* article to be a missing link between modern-day lions and prehistoric cave lions. (This in turn had been based upon a suggestion posed in an *In The Field* article from 2000 that these buffalo lions might represent an ancient lineage phylogenetically discrete from all other living lions, and may even constitute relict, Pleistocene-type lions.) And as they reputedly sported gun-metal coloration on their back and flanks, Forés wondered if the mngwa might be an unknown lion form similar in appearance to buffalo lions.

This cz@yahoogroups post initiated a more detailed response the next day from British paleontologist Darren Naish, who had been investigating buffalo lions in some detail (their name, incidentally, comes from claims that they are adept at bringing down very large prey, such as buffaloes). Not only are they reputedly maneless but also they are said to be solitary, not living in prides like typical lions. As he pointed out, however, their manelessness may not be a taxonomic trait but due merely to hormonal problems or genetic abnormalities. In any case, manelessness is relatively common among lions (the notorious man-eating lions of Tsavo famously lacked manes). Moreover, this in turn may explain why they are solitary—in many different species, freak individuals are often shunned by normal members of their species. He did not consider that such lions had anything to do with the mngwa.

There is one further, final mngwa identity still requiring

consideration, one that if ever shown to be correct would largely if not entirely dispel the mngwa's widely-assumed status as a cat of cryptozoology. Is it possible that there was no feline mngwa, certainly as far as the Lindi killings documented here were concerned? The following information, quoted from *Wild Cats of the World* (2002) by Mel and Fiona Sunquist, may have significant bearing upon this:

> ...witch doctors in the Singida area, in what was at that time Tanganyika, ran a lucrative extortion business in the early twentieth century by threatening to turn themselves into lions and kill people who did not pay them. Many people were murdered by young men dressed as lions, wearing lion paws as gloves on their hands and feet. These mjobo, or "lion men," reappeared in the same area in 1946, some twenty-five years later, and 103 deaths were attributed to their activities. Murders were made to look like the work of man-killing lions, and the common belief in were-lions was exploited by secret societies.

Could it be that the Lindi killings were also the work of these lion men? The timings are similar—two separate outbreaks, one during the early 20th century and a second over a decade later. Claims that the killings resembled maulings from some great cat are also consistent with this theory, and what better way to increase fear among the highly superstitious local people than to spread rumors that the assailant was not even a lion but rather a greatly-feared feline monster of traditional local folklore—the mngwa?

Moreover, to provide physical evidence for this claim, what easier way than to place clumps of grey fur in the hands of the murder victims, fur that was clearly not from a lion but matched the folklore descriptions of the mngwa? In reality, of course, the fur could have originated from anything, even a domestic dog, but as long as it didn't look like a lion's and did look like the fabled mngwa's, that was all that was required of it. Even the extra-large footprints would not be beyond the ingenuity of the lion-men to fake, thereby yielding further evidence of mngwa presence in the eyes of the frightened locals. Consequently, although I don't entirely deny the reality of the mngwa, I do consider this particular explanation a very plausible one, at least for the two outbreaks of mngwa-attributed Lindi killings.

Tigers in Africa?

Although Asia possesses a small number of lions, it is surely a bit much to suggest that Africa houses an entourage of tigers, isn't it?

German explorer Theodor von Heuglin and other 19th-century explorers of Abyssinia (now Ethiopia) learnt of an extremely dangerous wild beast allegedly inhabiting that country that was called the wobo by the Amhara people and the mendelit by the Tigre inhabitants. It was described as being larger than a lion, yellowish-brown or grey-brown in color, and decorated with black stripes. Writing in his book *Simba* (1962), C.A.W. Guggisberg added the following tantalizing snippet:

> King Theodoros [=Tewodros II of Abyssinia] told Heuglin that for a long time a skin of the "Wobo" had been hanging in the main Cathedral of Eifag, and many people claimed to have seen it. Could it have been a tiger skin which had somehow found its way into Abyssinia?

This is possible, but it does not explain the widespread native knowledge of such a creature, unless we assume that all of this evolved from knowledge of the skin at Eifag. Incidentally, for quite some time I was unable to find any first-hand mention of Eifag (let alone its cathedral) anywhere other than in one of von Heuglin's writings—the Cathedral of Eifag is specifically named on page 57 of his travelog *Reise in Nordostafrika, Vol. 2* (1877). Consequently, I'd begun to wonder whether "Eifag" was actually a Germanic rendering of some other, more familiar place-name but which Guggisberg had simply quoted verbatim from von Heuglin's tome.

Happily, however, after perusing various other 19th-century travelogs, German cryptozoologist Markus Hemmler was able to supply me with some relevant details. It turned out that Eifag (aka Ifag, nowadays Yifag) was (and still is) an important commercial emporium consisting of a cluster of Ethiopian villages and churches encircling a dormant volcano east of a hilly region named Tisba, situated between the rivers Arno and Reb. Moreover, its "cathedral" was merely a main church. Eifag was famously visited by French consul-explorer Guillaume Lejean, as chronicled in Richard Andree's book *Abessinien, das Alpenland Unter den Tropen und Seiner Grenzländer* (1869). Also worth noting is that the type specimen of the chestnut climbing mouse *Dendromus mystacalis*, formally described and named by von Heuglin in 1863, was actually procured by him in Eifag. So Eifag definitely exists!

As for the suggested tiger skin identity for the alleged wobo pelt, there is a fundamental problem with this. Namely, the wobo is not an isolated case because von Heuglin also noted that a similar beast appeared to occur in neighboring Sudan, where it was referred to as the abu sotan. Inhabiting rocky mountains near to the River Rahad, it was described as being marked with great black blotches or stripes.

Muddying the waters even further is a brief excerpt from the second edition of English traveller Mansfield Parkyns's book *Life in Abyssinia* (1868): "[A wobo] had been killed some years ago on the river Weney, and its skin presented to Oubi (king of Tigre); but I could never discover what became of it." Was this the same skin as the Eifag pelt noted by von Heuglin and Guggisberg, thereby indicating that it did originate in Ethiopia after all? Or was it a second, totally different one?

Also, is it possible, as von Heuglin suggested, that the wobo and abu sotan belong to a currently unknown species? Certainly it is very difficult to suggest a known African species that possesses stripes; ironically, of all African cats, it is the equally mysterious mngwa that in terms both of brindled pelage and of huge size comes closest to these. Curiouser and curiouser, as Alice would have said!

Speaking of tigerine beasts in Sudan, in a paper published by the journal *Spolia Zeylanica* in 1951, eminent Sri Lankan zoologist Paulus E.P. Deraniyagala actually described a new subspecies of tiger from Africa—*Panthera tigris sudanensis*, the Sudan tiger, based upon a pelt that he had seen in a bazaar in Cairo, Egypt, and which the seller claimed had originated from a tiger shot in Sudan. Although he did not purchase the pelt, Deraniyagala did photograph it, and according to Czech mammalogist Vratislav Mazák, who documented this curious affair in his book *Velké Kočky a Gepardi* [*Big Cats and Cheetahs*] (1980), it clearly resembled that of a Caspian tiger *P. t. virgata* and had probably been smuggled into Egypt from Turkey or Iran.

Back to bona fide striped mystery cats of Africa, and definitely in need of an explanation is an illustrated limestone relief from the tomb of Ti, a wealthy 5th-Dynasty Egyptian landowner and courtier (c.2490-2300 BC), at Saqqara—because among the many animals portrayed is a striped tiger-like cat but with a distinctly leonine tuft at the tip of its tail. Is it a freak lion whose juvenile spots have coalesced into stripes (as discussed previously), or an imported tiger poorly represented? Or could this enigmatic image be a portrait of a species still unknown to science? Whatever the answer, it is significant that all of the other animals present in this relief, including a lion and a leopard, are accurately depicted and readily recognizable.

And why, as noted by Nelson Mandela no less in his autobiography *Long Walk To Freedom* (1994), is there a word for "tiger" in South Africa's Xhosa language? Mandela revealed this fascinating little fact while arguing with various fellow prisoners on Robben Island who were insisting that they had seen tigers in Africa's jungles. Here is what he wrote:

Saqqara striped mystery cat (arrowed) depicted on the tomb of Ti (public domain)

> One subject we hearkened back to again and again was the question of whether there were tigers in Africa. Some argued that although it was popularly assumed that tigers lived in Africa, this was a myth and they were native to Asia and the Indian subcontinent. Africa had leopards in abundance, but no tigers. The other side argued that tigers were native to Africa and some still lived there. Some claimed to have seen with their own eyes this most powerful and beautiful of cats in the jungles of Africa. I maintained that while there were no tigers to be found in contemporary Africa, there was a Xhosa word for tiger, a word different from the one for leopard, and that if the word existed in our language, the creature must once have existed in Africa. Otherwise, why would there be a name for it?

In 2005, British scientist Tim Davenport made zoological headlines with his co-discovery of the kipunji *Rungwecebus kipunji*, a new species, and genus, of mangabey monkey in Tanzania's Rungwe (Rongwe) highlands that was well known to the local people ("kibunji" is its native name) but which had previously been dismissed by scientists as a wholly mythical spirit beast. Still dismissed today as such, conversely, is the so-called Rungwe tiger. According to the locals, it is a large striped animal, a description not matching that of any known species from this area of Tanzania. Davenport concedes that it could be a striped hyaena *Hyaena hyaena* or aardwolf *Proteles cristatus*—albeit way out of its known range—but does not discount that it might be a still-unknown species (*Guardian*, 7 December 2005). After all, he has only to look at the kibunji to know that such a prospect is far from unrealistic in this remote African locality.

And what are we to make of a lion-clawed, tiger-headed, leopard-rosetted felid on the rampage in the Mayanja district of Kenya during early 1974 and likened in some reports to a giant cheetah?

This anomalous amalgam allegedly devoured many domestic animals during its reign of terror, while eluding with ease the many hunting parties that went in search of it. Its identity was made particularly mystifying by the fact that leopards are very rare in this region, and the last reported lion here was slain more than 20 years earlier. Conversely, the Mayanja monster was never killed, unless it happened to have been one and the same individual as the Beast of Bungoma?

This was an equally ferocious felid that hit the headlines a little later in Kenya's Bungoma district, where it was held responsible for the deaths of hundreds of farm animals. A very large and savage leopard was ultimately trapped by forestry Rangers, but whether this was really the Beast is unknown.

Mountain Tigers and Water Lions

One of the greatest losses for the modern-day mammalogist must surely be the huge-fanged saber-toothed cats or machairodontids. The African Pleistocene possessed various saber-tooth genera, including *Megantereon*, *Dinofelis*, and *Homotherium*; in life they undoubtedly constituted a sight more awe-inspiring than even the most magnificent of today's lions, tigers, jaguars, and other large cats.

As it happens, there are many Africans alive today who may actually have firsthand knowledge of such experiences because it could be that a few saber-tooths still exist, as I comprehensively documented in my book *Still In Search Of Prehistoric Survivors* (2016).

Take, for example, the unidentified mountain tiger of Chad, concerning which I am exceedingly grateful to French cryptozoologist Michel Raynal for supplying me with much information from his files. The Zagaoua people of Ennedi, northern Chad, tell of a cat form referred to by this region's French-speaking people as the tigre de montagne (mountain tiger) that if real is definitely not known, at least in the living state, to modern science. For they describe it as being larger than a lion but lacking a tail, sporting red fur banded with white stripes, plus long hairs on its feet, and—most interesting of all—fangs that protrude from its mouth. Chad's mysterious mega-cat inhabits the Ennedi mountains and caves and is strong enough to bear away sizable antelopes.

Mountain tiger, based upon eyewitness accounts (Tim Morris)

A comparable creature may once have existed in Senegal too. According to naturalist Owen Burnham, who lived for much of his youth here, hunters in the Casamance area still readily recall a huge cat termed the wanjilanko. They state that it was striped, possessed very large teeth, and was so formidable that it could kill lions. Today, however, it has disappeared, together with the lions that also formerly inhabited this region (Burnham, pers. comm.).

This is not the only information on record concerning such cats either. In 1967, French ethnologist Jeanne-Françoise Vincent made a study of the Hadjeray tribe from Temki in Chad. Published in 1975 as *Le Pouvoir et le Sacré Chez les Hadjeray du Tchad*, it revealed that one of the Hadjeray's clans believes in the existence of a mystifying lion-like cat, dubbed the hadjel. Although it greatly resembles a lion in overall appearance and possesses an impressive mane, the hadjel is even bigger in size, but has only a short tail, variously likened by Hadjeray eyewitnesses to that of a hyaena or a small mare.

As for its fangs, these are said to be so long that the unfortunate creature has trouble opening its mouth. Indeed, despite the hadjel's greater size, the Hadjeray aver that it is less dangerous than the lion—because of the time it takes to open its mouth! Commenting that their description of hadjel corresponds exactly with that of a saber-tooth, Vincent speculated that perhaps a branch of these spectacular cats' supposedly extinct feline lineage does still survive in this isolated montane region of Africa.

In 1932, Lieutenant-Colonel André J.V. de Burthe d'Annelet published a travelog entitled *À Travers l'Afrique Française...*, in which he referred to the alleged existence in Chad's Ouada district of a very savage beast dubbed the yassou, which is apparently unknown to science. It is said to be feline, yet resembles a bear, and like a bear it is plantigrade (walks on the soles of its feet). True cats, conversely, are digitigrade, walking on their toes. However, there was once a most unusual type of cat whose hind feet were at least partially plantigrade. This was *Homotherium*, the scimitar cat, which happens to have been a saber-tooth. Coincidence?

A similar beast is also known to the Youlou tribe of the Central African Republic (CAR), which neighbors Chad. They call it the coq-ninji or coq-djingé, whereas this region's French speakers again use the term "tigre de montagne." In other, unidentified languages of the CAR, the same creature is apparently referred to as the gassingram or vassoko.

In 1937, Lucien Blancou, formerly Chief Game Inspector in what was then French Equatorial Africa (subsequently split into Chad, the CAR, the Republic of the Congo, and Gabon), was informed by the village chief of Ouanda-Djailé that the gassingram was reddish-brown in color, larger than a lion in size, and with equally over-sized footprints. Primarily nocturnal, with eyes that shone like lamps in the darkness, it bore its prey to mountainous caves. In view of its close correspondence with the tigre de montagne of Ennedi in Chad, the gassingram is surely one and the same species.

While in the company of an old game tracker in the Ouanda-Djailé area during the 1960s, French hunting guide Christian Le Noël heard a terrible roar near a large cavern that was totally unlike anything known to him, despite his very considerable experience with animal calls. Conversely, his companion had heard it before and identified it as that of the tigre de montagne; his subsequent description of this beast suggested a saber-toothed cat. And sure enough, when Le Noël returned to this area a year later armed with drawings of modern-day and fossil animal forms, his game tracker acquaintance positively identified the picture of the prehistoric Miocene-dated African saber-tooth *Machairodus* as the mountain tiger. In addition, Le Noël once saw a hippopotamus in southern Chad that had died of strange wounds that could only have been inflicted by a cat armed with exceptionally well-developed upper canine teeth.

Certain other reports from the CAR, but this time derived from localities to the north of its capital, Bangui, describe creatures that although comparable in some ways with those mountain-dwelling forms discussed here are nonetheless most probably of a quite separate species. This is because they display a most unexpected and singular behavioral attribute for cats—they are predominantly aquatic.

Amphibious Saber-Tooths?

The marked predominance of this feature is acknowledged in the various native names given to these animals by different CAR tribes. Thus the Banda speak of the mourou n'gou or muru-ngu (meaning "water leopard"); the Baya speak of the dilali ("water lion"); the Sangho of the ze-ti-ngu or nze ti gou ("water panther"); and the Zande of the mamaimé ("water lion") or (just to be different) ngoroli ("water elephant"). Judging from all but the last-mentioned term, these creatures are quite clearly feline, an assumption supported by the various descriptions of them on record.

According to an old Banda tribesman called Moussa, interviewed in 1934 by Blancou, the mourou n'gou was somewhat larger than a lion in size (Moussa estimating its length at about 12 feet), with an overall body shape and pelage background color reminiscent of a leopard's but additionally adorned with stripes. Curiously, its paw-print was described as "containing a circle in the middle."

Apparently, Moussa at some stage had observed one of these creatures emerging from the Koukourou river in close proximity to a soldier in a canoe. The mourou n'gou seized the hapless man and dragged him down into the water. As a result of this incident, the detachment to which the beast's victim had belonged decided not to cross the river at this point thereafter but instead at a new location some considerable distance to the east. In 1945, a native gunbearer called Mitikata drew a sketch of the mourou n'gou, which showed a small-headed, large-fanged creature about 8 foot long, with a plump, uniformly brown-colored body and a panther-like tail.

During December 1994-January 1995, Belgian cryptozoologist Eric Joye led a two-man expedition, dubbed "Operation Mourou N'gou," in search of this elusive creature. Although they failed to spy it themselves, Joye and his teammate, hunting guide Willy Blomme, succeeded in gathering some very interesting anecdotal evidence, probably the most extensive obtained since the material collected here during the 1930s by Lucien Blancou (*Cryptozoologia*, September 1994, September 1996).

Claiming to have narrowly avoided being propelled into the Bamingui River by such an animal as he sat fishing at the river's edge one afternoon in February 1985, a native guide called Marcel told Joye that the mourou n'gou hunts in pairs, one waiting in the river to seize any prey chased into the water by the other.

According to Marcel, the mourou n'gou compares with a leopard in shape and size, and its pelage is ochre, dappled with blue and white spots that are very distinct upon its back but less well-defined upon its flanks. It has a long tail, hairier than the leopard's, and its head is said to be a little like that of a civet (does this mean that, like a civet, it has a dark face mask?), but its teeth are very large and resemble those of a Big Cat such as the leopard or lion.

Marcel followed the mourou n'gou's trail, which was like a leopard's but bigger. Also, when it runs, it leaves behind the impression of claws, which is not usual for a leopard.

A second so-called water leopard is the nzemendim, spoken of by the natives in the N'velle distinct of Yaounde, Cameroon, and documented in *Gorillas Were My Neighbors* (1956), written by Fred G. Merfield and Harry Miller. The locals claimed that it was a type of leopard that lived in small rivers and frequently carried off women and children. After several failed attempts to spot this dangerous mystery beast, one of this book's authors believes that he finally achieved success:

> One morning, just as it was getting light, an animal swam rapidly upstream and landed on the opposite bank. The light was still poor, and all I could see was something furry, spotted and four-legged, so I fired. The animal spun round and dived back into the water. I thought it was lost, but when the sun rose and my men turned up, we found it dead and washed ashore a few hundred yards downstream. It was a big dog-otter, a very old fellow whose fur had gone grey and blotchy, giving the appearance of spots. The natives would not admit that this was their nzemendim, and having no more time to spare I gave up and went home.

Lucien Blancou learned a little concerning another type of supposed aquatic mystery cat, the dilali, which according to an interpreter at Bozoum in eastern Ubangi-Shari possessed the body of a horse and the claws of a lion, as well as large walrus-like tusks (this last-noted feature was mentioned to Blancou by a Zande native guard).

Then there is the nze ti gou, a nocturnal beast of leopard stature, having red fur marked with pale stripes or spots, living within hollows in large rivers, and emitting a thunderous noise. Such creatures are not restricted to the CAR and Cameroon either.

The Mbunda tribe of eastern Angola speak of the coje ya menia ("water lion"), which attracted the interest of Ilse von Nolde during her sojourn there in the early 1930s, and which she subsequently wrote about in an article published by the periodical *Deutsche Kolonialzeitung* in 1939. Like the nze ti gou, this beast is well known to the natives for its loud rumbling vocals and, although principally aquatic, sometimes ventures forth onto dry land. Like the other beasts mentioned here, it too is armed with large canine teeth or tusks and supposedly kills hippopotamuses with them, despite its own smaller body size. The observations and dealings of von Nolde with this region's natives convinced her that the coje ya menia was not based upon misidentified sightings of crocodiles or hippos, or of these species' spoor. On the contrary, the natives were very adept at identifying and interpreting spoor.

She was equally convinced of the sincerity of a Portuguese lorry driver who told her that in the company of some natives, he had actually tracked a coje ya menia that had chased after and killed a hippopotamus. Sure enough, the tracks of the two beasts had led to the mutilated carcass of a hippo, yet no part of it had been eaten.

The tracks of the coje ya menia were smaller than those of the hippo, and although somewhat reminiscent of an elephant's in shape, they additionally contained the impression of toes. Worth noting here is that as far back as 1947 in a *Kosmos* article, renowned German cryptozoologist Ingo Krumbiegel had suggested a surviving saber-tooth as a possible identity for this Löwe des Wassers ("water lion").

The Zande or Azande is a native African tribe indigenous to the northeastern part of the Democratic Republic of the Congo (DRC), the Republic of South Sudan, and the southeastern section of the CAR—which in combination is often referred to as Zandeland. The Zande people speak of several mysterious beasts that may possibly be cryptozoological in nature. One of the most intriguing of these is the mamaimé or water leopard—but one that sounds very different from both of the previous two versions. In an article published by the journal *Man* in September 1963, Oxford anthropologist E.E. Evans-Pritchard noted that the Zande's water leopard had been described to him by them as follows:

> The water leopard is a powerful kind of beast, dark and with a blackish skin and a head of hair shaggy to the neck. Its pads are very large and its palms hairless like those of a man. It has powerful teeth in its mouth. It seizes a person as does a crocodile. It appears in places of deep water. Its eyes are very large and red, like the seeds of the nzua vegetable [like tomatoes]. It lives in holes as crocodiles do, but where it resides there is water and many fish near it. This water never dries up, for it is its home.

Professor Evans-Pritchard was also told that this strange creature has hair like a man's that falls over its body, and he noted that a Major P.M. Larken had stated in an article published in 1926 that water leopards are said to live in deep pools within large rivers, and that big fissures in the banks of rivers have often been pointed out by locals as being these aquatic cryptids' homes. Evans-Pritchard doubted that the water leopard genuinely existed but was at a loss as to how to explain reports of it.

Moving to the DRC's southern regions, we encounter reports of yet another similar animal, known variously as the simba ya mai, ntambue ya mai, and ntambo wa luy, all of which translate as "water lion." More mystery beasts apparently belonging to this same type include the Kenyan ol-maima and the Sudanese nyo-kodoing.

Even though from description and native nomenclature a feline identity seems probable, and from their formidable upper canines a saber-tooth is ostensibly intimated, the existence of an aquatic version remains a very bizarre concept. Certainly there is no evidence from fossil saber-tooth finds to suggest that any extinct form ever became adapted for this particular lifestyle. So is it really possible that such a creature is the explanation for any or all of the various mystery beasts described here?

In his books *Les Derniers Dragons d'Afrique* (1978) and *Les Félins Encore Inconnus d'Afrique* (2007), Bernard Heuvelmans asserted that it was not only possible but was, to his mind, more than likely. The principal arguments that he put forward in support of this can be summarized as follows:

1) Despite popular belief, many cat species are not afraid of water.

2) The very effective use by walruses of their own huge upper canines for anchorage, dragging themselves on to land or ice floes, and ploughing up the sea bed sediment in search of modest-sized prey demonstrates that the saber-tooth's enlarged canines would be of great benefit for an aquatic existence.

3) Conversely, on land such teeth would surely have been a great handicap to the saber-tooth when attempting to tear off and devour pieces of meat from its prey.

4) Due to this, saber-tooths would have faired badly in competition with true felids of comparable size, e.g. lions and leopards, and ultimately would have been forced to move into ecological niches unoccupied by such cats, namely remote mountains and aquatic realms (which just so happen to be the very environments from which reports of alleged modern-day saber-tooths are emerging).

Consequently, Heuvelmans believed that saber-tooths could exist very readily in an aquatic environment simply by paralleling the walrus's lifestyle. Moreover, if challenged by potential rivals such as elephants or hippos, the saber-tooth would be more than amply equipped to dispose of them and could even drink the blood spurting from any vanquished opponent's neck (some authorities believe that the fossil terrestrial saber-tooths were themselves sanguinivorous). Its carcass could serve as a food source too. For if dragged underwater by the saber-tooth and allowed to decompose, the carcass's meat would eventually soften, enabling it to be torn off in pieces and devoured more readily by the saber-tooth, thereby cancelling out the problems that the felid might otherwise experience due to its cumbersome canines

Once again, this suggestion is supported by walrus activity. Some walruses have been reported killing seals and small whales, and lumps of blubber and seal remains (soft and/or from decomposing beached carcasses, hence easy to tear) have been found in the stomach of some walruses.

Nor have we come to the end of walrus parallels. Any saber-tooth modified via evolution for an aquatic existence that compared closely with that of a walrus behaviorally would very likely follow a similar course morphologically too (convergent evolution). In addition to eyewitness reports of the various mystery beasts mentioned here, is there any other evidence for a walrus-like creature in tropical Africa?

The answer is both affirmative and rather stunning. A cave painting discovered at Brackfontein Ridge in South Africa's Orange Free State, and depicted in *Rock-Paintings in South Africa* (1930) by George William and Dorothea Bleek, portrays an animal that bears a startling resemblance to a walrus—from its rounded head displaying two very large downward-curving tusks to its elongated, tapering body and paddle-like limbs. It differs primarily from the walrus in possessing a long tail, because over its 23-million-year evolution from original terrestrial ancestors, the walrus has lost its tail.

Conversely, as the known fossil saber-tooths of Africa's early Pleistocene (2 million years ago) displayed terrestrial-related morphology, any amphibious development must have occurred since then. Yet it is rather unlikely that as drastic a change as the complete loss of a tail via evolution would have occurred in so short (geologically-speaking) a space of time. Consequently, the painting is precisely what we would expect a recently-evolved amphibious saber-tooth to look like: a more streamlined version of its terrestrial counterpart with limbs modified for an aquatic life but with the tail not yet lost. Also worth noting is that such a saber-tooth may be expected to be larger than land-living species because the accompanying weight increase would be buoyed by its surrounding liquid environment. This would explain the notable size of some of the mystery beasts reported in this section.

Walrus-like mystery beast in Brackfontein Ridge cave painting (public domain)

Many years ago, an Ituri Forest pygmy identified a picture of a walrus shown to him by aptly-named big-game hunter John A. Hunter as a savage, nocturnal beast that lived in the deepest parts of the forest. Not surprisingly, Hunter dismissed this claim of a "jungle walrus" as nothing more than the pygmy's desire to please him, as he noted in his book *Hunter* (1952). Nevertheless, in view of the Brackfontein painting and this chapter's unidentified animals, it may be prudent not to discount it totally.

Also worth keeping in mind is a certain Ngona Horn folktale of the Wahungwe people from Zimbabwe, documented in *African Genesis* by Professor Leo Frobenius and Douglas Fox. For it refers to some very odd felids as "lions under the water" and is itself entitled "The Water Lions." Perhaps they are comparable with nearby Angola's coje ya menia?

Clearly, the concept of aquatic felids is not at all alien to native Africans. And the close correspondence in such creatures' descriptions over so far-reaching an expanse of Africa, cutting across many different tribal cultures and histories too, is surely indicative that there is something more substantial and fundamental to their accounts than mere folktale and superstition.

The mysterious Kenyan dingonek—an aquatic beast of generally feline appearance, familiar to the Wa-Ndorobo tribe and allegedly shot at by adventurer John Alfred Jordan—is another possible member of the aquatic saber-tooth association. Its fangs are huge; its body is long and as wide as a hippopotamus's; its tail is lengthy and broad, and swings gently against the river current when the animal is in the water; it has four short legs, and its feet are as big as a hippo's, but they bear large claws like those of a crocodile.

True, it possesses one feature that initially seems to exclude it from a felid identity—namely, a body covered in scales. Heuvelmans, however, considered it more likely that these are optical illusions produced by light shining upon a particularly reflective pelage, and one that also bears tufts of hair that have adhered together as a result of the animal having been submerged in water, thereby creating the effect of scales. Having said that, in his description of an alleged first-hand encounter with a dingonek in 1905, documented in his book *Elephants and Ivory: True Tales of Hunting and Adventure* (1956), Jordan seemed very emphatic that this creature was genuinely scaled:

> I slid down the bank and got in the cover of the bushes. There it was.
>
> It was in midstream, about thirty feet from me, a beast-fish, a creature from your nightmares. It was fifteen to eighteen feet in length, with a massive head, not a head like a crocodile's, but flat-skulled and round. It had two yellow fangs dropping from its upper jaw, and its back was as broad as a hippo's, but it was scaled in beautifully overlapping plates, as smooth and as intricate as those I've seen on an old Arabian cuirass. The sunlight fell on those wet scales and was dappled by the leaves, and made them seem as brilliantly colored as a leopard's coat. It had something of every animal in it. It was impossible.
>
> There was a broad tail, and this was swinging gently against the current, keeping it midstream, keeping it stationary, whatever it was.
>
> At last I took aim on it...I aimed the .303 at the base of the neck and gave it one solid round.
>
> I saw the bullet hit, and heard it hit the way you do at short range. The beast turned in a great flurry of yellow water until it was facing the bank and my cover. It leaped into the air until it was standing, or so it seemed, its pale belly scales vivid, ten or twelve feet on end.
>
> [Jordan and his native helpers then fled through the forest for 300 yards before halting, but after a time they cautiously returned to the river] It had gone, but its spoor was all over the soft mud, huge prints about the size of a hippo's, but clawed.
>
> Some Wanderobo told me that they knew about this thing, they called it a dingonek. The Kavirondo knew of it too. They had seen more than one of them and made a god out of it whom they called "Luquata." They were worried when they heard that a white man had shot at Lukuata. They said that now they would all die of sleeping sickness, and it is true that there was an epidemic of it among the Kavirondo that year.

Jordan's noting that it was considered bad luck to kill a dingonek echoes a belief that is also prevalent concerning DR Congo's equivalent mystery beast, the ntambo wa luy or simba ya mai, as recorded in Charles Mahauden's book *Kisongokimo* (1965). Such taboos have undoubtedly assisted in saving these potentially dangerous creatures from extirpation.

At much the same time as Jordan's sighting, another hunter spied a dingonek floating on a log down the Mara River (also running into Lake Victoria) while in high flood, but it quickly slid off and into the water. Near Kenya's Amala River, the Masai call it the ol-maima and sometimes see it lying in the sun on the sand by the riverside; but if disturbed, it slips into the water at once, submerging until only its head remains above the surface.

Dingonek illustration, *Wide World Magazine*, 1917 (public domain)

Jordan also documented his dingonek encounter in his later book, *The Elephant Stone* (1959), and here he emphasized its fangs and scales:

> Fifteen to eighteen feet long, with a massive head shaped like that of an otter, two large fangs, as thick as walrus teeth, descending from the upper jaw; its back as wide as a male hippo's, yet scaled distinctly, like an armadillo; and I could see also why my Lumbwa [native helpers] had spoken of a leopard's body, for the light reflected on the scales in that cat's colours. Idly it switched a broad tail...

In a much earlier account of his dingonek sighting that he provided to a London *Daily Mail* newspaper reporter almost 40 years before his own books were published, Jordan specifically stated that this creature's two fangs "were like those of a walrus protruding from its mouth" and that its body "was shaped like a hippopotamus but scaled." However, after stating that he fired at it, Jordan then claimed: "My further observations were cut short by the animal charging us," and went on to say that he returned to the area the next day (*Daily Mail*, 16 December 1919).

Needless to say, these claims directly contradicted Jordan's accounts in his two books, in which no mention was made of the dingonek charging him, and in which he and his helpers returned not long afterwards rather than the next day. Such notable discrepancies cannot help but make me wonder just how reliable the remainder of his dingonek testimony was.

Another aquatic mystery beast, the chipekwe of Zambia's Lake Bangweulu, has also been put forward as a possible water-dwelling saber-tooth. Unlike the dingonek, however, most reports of the chipekwe imply either some form of reptilian beast or a herbivorous mammal, rather than a carnivoran.

Digging Up a Burrowing Cat

I am greatly indebted to English cryptozoological researcher Richard Muirhead for bringing to my attention an extremely curious report published on 26 January 1925 by the *Leeds Mercury*. It concerns what was referred to in the report as a burrowing cat, but what it truly was could well be another matter entirely. Here is the report in question:

> **A "Burrowing" Cat**
>
> Captain Buchanan was engaged on scientific work, for Lord Rothschild and the British Museum, and brought back some remarkable relics of his journey through the heart of the Sahara.
>
> His collection was the first brought from there.
>
> One of the most valuable specimens was the skin of a "burrowing cat," the only specimen in any collection in the world.
>
> This animal greatly resembles a cat, but is able to burrow like a rabbit. It is beautifully marked, has a fine coat, and lynx-like ears.
>
> Captain Buchanan started from Lagos, Nigeria, travelled up country about 700 miles to Cano, and then struck across the vast desert to Algiers.

Captain Buchanan was Captain Angus Buchanan, the famous British explorer who (with his cameraman) was the first white explorer to cross the Sahara by camel. So, assuming that the above report is genuine and not journalistic hokum (but its specific naming of Buchanan, his zoologist sponsor Lord Walter Rothschild, and the British Museum suggests that it is indeed genuine), what could this odd-sounding creature be?

When reading the report, I immediately thought of the sand cat *Felis margarita*, a small species that does inhabit the Sahara and is in fact the world's only known felid adapted for desert life. Some (although not all) specimens are handsomely marked with stripes and spots, and its ears are very pointed and therefore somewhat reminiscent of a lynx's. Moreover, it does indeed retreat into burrows when the prevailing temperature is too extreme.

However, the sand cat's existence was known to science long before 1925 (it was formally named in 1858), with many specimens already preserved in museums worldwide. So if it were a sand cat, the report's claim that Buchanan's was the first specimen in any collection is very mystifying.

Madagascan Mystery Cats

Southeast of the African mainland lies the island of Madagascar—a zoological time-capsule. For it is the home of a vast variety of creatures extinct elsewhere or totally unique, a wonderland of lemurs and tenrecs, falanoucs and vanga-shrikes. It has no native canids or felids, instead the euplerids, i.e. Malagasy civets and mongooses, reign supreme here. Among this heterogeneous assemblage, the largest species—and the creature that assumes on Madagascar the ecological roles occupied elsewhere by sizable felid species—is the fossa *Cryptoprocta ferox* (not to be confused with the Madagascan civet or fanaloka, whose scientific name is *Fossa fossa*). Despite its euplerid affinities, the puma-sized fossa is strikingly cat-like in appearance and is especially comparable to the Neotropical jaguarundi.

However, Madagascar may also possess some uncategorized true felids. In a report published by the *Chasseur Français* in October 1939, Paul Cazard recalled that while in Madagascar he had been informed by a civil engineer named Belime that native tales originating from areas of the island still unexplored by Westerners told of giant lions that lived in caves and which ravaged the island's other fauna as well as the inhabitants of these regions' native villages.

Cazard contemplated whether these lions of the rocks could possibly be living saber-tooths, and wondered if it would be feasible for an expedition to be mounted to seek out these mighty beasts. Feasible or not, no such expedition has set out on their trail to date, so their identity remains unknown. Needless to say, zoogeographically-speaking it would be a great jolt to scientific conceptions if a bona fide cat form were to be discovered here. Yet it would certainly not be without precedent, as the discovery of so many hitherto unknown and highly unexpected animals within the 20th and 21st centuries can readily verify.

And this is where my coverage of Madagascan mystery cats came to an end in this book's original 1989 edition. Since then, however, I have uncovered details concerning some very intriguing additional examples, so here they are.

Domestic cats *Felis catus* had been introduced into Madagascar by the 17th century, and many have since run wild, yielding widely-distributed feral populations across this extremely large island. However, in a cz@yahoogroups.com posting of 19 May 2003, I recalled that back in 1967, in his book *The Life, History, and Magic of the Cat*, Fernand Mery had included the following tantalizing snippet concerning a felid specimen procured in Madagascar that may constitute something much more significant than a mere feral domestic:

> The Malagasy Academy possesses a specimen of a magnificent tabby cat, larger than a domestic cat. Details of its capture on Madagascar are uncertain, but of interest is that in the local Malagasy language, pisu = domestic cat, with kary used to denote 'wild cats', even though wildcats do not officially exist on the island.

Mery considered that this lent support for the probable existence of wildcats on Madagascar. Interestingly, in a letter to *Fortean Times* (November 2003), Geoff Hosey from the Bolton Institute in Manchester, England, noted that Mery's account appears to have been lifted almost verbatim from an earlier work, Raymond Decary's book *La Faune Malgache* (1950). Decary had also alluded to wildcats appearing in various Malagasy folktales, thereby providing further evidence that such cats do indeed exist in Madagascar. Moreover, Hosey included in his published letter a very intriguing color photograph he took in August 1998 of a cat curled up asleep that may have been merely a feral domestic cat but which in his opinion looked very like an African wildcat *Felis lybica*. The cat was in an unlabelled cage at Parc Tsimbazaza, the zoo that occupies the grounds of the old Malagasy Academy. Unfortunately, however, due to its curled-up position, the cat presented insufficient morphological details for a precise identification of it to be made from the photo alone.

My cz@yahoogroups.com posting of 19 May 2003 had been in response to a previous one that same day by British paleontologist Darren Naish, who, a little earlier in May 2003, had unexpectedly obtained some interesting information while watching a television program, information that bestowed added significance to Mery's statement.

The program was a documentary in *National Geographic*'s "Out There" series, during which, while conducting studies in northwestern Madagascar's Ankarafantsika National Park, Tennessee University fossa researcher Luke Dollar trapped what looked like a wildcat—the second such creature that he had caught there. Moreover, instead of resembling a feral domestic cat, it seemed exactly like the African wildcat. In the program, Dollar hinted that it may be a valid new record for Madagascar, or even a bona fide new species.

A blood sample from this intriguing specimen, a pregnant juvenile, was taken for examination; how remarkable it would be if Mery's (and Decary's) belief in Madagascan wildcats had finally been justified. Sadly, however, although I emailed Dollar concerning it in February 2012, I never received any response from him, so I have no idea if any information of significance was obtained from this sample (but as I have not uncovered any follow-up details regarding it online or elsewhere, I am assuming that nothing was).

Meanwhile, in February 2020 a *Conservation Genetics* paper finally revealed the precise nature and origin of Madagascar's tabby-striped feral domestic cats. It presented the findings of a team of researchers including Missouri University cat genomics expert Leslie Lyons who had been conducting comparative DNA analyses using blood samples from specimens of these cats and from other domestics around the world, which revealed the closest match with the Madagascan ferals to be domestics from Arabian Sea locales. Consequently, the team proposed that perhaps as far back in time as a thousand years ago, some such Arabian domestics had made their way to Madagascar by stowing away on Arab trade ships, then disembarking onto the island and over time establishing thriving populations here.

Nor does the fascinating saga of mysterious felids on Madagascar end there, as my continuing researches have duly discovered. In November 2013, a remarkable paper authored by Massachusetts University anthropologist Cortni Borgerson was published in the journal *Madagascar Conservation & Development*, concerning a Madagascan mystery beast hitherto unknown to me. It was referred to locally as the fitoaty, and native descriptions of it given to Borgerson and her assistants suggested a gracile, entirely black-furred felid (as opposed to any form of euplerid), but larger and leaner than feral domestics and confined to the rainforests of northeastern Madagascar's little-studied Masoala peninsula.

During 2011, Borgerson was fortunate enough to observe a fitoaty personally, when she saw what she described in her paper as "a medium-sized melanistic carnivoran crossing a village trail just outside the Masoala National Park boundary. The sighting occurred at approximately 15:00h, in a transitional area of primary and secondary forest." She tentatively classified the fitoaty as *Felis* sp., and stated that trapping and genetic testing of this unidentified felid was needed to assess its taxonomic identity, distribution range, and potential impact upon local ecosystems.

In December 2015, a second paper concerning the fitoaty appeared, authored by a seven-strong team of researchers that included two from Madagascar's Wildlife Conservation Society, and published in the *Journal of Mammalogy*. It presented not only the first population assessment of the fitoaty, or black forest cat as it was now being called colloquially, but also some excellent full-color and black-and-white photographs of fitoaty specimens obtained via camera-trapping methods.

Interestingly, the team discovered that there was minimal interaction between the fitoaty and feral domestics in the wild. Nevertheless, based upon their field research they suggested that this mystifying melanistic was "a phenotypically-different form of the feral cat [rather than either an African wildcat or any other felid species, known or unknown], but additional research is needed." In view of the successful new findings concerning the genetic identity and origin of Madagascar's typical feral domestics, I now look forward to equivalent fitoaty studies to determine conclusively the precise taxonomic and genetic nature of this unexpected "new" member of Madagascar's mammalian fauna.

Finally, just in case you are wondering, the fitoaty's name is Malagasy for "seven livers," which stems from a somewhat strange native belief concerning this animal's internal anatomy. Moreover, its flesh is claimed by locals to be poisonous, and therefore is never eaten by them.

I began this chapter with a quotation from Pliny the Elder. I end it now with a riddle from another philosopher (one whose words seem to have lived on long after the memory of his name ceased to exist):

> What is the world's most cunning animal?
> That which no man has ever seen.

I have a strong feeling that Africa may still house quite a few comparably cunning animals.

Chapter 5
North America: Panthers Aplenty and Bobcats of Blue

Velveteen silhouette, silent and sinister;
Satin-furred midnight on ebony paws.
Eyes hewn from emeralds, seeking a sacrifice;
Death to deliver with ivory claws.
— Patrick Krushenk, "The Panther"

Although North America also houses the Canadian lynx *Lynx canadensis*, bobcat *L. rufus*, and a few rarer and more exotic species, the feline personification of this continent must surely be the puma *Puma concolor*. Not only is it undoubtedly the most familiar of America's cat species, demonstrated by its long list of common names (including cougar, mountain lion, catamount, painter, and panther), but it is also America's most widely distributed. Until the 20th century and the severe persecution that this spectacular New World species suffered during it, the puma's range had extended from British Columbia to Patagonia—over 8700 miles as the puma runs! Nowadays, however, it has been exterminated from much of South America's more densely settled areas as well as from many of its former North American haunts.

Eastern Cougar

In past times, many different puma subspecies were recognized, whereas today they have all been lumped together into just two. At the time of this book's first edition in 1989, however, both the Florida cougar (aka Florida panther) and the Eastern cougar (aka Eastern panther) were deemed to be distinct subspecies (*P. c. coryi* and *P. c. couguar* respectively). Yet although this is no longer the case as far as most felid researchers are concerned, they remain geographically distinct, so for the sake of convenience here I shall treat them as discrete entities on that latter basis.

Florida cougar (public domain)

Both of these puma forms have suffered greatly in modern times. Having sunk as low as a mere 20 individuals by the 1970s, the Florida cougar is now represented by approximately 230 specimens, lingering predominantly within the Everglades, and is currently the only confirmed puma population in the eastern USA

This is because the plight of the Eastern cougar has been even more severe. It once ranged from the eastern US to provincial Canada west to the edge of Alberta and the Plains, but by the early part of the 20th century most zoologists believed that it had been hunted into extinction in the USA and virtually eliminated in Canada too. Yet sightings and occasional killings of pumas in eastern North America continued to be reported, attracting in particular the attention of Robert Downing and Bruce Wright. Their writings are essential reading on the Eastern cougar and proved invaluable during my preparation of this section.

Comparable reports are still being made, and due to the Eastern cougar's perpetuating suspension between existence and extinction in the USA, it has ultimately gained the status of a cryptozoological cat—hence its inclusion here.

Of the various accounts on record for the first two decades of the 20th century, the following couple are particularly important because they are supported by preserved specimens. In 1903, as documented by F.B. Golley in his book *Mammals of Georgia* (1962), a mounted puma from Georgia's Bulloch County was displayed (a photograph taken of it was still held at the University of Georgia at the time of my book's original 1989 publication). Two years later, a puma was collected near Vidalia, Louisiana, and housed in the US National Museum. Other notable reports of eastern specimens killed during this period (but not preserved afterwards) were documented from Maryland, Maine, Pennsylvania, and Wisconsin.

During the 1920s, a puma cub was killed in Maryland's Garrett County and is now ensconced at the US National Museum, whereas in 1927 an adult was dispatched at Shutesbury, Massachusetts. Reputable reports of killed pumas in the east during the 1920s-1930s also emerged from Georgia, Louisiana, North Carolina, Tennessee, and Maine.

For many years, it was believed that the last "official" killing of a puma in Missouri took place during 1927, then on 21 January 2017 a vehicle struck and killed one on I-70 in Warren County, but whether it was a genuine Eastern cougar or either a western migrant or merely an escapee/release from captivity remains undisclosed (*Springfield News-Leader*, 1 February 2017). Meanwhile, in January 1938 a puma scare occurred near Lamar when strange sounds believed to be those of a "painter" (to use the Ozark vernacular) were reported by several local farmers. Nevertheless, despite many searches and forays, the creature responsible remained aloof and unidentified. A few coyotes were flushed from cover, but no sign of a puma (or even a bobcat) was discovered.

Moving momentarily to Canada, in 1931 a puma was shot at Mundleville in New Brunswick; the last specimen shot in this area and actually preserved dated back to 1881. Later in the 1930s, New Brunswick was involved in another Eastern cougar incident, although on this occasion the site of procurement was in the USA In January 1938 (perhaps a couple of years earlier according to one opinion), a puma was trapped east of Little St John Lake, Maine, by Rosarie Morin. Preserved as a mounted exhibit, it was ultimately sold in 1960 to New Brunswick's Northeastern Wildlife Station, whose director was Bruce Wright. All three of these specimens were documented by Wright in a *Journal of Mammalogy* article for May 1961.

One of the most compelling of eastern US accounts from the 1940s is the report by Fred Barkalow that a 109-pound specimen was killed in northern Alabama's TVA lakes region in 1942, and that during this same year several other pumas were flushed from cover while the lake bed was being cleared for the construction of Fontana Dam. Also worthy of note is an account published on 13 March 1954 by Herbert Sass in the *Saturday Evening Post* concerning a puma hit and killed by a truck driven by Alan Broun Jnr in South Carolina's Georgetown County sometime during 1942-1943.

As far as pumas in Arkansas were concerned during this period, the official *Arkansas Guide Book* published in 1941 had stated that "only a few remain, and these in the most remote sections." However, Ronald Nowak noted the killing of one specimen near Mena, Polk County, in 1948.

On 29 March 1947, in the company of two colleagues and traveling on skis and snowshoes on the border between Albert and St John counties in New Brunswick, Bruce Wright discovered a set of unmistakeable puma tracks, clearly belonging to a large male puma. Investigating these, Wright observed that they were joined just 200 yards further along by two more sets of tracks belonging to a female with a one-third-grown cub. In short, they were the spoor of an entire puma family. Wright took several photographs of them, and on 29 July 1947 Paul F. Elson discovered large cat tracks in this same area. On this occasion, casts as well as photos were taken, and Wright sent them all to the US National Museum, where they were confirmed to be puma. Wright was most excited by these finds—and with good reason because they constituted the first puma tracks unequivocally recorded in eastern Canada for more than a century. On 23 March 1963, Wright published his discoveries in the then-weekly British magazine *Animals*.

Back in the USA, in 1949, just over a decade after the trapping of the Little St John Lake individual, Maine produced a further specimen in the shape of a skull submitted to Safari Club International. The 1940s also saw the publication of Stanley P. Young and Edward A. Goldman's definitive book *The Puma: Mysterious American Cat* (1946). Judging from its hide-and-seek (or should that be cat-and-mouse?) tactics in the eastern US, their book had been very adroitly titled.

By the 1950s, scores of alleged puma sightings were emerging each year from many of these states. Indeed, in Vermont an entire association was founded by persons willing to testify that they had observed pumas there, whereas puma pandemonium in Minnesota had reached such proportions in one county that, as noted in Sass's *Saturday Evening Post* article, a sheriff posse and a National Guard Unit were called out to investigate reports (officially the puma has been extinct in this State since 1875).

Several killings of pumas were also reported during this decade. These include at least three in North Carolina (1950, 1952, and 1959); a roadside victim at Charleston, South Carolina, in 1952; and a specimen from Alabama's Tuscaloosa in 1953.

Moving into the 1960s, at least four specimens in the eastern US (in Louisiana, Arkansas, Oklahoma, and Pennsylvania respectively) were killed and preserved. However, Robert Downing noted in his extensive *Cryptozoology* article of 1984 concerning the Eastern cougar that one of these—now housed at the Carnegie Museum—possessed deformities suggestive of a former existence in captivity. Moreover, according to a *Latrobe Bulletin* article of 12 August 1987 by T. Fegely, it was allegedly traced back to an Ohio zoo.

The year 1971 saw the publication of Bruce Wright's major work *The Eastern Panther*. Also in that year, while deer-hunting north of Crossville, eastern Tennessee, a Mr. Buckner shot a puma and arranged for its body (including the skull) to be preserved as a permanent exhibit. It was partly as a result of this latest specimen to be obtained that in 1973 the US Department of the Interior changed its official classification of the Eastern cougar from extinct to endangered. Back in the land of the living—and officially too—at least for a time.

In April 1974, a puma was photographed in the Adirondacks by New York teacher Alex McKay (*The News*, April 1975). A year later, as recorded by Downing, the Arkansas Game and Fish Commission offices gained a mounted taxidermy puma, a 118-pound male that had been killed in Logan County during 1975. Additionally, reports of killings unsubstantiated by physical evidence emanated from several other states.

In Canada, a puma sighting on the Bruce Peninsula in Ontario was recorded by Provincial Police Constable Art King in 1978, followed a month later by a similar sighting made by writer John Kerr on the banks of the Saugeen River; casts taken of spoor left behind were confirmed to be those of a puma by Ontario's Ministry of Natural Resources (*Fortean Times*, winter 1979 and fall 1980).

On 10 June 1979, Charles and Helen Markses, co-managers of the Lake Estates trailer court in Ohio's Westerville, paid close attention to more than 200 large paw prints that had been discovered in a grassy muddy field and alongside a small lake nearby. Some of the prints showed claws, others did not. The Markses took casts of the most dramatic ones—those with claws—and showed them to the police. They were identified as dog spoor, but the Markses were sure that they were puma prints. Within hours, three boys who knew nothing concerning these spoor reported seeing a very large tan-colored felid in a tree behind the trailer court. It jumped down from the tree and fled but was later seen again by the boys on a roadway close by. Only after this second appearance of the mystery cat did the boys learn of the prints (*Fortean Times*, fall 1980).

The early 1980s were memorable on the Massachusetts mystery carnivore front for the Beast of Truro, which many locals believed to have been either a wild puma or an escaped pet puma. In September 1981, several mutilated domestic cat carcasses were found in the vicinity of Truro, a town chiefly composed of scrub pine forests and sand dunes, ensconced within the National Seashore Park at the tip of Cape Cod.

During the next few months, further livestock killings were reported, and two hogs belonging to police officer David Costa were severely clawed. At first these attacks had been attributed to feral dogs, but this theory was soon abandoned when piercing screams began to ring out at night and sightings of puma-like beasts were reported. Also worth noting is that Costa's hogs had been clawed horizontally, implying that their attacker had jumped onto their backs.

Moreover, while walking along a path near the parking lot in the Cape Cod National Seashore Park one day in late September/early October 1981, William and Marsha Medeiros were convinced that they saw a puma about 50 feet further along the path. It weighed

60-80 pounds, had short ears, and a long slender tail held low but curling upwards at its tip (a puma characteristic). Surprisingly, when the animal saw its human observers, it showed no alarm, simply sauntering unconcerned into the nearby woods within which it soon disappeared from view. Similarly, on 7 February 1982, Truro police officer Rodney Allen claimed that a cougar crossed in front of his cruiser.

Certainly there is more than enough undeveloped woodland around Truro to conceal such a cat, and a pet puma identity could explain the unfrightened attitude of the animal sighted by the Medeiroses (wild pumas are notoriously shy of humans). Unafraid or not, extensive searches of the entire area by Park Rangers failed to uncover any positive signs to verify the presence there of a large felid. No more reports emerged from Truro either. Another mystery animal had made its escape (*ISC Newsletter*, spring 1982).

The Pennsylvania Association for the Study of the Unexplained (PASU) had been investigating suspected puma sightings in this state for a number of years, and in February 1987 a selection of some of the more recent ones were published by *PA Woods & Waters* in the form of a report written by PASU Director Stan Gordon. Of particular interest was the discovery made by a PASU team on 25 July 1984 of a short trail of paw prints near to South Buffalo Township in Armstrong County. A cast was made of the best spoor, which bore three lobes on the plantar pad's posterior edge (a felid characteristic). No dimension for the spoor was given, but according to Gordon, several (unnamed) wildlife experts identified it as a puma print

Subsequently, the dog of John Galinat, from Chaplin, Connecticut, treed a small-headed, large-bodied felid of roughly 40 pounds in weight with spots on its body and a lengthy ringed tail. As luck would have it, Galinat had a film-loaded camera with him and took some photographs of this unexpected cat, which depict an animal resembling a young puma (immature pumas do have spotted pelages and rings round their tails).

By fall 1987, John Lutz's Eastern Puma Research Network based in Maryland had recorded 70 puma sightings so far that year, plus 97 in 1986, with West Virginia supplying the largest number, and further reports stretched into 1988. Moreover, at the 1988 International Society of Cryptozoology's annual members meeting, held at Maryland University, Robert Downing presented an Eastern cougar update. And on 3-5 June 1994, Gannon University in Erie, Pennsylvania, hosted an official Eastern Cougar Conference, with papers presented on all aspects of this elusive felid as well as the Florida cougar and even mystery black felids.

During the 31 years that have passed since this book's first edition appeared in print, the cryptozoological situation regarding the Eastern cougar has remained the same, inasmuch as alleged sightings of pumas in eastern North America have continued to emerge on a regular basis, but its survival here has not been officially substantiated. On the contrary, in 2011, a US Fish and Wildlife Service evaluation unofficially declared the Eastern cougar extinct, followed by an official statement to this effect by the same authority in 2018, which was accompanied by the removal of this puma form from its list of endangered species. Conversely, Canada has made no official statement on the subject so far.

Yet from the accounts recorded here and numerous more recent ones in the literature, there can be no doubt that pumas do still occur not just in eastern Canada but also in several regions of the eastern USA. Less evident, however, is whether these constitute remnant breeding populations of the genuine Eastern cougar, or whether they are isolated stragglers, perhaps even wanderers from more westerly locations, or escapees/releases from captivity. In addition to the Carnegie Museum specimen already noted, other escapee/released examples may include an individual apparently killed and photographed near Kentucky's Central City in 1960 and rumored to have escaped from a roadside zoo; plus the killing of one puma and the capture of another in West Virginia during 1976—Robert Downing noted that these two were shown to possess behavior and parasite loads indicative of recent captivity.

Misidentification of other species is another important consideration. Downing, for example, recorded that certain "puma" carcasses and skeletons proved to be those of dogs or bobcats. Also, there is always the possibility of deliberate hoaxes being responsible for some reports, as exemplified by three episodes from the 1950s in West Virginia that proved to be hoaxes perpetrated with Western and Mexican pumas, plus a newspaper April's Fool hoax in Tennessee.

Nevertheless, it still seems very unlikely that all Eastern cougar reports are due to western invaders, escapees/releases, misidentifications, and hoaxes. When dealing with a species as elusive as the puma, one should never be too hasty in declaring it extinct.

American Mystery Black Panthers

Creatures closely resembling black panthers (i.e. melanistic leopards) but appearing in totally unexpected locations are not restricted to Britain and mainland Europe. A vast dossier could be compiled from all of the reports of such creatures that have been filed in North America too. Moreover, regardless of the state in which they have been recorded, the eyewitness descriptions of these felids' morphology is much the same. Namely, a powerful jet-black creature of pantheresque outline and build, with a total length and height sometimes said to equal those of a puma or jaguar. The head is relatively small with short pricked ears and bright greenish-yellow or orange (occasionally even red) eyes, the limbs are fairly (although not markedly) long, and the tail is lengthy and slender. Spoor left behind is generally cat-like in outline, but sometimes claw impressions are present. Yet many eyewitnesses have stated categorically that the beast that they have seen was not a dog but instead a large, unmistakably feline animal.

As a result of such marked morphological uniformity, in the following necessarily brief selection of reports (but see also Appendix 2 for an additional, previously-unpublished selection, exclusively excerpted from my *ShukerNature* blog) it should be understood that unless stated otherwise, the creatures seen fit the standard description. Also, as it has not been demonstrated conclusively so far that any of these mystery cats are bona fide black panthers, quotation marks around that term should be understood throughout its use in these reports.

American mystery black panthers have been reported for many years (William M. Rebsamen)

A leading investigator of these unidentified American felids is veteran US cryptozoologist Loren Coleman, whose writings (especially the various editions of his classic book *Mysterious America*) include very detailed documentation of this subject and contain full coverage of certain accounts mentioned here. Also of great worth is Mick Mayes's recent book *Shadow Cats: The Black Panthers of North America* (2018), which is by far the most extensive book-length treatment of such cats, with particular attention paid to examples that have been reported from Texas and other southern US states.

Local inhabitants in several US regions claim to have seen black panthers in their vicinity for decades. However, not until the end of World War II did these animals gain widespread attention from the general public and the media, thereby making this period a convenient point at which to begin our coverage.

On 4 September 1946, Harry McClain of Lebanon, Indiana, claimed that a black panther that had been terrorizing the area for some time had now been shot by his assistant Roy Graham, but that its body had fallen into a creek and could not be recovered. Just a fortnight later, two black panthers made an appearance near Oquawka, Illinois, but a 150-man search three days later (assisted by an aerial reconnaissance of the region via three airplanes) found nothing. This incident was documented a year later in *Doubt*, one of the very first Fortean magazines.

Into the 1950s, and I am very grateful to American correspondent David McAvoy for bringing to my attention a most unusual, memorable example of evidence for black pantheresque cats in America that dates from that decade—namely, a remarkable painting on display at the Memphis Brooks Museum of Art in Tennessee. Entitled "Story Told By My Mother," it was produced in 1955 by highly-acclaimed Arkansas-born artist Carroll Cloar (1913-1993), and depicts a snow scene in which a woman is stepping briskly away from a very large black panther-like cat standing at the edge of some trees. David informed me that it was inspired by tales that Cloar had heard from his mother concerning so-called black panthers that had once roamed Arkansas. Moreover, David himself hails from Arkansas, and he mentioned that he has heard such stories for as long as he can remember (*ShukerNature*, 15 April 2016).

As reported by Loren Coleman in a series of early 1970s *Fate* articles, the state of Illinois is a rich source of black panther incidents, of which one in particular could have provided an answer to the panther phenomenon. On 25 October 1955, a black panther was unexpectedly encountered near Decatur by game warden Paul G. Myers, who promptly shot it—but not fatally because the panther escaped, and its body was never found. A large cat, described as very dark brown, possibly black, was the subject of many reports from at least three different localities in Illinois's Champaign County during 1963, and eluded even a 300-man search posse. Its tracks, however, were discovered and were described as being between a dog's and a puma's in size. In 1965, panther accounts re-emerged from Decatur's environs.

Two adult panthers were observed in 1964 near Ventura, in California's Conejo Valley, whereas on 12 December 1967 a specimen was sighted there by Henry Madrid and three companions. Due to the car of the investigating sheriff deputies becoming bogged down (literally) in a marshy field, however, the panther was readily able to avoid further enquiries.

The year 1970 saw more Illinois panther sightings, subsequently documented by Loren Coleman, including a bizarre incident that occurred on the evening of 10 April. Mike Busby, while investigating the reason for his car's sudden engine trouble a mile south of Olive Branch, Alexander County, was attacked by a macabre beast resembling a black panther but walking on its hind legs! It was frightened off by an approaching truck, and Busby later received medical attention for his wounds.

This is not the only record of a bipedal panther-like beast either: a comparable individual was recorded in Abesville, Missouri, in June 1945; a second one in South Carolina's White Oak Swamp during fall 1948; and a third in Queens County in Canada's New Brunswick on 22 November 1951, which was documented by none other than Eastern cougar investigator Bruce Wright in his book *The Ghost of North America* (1959).

Also noteworthy is that in December 1970, a large black panther and cub were sighted twice in a field near Decatur by Clarence Runyon and family. Three years on and over 1850 miles further west, a large cat described as a 5-foot-long black panther or very dark puma was the quarry of a posse of police and County Animal Control officers in the eastern foothills of California's San Jose Hills, south of San Francisco. Local ranchers believed that it had been roaming the area for 2-3 months, and allegedly had even been treed on one occasion by a collie belonging to Navy Lieutenant Commander Thomas Mantei.

On to spring 1977 and to Ohio's Allen County, the scene of many black panther sightings and ferocious sheep killings attributed to them. Spoor discovered in the vicinity of these latter killings were found to be cat-like in general shape but sporting the clear impression of claws (*Fate*, November 1977).

Summer 1977 saw the debut of New York's media-dubbed "Van Etten swamp monster," which was investigated in detail by Loren Coleman. It apparently inhabited the two-mile-long swamp lying between Van Etten and Spencer, was held responsible for the deaths of several domestic animals, and was documented extensively by Elmira's *Daily* and *Sunday Telegram* newspapers. During July and August, many sightings of a black panther in the area were reported by local inhabitants, and State Conservation Officer Charles Winant recalled rumors that such an animal had been there since 1971. On 25 July, news emerged of the discovery of strange claw-lacking spoor by Van Etten gardener John Palomaki, and Cornell University mammalogist Milo Richmond commented that the absence of claw marks was indicative of a felid identity, although further evidence would be needed for confirmation of this.

Van Etten swamp monster, based upon eyewitness accounts (Fortean Picture Library)

During the next few weeks, however, various scientists questioned the prints' authenticity, arguing that they did not match those of any North American mammal but could have been made by a human. Yet the argument ultimately proved to be academic because following a brief reappearance in October and again in December 1977, the Van Etten swamp monster was seen no more.

The 1980s saw mystery panthers come and go too. Summer 1984, for example, marked the descent of panther pandemonium upon several lower Michigan localities, beginning in the village of Manchester with a good black panther sighting by patrolman Kevin Deacons, although by the end of the year attention had moved to Cincinnati in Ohio. The usual pattern was prevalent: many sightings, many searches, nothing caught. The closest came during the Michigan outbreak, when some discovered paw prints were examined by Stephen P. Schiffer of Michigan University, who confirmed the presence of a black leopard; others went even further by speculating that more than one cat could be involved (*Fate*, January and May 1985).

During May 1987, a mysterious cat described as a "black cougar" and displaying a predilection for posing on peoples' porches in Nauvoo hit the headlines briefly before departing for pastures (or porches) new (*Tuscaloosa News*, 29 May 1987). It was rapidly replaced by the equally elusive black panther of New Jersey's southern regions, as now revealed.

For some months in 1987, livestock killings and eerie screams piercing the night had been reported by the inhabitants of Salem County's rurally-sited Pittsgrove Township and Cumberland County's Vineland. In June, a Pittsgrove Township resident informed police that he had seen a large, long-tailed cat resembling a black panther, and a few days later a paw print exceeding 4 inches long was discovered near to a trap set in Vineland woods by officials of the State Division of Fish, Game and Wildlife. An unnamed wildlife authority identified it as that of a bear, but Vineland Police Captain James Forcinito stated that he would not be surprised if the animal in question turned out to be a black cougar, remarking that reports had been received for years by police regarding such animals in the nearby woods. Yet fur samples ultimately obtained were identified by scientists as canine, and by now some authorities were suggesting mystifyingly that many (if not all) of the slaughtered livestock had died from natural causes (*Creature Chronicles*, February 1988)!

So in best official traditions, a black panther/puma identity was discounted in favor of a feral domestic cat. Undoubtedly it is true that an element of panther paranoia had entered into the proceedings, with such felids being seen everywhere. For instance, one lady reported a panther prowling behind her home, only to discover that it was her own pet cat! Yet what of the 4-inch spoor? No domestic cat could have made that. Once again, a cryptozoological case had been concluded by a convenient but clearly incorrect solution having been put forward.

During the late 1980s, Gerald Cameron Jr. of Savannah, Georgia, informed then-university zoology student Victor Albert (see also Chapter 7) that he had seen a black panther on three separate occasions locally (twice in the Dean Forest Road area, and once near Arnesville). He described them as being 4-4.5 feet in head-and-body length (plus a long smooth tail), 2.5-3 feet in shoulder height, and weighing about 100-140 pounds, with feline outline and pure black pelage. Apparently they inhabit the wilder, more remote high woodlands, and according to various friends of Cameron they range down from South Carolina's Charleston to Florida's Tampa region, and come to the edges of Okefenokee Swamp to hunt. They are extremely shy of humans, keeping well away from them, but as a result of their noticeable coloration are frequently seen. Indeed, although happy to answer all of Albert's questions concerning these animals, Cameron remained perplexed as to why he should be so interested in such common animals that everyone knew about! (Albert, pers. comm., 1987)

As with British and mainland European mystery black panthers, during the 30-odd years that have passed since this book's original 1989 publication many additional reports of American equivalents have been filed, swelling the archives immensely, but all of such similar content that to document them here would be repetitious in the extreme, and would serve no useful purpose anyway, as we already have sufficient data to analyze in search of an identity for such cats. Suffice it to say, therefore, that, indisputably, creatures closely resembling black panthers are being seen in North America—but what might they be?

Too many eyewitnesses have been adamant that the beasts they had seen were cat-like and not dog-like for a canine identity to be generally tenable in any but the most fleeting of sightings or those made in poor viewing conditions. Also, it must be remembered that clawed spoor cannot be used to distinguish unequivocally between felids and canids. Thus, with one notable exception, we should look to the cat family for these animals' identity—or identities, because it is conceivable that more than one is involved here.

(1) Fisher?

This is the notable exception to the "felids only" list of possible panther identities. Inhabiting Canada and northeastern USA (although reports as far south as Florida have been recorded), the fisher *Pekania pennanti* is a mustelid. In other words, it is related to weasels and especially the martens (with whom it was traditionally classed, within the genus *Martes*, until 2008). It attains a total length of up to 3.5 feet, of which 16 inches constitutes its fairly bushy but tapering tail, yet it appears even larger, because of its dense pelage. Indeed, the fisher has great potential for being confused with a modest-sized panther-like cat.

Its body is long and slim, its head is rather cat-like in shape, and its ears are small and pricked. In color it varies from black to dark brown. Of particular pertinence to the panther question is the fisher's overall mien, because this is strikingly cat-like: apparent in its fluid movements, its precise manner of stalking its prey, and also in its attitude of intense concentration.

Fisher (public domain)

Indeed, the correspondences between fisher and felid are so marked that Native American nations living within its range of distribution refer to it in their languages as "the black cat." Similarly, a friend of English archaeologist A. Charles Thomas obtained a clear sighting of a fisher while in Maine during the early 1980s and noted in particular its "general cat-like appearance" and "sinuous" mode of movement (Thomas, pers. comm., 1984).

Consequently it is by no means unlikely that some panther reports may actually be based upon fisher sightings. Indeed, from the description given by K.E. Ives, *INFO Journal* editor Paul Kelsey suggested this identity for the black pantheresque mystery beast spied by Ives on the shore of New York State's Oneida Lake sometime prior to 1982 (*INFO Journal*, October 1982); the fisher's distribution does extend as far south as this area.

(2) Feral Domestic Cat?

This potential identity has already been covered in Chapter 1 relative to British mystery panthers, but it is equally applicable to the North American situation and is also an option favored for at least some, although by no means all, reported American mystery cats by American cryptozoologist Chad Arment in his book *Varmints: Mystery Carnivores of North America* (2010). Moreover, although it may seem implausible that domestic cats could be mistaken in the field for much larger panther-like felids, such misidentifications have indeed occurred.

For example, during the 1970s a "black panther" was sighted and shot near to Cosby, Tennessee, and a photograph of the dead animal was later published in a Newport newspaper. However, while still fresh its carcass had been examined by Park Ranger J.R. Buchanan and Michael Pelton, and it was identified as nothing more than a large domestic cat. Similarly, a "black panther" killed in 1979 near Central, South Carolina, also turned out to be a domestic cat (an identity rumored in relation to the San Jose panther of 1973 too). Both cases were documented in Downing's 1984 *Cryptozoology* paper. Nevertheless, it is clearly ludicrous to suggest that all panther sightings can be explained in this way.

(3) Jaguarundi?

The jaguarundi *Puma yagouaroundi* is a slender felid slightly larger than a domestic cat with a small head, short legs, and a very long tail. Predominantly thought of as an exclusively Central and South American species, during the early part of the 20th century it was still common in southern Texas too, although an individual killed by a car in Cameron County during 1986 was the first confirmed record of the jaguarundi in Texas since 1952 (*Oryx*, April 1987).

It exists in two color forms: a red morph referred to as the eyra, and a grey-black morph that is the jaguarundi proper. Assuming that it still survives there, this species may be responsible for certain briefly-spied "panthers" recorded from Texas, and also from Florida; the jaguarundi was introduced here during the 1940s and is now present in large numbers.

Grey-black morph of the jaguarundi (Bodlina/ Wikipedia)

(4) Bobcat?

Also worth a consideration relative to reports of black panthers in Florida is whether at least some of them could have been melanistic bobcats. For in stark contrast to the typical reddish-brown, spotted pelage of this species, some all-black individuals have been recorded in Florida, as documented by several authors, including John Paradiso in a *Florida Scientist* paper in 1973. Moreover, the tails of various of these specimens have been extremely long in comparison to those of normally-colored bobcats, thereby enhancing a pantheresque appearance.

Speaking of long-tailed bobcats, in his book *Cryptozoology: Science and Speculation* (2004), Chad Arment devoted a chapter to intriguing reports of mysterious American "wild cats" (the New World does not of course harbor any native representatives of the bona fide wildcat *Felis silvestris*) characterized by their long tails but which do not appear to be feral domestics, leading some to speculate whether they could be bobcat x feral domestic cat hybrids.

5) Puma?

The puma is panther-sized and is superficially similar to this felid in general outline too. Also, it exhibits a wide variety of coat colorations, divisible into two morphs: tan-brown and grey-blue. So could North America's black panthers actually be black pumas? In fact, although melanism has been recorded frequently from a number of different cat species, the puma is not one of them. For whereas unconfirmed reports of black pumas in Florida have surfaced from time to time, only one generally accepted case is on record from anywhere in the New World.

Namely, an individual shot by professional hunter William Thomson in the Carandahy River section of Brazil during 1843, but sadly its skin was not preserved.

In addition, I have seen various online mentions of an enigmatic taxidermy cat dubbed the "Cherokee cougar" that has been claimed to be a black puma. Measuring 6 foot 2 inches long, and variously said to have been shot in Tennessee or Montana, it has been denounced by skeptics as a normal puma that had been dyed black, or even some entirely different, non-feline species, such as a bear. However, hair samples from it that were tested by researchers from East Tennessee State University's zoology department confirmed that they had not been dyed, and DNA samples verified that it was a puma. Nevertheless, photographs of it apparently suggest that it is dark brown rather than truly black. Unfortunately, no primary source for this potentially significant specimen is provided by any online documentation that I have seen.

Black puma—likely morphology, created by digital photo-manipulation (Karl Shuker)

On 18 June 2016, however, mystery cat investigator Ben Willis informed me that he had conducted some minor research on it 10 years previously, and he kindly emailed to me the following report of his findings that he had posted back then on his e-mail group:

> Saturday, July 23, 2005 1:20 AM
>
> Sometime early last year [2004], a member of our British big cats group was contacted by Jerry Coleman. Mr. Coleman is a resident of Tennessee, and the author of several books on cryptozoology and the paranormal.
>
> Our member was informed that Mr. Coleman had possibly acquired the carcass of a melanistic puma, killed in Tennessee some years ago. The carcass was said to be owned by a private party in North Carolina, and hair samples had been sent to the East Tennessee State University for analysis. The preliminary report stated that there was no evidence that the hair had been dyed, and the muzzle of the animal appeared too extended for a common cougar. Mr. Coleman further speculated that this might even be a cross between a puma and the legendary onza [see Chapter 6]. The report sounded promising, and my first thoughts were the purported Costa Rica black cougar killed in 1959 [see Chapter 6 again], which also had an

> elongated head, more similar to the Temminck's golden cat.
>
> After much discussion of the subject, a few weeks later, our member informed us that Mr. Coleman had released further information and a photograph of the mysterious cat. Rather than a frozen carcass, it was instead a taxidermy mount located in what was described as an Indian trading post inside the reservation at Cherokee, North Carolina. I immediately remembered my brother telling me of seeing a mounted cougar specimen in one souvenir shop just off main street in Cherokee. He said it appeared that someone had taken a number [of] bear-skin remnants, and fashioned them over a plastic cougar mould. Our initial plans were to feature a web-page and article on this mystery on our British web-site, but since I was certain it was the same mount, I at least wanted a final report on the hair samples.
>
> Now, more than a year later, this mysterious cat is still being discussed on virtually every cryptozoology site on the Internet. To my knowledge, no analysis was returned on any hair samples to date. The only recent development is that the mounted animal has somehow vanished from the shelf in Cherokee, and no one seems to know its whereabouts. Mr. Coleman has published one additional photo depicting a clear frontal view of the animal's face. In my opinion, the nose, which cannot be distorted by the taxidermy process, is undoubtedly bear. The anterior nasal openings, or nostrils, have no resemblance whatsoever to any feline.

Ben kindly sent me a scan of the photograph, and although the cat's face is certainly feline, its nose does appear very odd and disproportionately large. Moreover, its face's fur color does seem to be brown, although if flash photography was used, this coloration may be a resulting artifact.

In 1998, American mystery cat investigator Keith Foster of Holcolm, Kansas, informed me that what had been reported to him by the person concerned as being a "glossy black puma" had been shot and killed in Oklahoma several years previously after it had been killing sheep on his father's farm. Afterwards, this person (a church pastor who was known to Keith) contacted the authorities, and the Oklahoma Department of Wildlife duly confiscated the cat's carcass. Nothing more was heard about it after that, but Keith vowed to trace it. Unlike so many cryptozoological cases featuring a missing specimen, moreover, Keith did succeed in doing so, but was disappointed to discover that it wasn't glossy black in color after all, merely grey (Foster, pers. comm., 1998).

Melanism in wild felids is usually associated with tropical climate and dense forests (e.g. black panthers in southeastern Asia, black jaguars in Amazonia), where a black pelage serves as effective camouflage and as an efficient radiator of body heat. Hence there would seem little advantage in (and thus one would not expect) the evolution of a black puma morph within most of the puma's North American range.

However, Florida actually contains sub-tropical forest and is therefore more likely than anywhere else in North America to give rise to black felids, as demonstrated by its black bobcats. Although it is extremely unlikely that a melanistic puma morph could arise from a few escapee pumas (as would be needed, for example, to reconcile Britain's pantheresque mystery cats with pumas), it is far less implausible that such a morph could arise in its native land, and particularly in an environment such as Florida's subtropical zone, which would seem well-suited for such felids. So perhaps black pumas may indeed be responsible for mystery black panther reports in Florida.

But what of other North American regions? Very dark brown or very dark grey pumas could explain some black panthers reported from these areas, especially any that were associated with eerie screams. For although leopards (including melanistic ones) can indeed give voice to screeching caterwaulings, the puma is indisputably the felid par excellence for producing blood-curdling screams and shrieks. At one time, scientists considered that its renowned scream might be nothing more than a myth. However, a director of Cincinnati Zoo personally saw and heard one of the Zoo's captive pumas give vent to a fine example of this vocal variation, and Robert Beam of the Chicago Zoological Park announced that the Park had never owned a female puma that had not screamed!

Since then, the puma's scream has been classed as fact rather than fable—and in some instances it has even been described as a roar. Yet in truth, the puma's Small Cat throat structure prevents it from producing a true roar.

An intriguing panther theory that involved pumas but not genuinely black pumas was contemplated by Bruce Wright: could normal-colored pumas appear black when wet? To investigate this possibility, Wright took the fresh hide of a newly-killed puma from Vancouver Island, suspended it by its edges, filled it with water, and left it overnight. When he examined it the following morning, however, despite viewing and photographing it from every conceivable angle he was unable to make it appear black in color (*Pursuit*, January 1972).

Wright also considered the prospect that backlighting of normal pumas could create the illusion of black fur but when checked this proved untenable too. Thus it would seem that if pumas are involved in the American mystery black panther phenomenon, they are bona fide black pumas (or at least very dark brown or very dark grey ones).

Such cats can even explain those bizarre bipedal panthers—a little-known facet of puma behavior is that it occasionally stands erect on its hind legs, as noted by Wright in his book *The Ghost of North America.*

(For more information and thoughts concerning black pumas, see Chapter 6 and also Appendices 2 and 3.)

(6) Melanistic Leopards or Jaguars?

In *Mysterious America* and other writings, Loren Coleman has noted that America's mystery black panthers often appear more aggressive than the extremely shy puma. With respect to morphology, there is no doubt that bona fide black panthers (i.e. melanistic leopards) provide the closest correspondence between America's mystery panthers and any known modern-day felid. And as leopards are indeed more belligerent and less afraid than pumas when confronted by humans, this identity is supported on behavioral grounds too. Of especial interest here is that, as noted in *Mysterious America*, a mystery black panther that was reported in northern California's Mount Diablo would keep its kill (usually a deer) up a tree, which is behavior typical of leopards.

Consequently the exotic escapee/release theory comes into force once more. This subject has already been discussed in detail relative to black panthers in Chapter 1, and the points raised there are equally valid in an American context.

A second possibility involving a melanistic Big Cat is that America's mystery panthers could in fact be black jaguars. From an escapee/release point of view this is less likely, if only because there are far fewer black jaguars than black leopards in captivity to begin with. Counterbalancing this, however, is the fact that at least until fairly recently, the jaguar was actually native to certain regions of the USA, where it was formerly believed to constitute a distinct subspecies known as the Arizona jaguar *Panthera onca arizonensis* (currently, however, no valid jaguar subspecies are recognized by the IUCN's Cat Specialist Group).

In the mid 1800s, this felid was known from southern Texas, most of Arizona, parts of New Mexico and California (where it was referred to as the hut'-te-kul—"big spotted lion"), and possibly Colorado, North Carolina, Florida, and Louisiana too. However, it was never a common felid, and excessive hunting resulted in this erstwhile subspecies' apparent extinction by 1905. Yet many reports of jaguars, sometimes supported by actual specimens, have continued to emerge from certain of its former US haunts since that date.

If jaguars do still exist in the USA, it is possible that some black specimens are amongst their number. Of great relevance to this is the fact that melanism in the jaguar is a genetically dominant trait (rather than induced via the recessive non-agouti mutant allele responsible for melanistic individuals of most other felids). More specifically, in a March 2003 *Current Biology* paper, a team of molecular biology researchers that included Stephen J. O'Brien, famed for his work on cat genetics, revealed that deletions in the receptor gene MC1R were associated with jaguar melanism (and also with the dark morph of the jaguarundi). Consequently, one might expect black jaguars to be relatively common.

Black jaguar with normal spotted jaguar (public domain)

In Chapter 1, I noted that some cryptozoologists have little faith in the escapee/release theory as a solution to mystery animal reports, in which case the idea of at least some of America's mystery black panthers being melanistic leopard escapees/releases would find little favor with them. Equally, however, on account of the jaguar's extreme rarity in the USA (assuming that it exists there at all nowadays), the possibility that melanistic jaguars are present in numbers sufficiently large to explain mystery black panther sightings would seem on the face of things rather remote.

In any event, to my knowledge, no such panther sighting documented prior to this book's 1989 publication contained any mention of the eyewitness having observed spots or rosettes on the creature itself (yet melanistic leopards and jaguars famously possess cryptic markings often visible at close range against their dark pelage). One conceivable reason for this is that the observers concerned have chosen not to get near enough to the felids to be able to see such markings! Yet without such information, the escapee/release black panther and escapee/release or native jaguar identities for America's mystery panthers are somewhat intangible.

The second American mystery panther report that was kindly brought to my attention by Victor Albert (pers. comm., 1987) is thus of particular interest and importance, but it had not previously been published anywhere prior to its inclusion within this book's 1989 edition. As in the first account (related earlier by Gerald Cameron), the location was Savannah, Georgia, and one of the principal witnesses was Karen Lingle Purser, one of Albert's cousins, who lived in the area on a 500-acre farm consisting of about 200 acres of pasture and 300 acres of woodlands (some dry, some swampy).

During the 1960s Purser was aware that these woods contained a very large cat, which she often heard at night giving voice to loud, fairly harsh cries. Intriguingly, by imitating these sounds she was sometimes able to lure it out of the woods and into the pasture close to the farm from where she could view it (she saw it 6-7 times via this method). Apparently it had been killing some of the farm goats, however, so eventually the decision was made to dispatch it.

While her father and others (including Cameron's father) hid in the pasture one night in or around 1968, Purser imitated the beast's cries in the hope of enticing it out of the woods. This she did, and when the animal had approached to within 10 feet or so from her lookout point at a window of the farmhouse, she shone a spotlight directly onto it. She described the animal as being big and black but stated that medium-sized spots could also be seen upon it. She compared its size with that of an Alsatian dog, and commented that it looked just like one of the black panthers that she'd seen at the National Zoo in Washington, D.C. She was adamant that it was a large cat, that it definitely did not resemble a dog. On this occasion, the coat appeared bluish-black; on some of the previous sightings it had seemed brownish-black (different lighting conditions, and/or different animals?).

Purser's father and Cameron Sr. shot a raccoon and hung it on a tree in the pasture, and when the pantheresque mystery cat drew near to the bait, it was shot dead. Very regrettably, the current whereabouts of this potentially important felid's remains are not known. The cat may have been taken to a nearby patch of ground where dead livestock were left; alternatively it may have been transported away by some unnamed wildlife investigators who subsequently showed an interest in it.

From this assessment, it is evident that America's panthers may actually constitute several different mammalian forms, a cryptozoological jigsaw puzzle composed of several different pieces, thereby reiterating the composite identity theory that I proposed in Chapter 1 for the British mystery cat situation. This notion was also favored by Michael Mayes in his recent book *Shadow Cats*. Nevertheless, there is little doubt that a hard core of sightings really do involve very large felids belonging to one or more of the species covered here, but of melanistic rather than normal coloration.

One further identity has been suggested, again involving a large felid, but one that is very different indeed from those discussed thus far, and which may also be responsible for sightings of the next type of crypto-felid to be examined here: America's maned mystery cats. So at the end of that section is where this final mystery panther identity will be examined.

Maned Mystery Cats

Black panthers are not the only out-of-place cats being sighted across North America. According to many documented eyewitness accounts from the eastern USA (plus an occasional report from the western states and from Canada too), a number of fully-maned lions, not to mention lionesses and lion cubs, are also on the loose here. As with the panthers, the principal sources of information concerning these felids are the writings and researches of Loren Coleman, who has been instrumental in bringing these intriguing mystery cats to the general attention of cryptozoologists in various books and articles (particularly *Fortean Times*). Jim Brandon's classic book *Weird America* (1978) also documented some examples.

Possibly the most famous of the early US lion episodes concerned a lioness-like creature nicknamed Nellie, which originally made headlines in central Illinois during the early summer of 1917 by attacking a butler picking flowers. Despite huge searches being instigated straight away in the Sangamon River region's woodlands nearby, Nellie was never captured, but an interesting discovery was made: she appeared to have a maned mate. For a beast whose description closely matched that of a male African lion was sighted during the hunt and several times afterwards—a "large yellow, long-haired beast," according to eyewitness James Rutherford, who spotted it near a gravel pit on 31 July 1917. Like Nellie, however, this mystery cat was never caught.

A similarly shaggy felid pursued a party of four adults and two children at Indiana's Elkhorn Falls in the early evening of 5 August 1948; happily they escaped unscathed. The eyewitnesses claimed that it resembled a lion, with a long tail and bushy hair around its neck.

Two days later, a comparably leonine beast was spied nearby and at close range by two farm boys—and in the company of a second strange cat that apparently resembled a black panther. One of the boys fired his rifle at them, whereupon they turned away along a lane. The very next day, two creatures identical to these were sighted by some farmers northeast of Abington, and the following morning in Wayne County, after which they were heard of no more. And in November 1950, an unseen creature that supposedly roared like an African lion was blamed for the killing and consumption of 42 pigs, 4 calves, 4 lambs, and 12 chickens in Peoria County—all in a single night!

Ceresco, Nebraska, was the scene of a prolonged but unsuccessful lion hunt just 12 months later, which had been sparked off by a number of reports of a mysterious maned cat in the area. Over the next few years, further sightings were made at various nearby locations, including Rising City and Surprise (the latter hosting an alleged sighting featuring a lion and lioness together).

American maned mystery cat with American mystery black panther (William M. Rebsamen)

Mystery lions are not exclusive to the USA either, as demonstrated by the observation of a five-foot-long maned felid standing at least three feet tall and possessing a long tail that terminated in a tufted tip, reported by Leo Dallaire. He spied it on his farm near Kapuskasing in Ontario, Canada, during June 1960.

A beast sighted by an Ohio tourist in Oklahoma's Big Cabin country, and described as being an unmistakeable African lion *Panthera leo,* initiated a major lion hunt in mid-March 1961. During the same period, an animal identified as a lion was also spotted by two nurses amidst shrubbery in the grounds of Oklahoma State Hospital. Local inhabitants affirmed that it had been in the area for the past two months, claimed to have heard it roar, and accused it of having eaten many of their chickens. Rogers County Sheriff Amos Ward and Tulsa Zoo Director Hugh Davis both felt that the mystery beast did indeed exist, and a rumor (albeit unsubstantiated) of a circus truck having overturned in that vicinity some time earlier was resurrected as an explanation for its origin (*Newsweek*, 27 March 1961). Like its predecessors, however, this crypto-cat was never caught and reports petered out.

During the mid-1960s a young African lion was allegedly killed by two deer hunters near Georgia's Blue Ridge. Its origin is unknown, but it was rumored to have been a pet that had become unmanageable. According to Downing's *Cryptozoology* article from 1984, a photo of this cat may still exist.

In May 1970, Illinois re-emerged as a center of lion sightings when yet another large-scale lion-hunt was launched. Its quarry was an eight-foot-long beast with a mane and long tail, observed at Parthenon Sod Farm near Roscoe by the farm owner, George Kapotas, together with Tom Terry and five of his fellow workers. The creature was not found. Nevertheless, a week later a number of livestock on Lyle Imig's nearby farm hurtled through two barbed-wire fences as a result of an encounter with a mystery beast—which, although remaining unseen itself, left behind some formidable tracks. These were described as being 5 inches long and 4.75 inches wide, with an inter-print length of 40 inches.

On or around 1 March the following year, a most peculiar cat was sighted skulking near the home of Howard Baldrige, near Centralia, Illinois. Baldrige described it as resembling a sort of shrunken lion. About twice the size of a large domestic cat and yellow in color, it had a long tail, very short legs, and a face "like something on television." Make of that what you will!

In 1976, and at a distance of 50-100 feet, a much larger mystery cat with a mane around its neck was sighted and shot at unsuccessfully by farmer J.H. Holyoak in his pasture in Georgia's Berrien County. The same year saw yet another police lion-hunt too, its quarry having a "shaggy black mane, light brown body and a black tuft at the end of a long tail," according to several people in Tacoma, Washington. The best that could be found, however, was a collie-Alsatian mongrel dog called Jake at the city dump.

In the late fall of 1977, a two-month-old lion cub was discovered by Police Lieutenant Ronald King in Muscatine, Iowa, but despite his putting out a teletype report enquiring whether anyone had lost a lion, the cub's origin remained obscure (*Res Bureaux Bulletin*, 1977). On 23 January 1978, a woman in Loxahatchee, Florida, was taken aback to see a lion just outside her window. Following the by-now-familiar pattern, a police search was duly instigated and returned home empty-handed. Moreover, the nearby Lion County Safari Park and two residents known to keep lions as pets all affirmed that none of theirs was missing (*Palm Beach Post*, 24 January 1978).

In August 1979, a lion cub appeared outside a back door in Ohio's West Chester. Although happy to be fed cat food by the family's children, the unexpected visitor was eventually traced back to an amusement company from which it had escaped (*Res Bureaux Bulletin*, August 1979). On 10 November 1979, a growling 300-400-pound lion prowling through Fremont, California, was sighted by several people and encountered at close range in the Alameda County Flood Channel by Police Officer William Fontes, but it succeeded in eluding the large-scale search party attempting to track it down.

The late 1970s also saw a supposed puma that had allegedly been killed in the North Carolina mountains; it turned out to be an African lioness whose body had been rescued from a dumpster. And in 1982, a skeleton found in a ditch, again in North Carolina, proved to be that of a young African lion. Both cases featured in Downing's 1984 *Cryptozoology* paper.

A sighting of a maned lion was made during July 1984 in a suburb of Cleveland, Ohio. Another was recorded near Texas's Fort Worth Zoo during late February 1985 (both were recorded in Coleman's book *Curious Encounters*, 1985). Needless to say, the zoo itself was checked, but no lions were missing, and although it was later seen by two police patrolmen who identified it unhesitatingly as a lion, it eluded all attempts at capture and was not sighted again.

More than 30 years later, maned mystery cats are still being reported and sighted across North America. One notable case on file from West Virginia in 2007 was reported widely by the media, together with another one the following year from Virginia and one that same year from Colorado too. There were also some highly confusing reports of unidentified "lions" from Georgia in 2009 that may, or may not, have been maned cats; plus a suspiciously canine maned mystery "lion" videoed in Los Angeles, California, during 2014, and another maned lion videoed in Milwaukee, Wisconsin a year later.

What could they be? There would appear to be three identities on offer for North America's maned mystery cats:

(1) Large Dogs?

This is a very popular "official" explanation for reports of mystery cats. Yet although various dog breeds can appear surprisingly panther-like or puma-like under certain viewing conditions, very few dogs (regardless of such conditions) can be mistaken for a full-sized, fully-maned lion.

Probably the closest approximation to a leonine dog is the Chinese chow—and by chance there is at least one episode on record involving a maned mystery cat in which a specific chow was put forward as the felid in question. This was during the case of the Fremont lion of November 1979. After Fontes had made his sighting of the mystery beast in the flood channel, a local inhabitant came forward to claim that his own 40-pound chow puppy was the animal responsible—to which Fontes soon retorted: "The puppy in no way resembled the 300 to 400 pound animal observed in the flood control channel." Considering that not even the very largest of adult chows can attain such dimensions, it is hardly surprising that Fontes should so readily dismiss a chow puppy from contention.

Even more remote is the possibility that anyone could confuse a lion with a Brittany spaniel, yet this was a highly regarded "official" identity for the lion spied near Fort Worth Zoo in 1985. Its police eyewitnesses, however, begged to differ—and who can blame them? Anything less like a lion than a white-and-orange/chestnut/black dog of maximum height 20 inches and bearing a pair of fluffy pendulous ears would be hard to imagine—except, perhaps, for mongrel Jake! In any case, whereas many maned mystery cats have been reported roaring, I have yet to read of a case in which one of these animals barked!

Sometimes, a dog suffering from a severe case of mange has appeared superficially lion-like, having lost much of its hair except for a fringe around its neck and at the tip of its tail. Obviously, however, by being a dog its dimensions were much less than those of a real lion.

(2) African Lion Escapees/Releases?

Despite being anathema to some mystery animal aficionados, the exotic escapee/release theory is a far more reasonable solution to North America's maned feline mysteries than large dogs. Lions are indisputably very popular animals, not only in public zoos and circuses but also in private animal collections and even as household pets. That escapes (or releases) of lions from captivity in North America do indeed occur has already been demonstrated by the discovery of the various dead specimens recorded here, and it seems highly unlikely that a lion or lioness would have much trouble in surviving in much of this continent's countryside. Moreover, we have already learned that cubs are not averse to escaping.

Of great pertinence is the horrific reality that people in Texas and elsewhere in the USA buy African lions from zoos, circuses, and pet owners for the express purpose of inviting would-be big-game hunters to come and shoot them, usually (although apparently not always) within fenced enclosures. Clearly then the possibility of escapees must be a very real one. During April 1988, for instance, news emerged that two African lion carcasses had been discovered by the US Army Corps of Engineers on a reservoir site near Wallisville, Texas; lion hunts had indeed been occurring on that site. Yet as lion hunting was technically not illegal there, US Federal officials feared that they could do little to prevent it (*Times*, London, 12 June 1988). As such activity is not limited to Texas either, we surely need not look much further for explanations of American maned lions.

Nevertheless, a third explanation is on offer, one involving lions that in America are not out-of-place but out-of-time, i.e. involving a putative prehistoric survivor. Moreover, this is also the final identity for American mystery black panthers that I promised to discuss at the end of the maned mystery cats section of this chapter.

(3) Surviving American Lions?

During the Pleistocene epoch, leonine cats existed not only throughout much of the Old World but also in the New World (as a result of the former presence of a Bering Sea land bridge connecting northeastern Asia with Alaska), thriving in North America, Central America, and as far south as Argentine Patagonia in South America. This New World felid, which was truly enormous in size, is most commonly dubbed the American lion (but should not be confused with America's mountain lion or puma *Puma concolor*, which is a much smaller, very different species of felid).

Although originally categorized as a gigantic jaguar (*American Museum Novitates*, 1941), until recently the American lion had traditionally been classed as a leonine subspecies, but separate from all of its Old World counterparts. Deemed to be a sister lineage to the Eurasian cave lion *P. (l.) spelaea*, it has been named *P. leo atrox*. During the past decade or so, however, many (but not all) paleontologists have considered it sufficiently distinct from all lions to be reclassified as a separate *Panthera* species in its own right, thus becoming *P. atrox* (e.g. *Molecular Ecology*, 2009). It appeared to die out at the end of the Pleistocene; no fossil evidence dating from the Holocene is known. However, developing a suggestion first put forward by fellow American cryptozoologist Mark A. Hall, Loren Coleman has postulated (originally in *Fortean Times*, summer 1980) that the unexpected maned felids being sighted across North America today may actually be surviving male individuals of *P. (l.) atrox*. Furthermore, he has proposed that the equally mystifying black panthers frequently sighted in many parts of North America but never formally identified could be *P. (l.) atrox* females.

In support of this theory, Coleman has commented in *Mysterious America* and elsewhere that the behavior of these two types of American mystery cat compares closely with that of genuine lions elsewhere—with the maned beasts being proud but cautious, like typical male lions, whereas the pantheresque felids are less retiring and more aggressive, like typical lionesses. He has added that the classification of these two unidentified felid forms as the two sexes of a single species can also explain those very occasional sightings on record of maned cats and black panther-like cats having been seen together.

Altogether a quite fascinating theory whose scientific development could constitute a stimulating intellectual exercise. Sadly, however, in practical terms it suffers from a number of fundamental problems, as brought to public attention by British researcher Mike Grayson in a *Fortean Times* communication (winter 1982).

First and foremost of these is the radical difference in pelage coloration between the maned and the black panther-like cats. As Grayson has pointed out, if we assume that these do indeed constitute the two sexes of the same species, the exhibition of normal leonine coloration by the male but very pronounced melanistic tendencies by the female would constitute an example of extreme sexual dimorphism totally without parallel among other mammalian species. The male-only mane of the lion is proof that this species is indeed capable of evolving a marked degree of sexual dimorphism, but even allowing for the fact that some lions have dark belly and neck manes, it is still exceedingly improbable that such dimorphism could aspire to the exceptional level required by the Coleman-Hall theory.

Equally contentious is the total absence of sightings in modern-day North America of mystery lion prides. As Coleman himself has noted, lions are social, and, according to felid specialist Helmut Hemmer in a *Carnivore* paper published in 1978, the high degree of cephalisation—brain development—displayed by the American lion makes it possible that this felid was too. So if Coleman's *P. (l.) atrox* theory is true, why no records of prides? And, indeed, why, out of the plentiful maned mystery cat records he and others have documented, are there so very few that involve one of these felids and a pantheresque cat being seen together? Having said all of that, however, since its redesignation as a species separate from Old World lions the American lion has been looked upon by some researchers as more likely to have been solitary, which would thus seem to favor Coleman's theory after all—except for one dilemma.

In the Old World lion, which is a polygamous social species, the darkness of a given male individual's mane serves as a significant visual clue regarding that male individual's quality in relation to potential mates and rivals—the darker the mane, the more dominant the male. As stated in an *American Scientist* article (May-June 2005) by Peyton M. West, a member of the research team that uncovered the hitherto-obscure purpose of the lion's mane:

> Female lions live in prides consisting of related females and their dependent offspring. As the cubs grow, young females typically join their mother's pride, and young males form "coalitions" and disperse to look for their own pride. This creates a system in which a small group of males can monopolize many females, leading to severe reproductive competition. Predictably, males compete intensely for mates.

Yet although the mane's very visible demonstration of a male's fitness will therefore be very beneficial in attracting mates and repelling rivals, it comes at a high cost to the lion's health, inasmuch as a dark mane inhibits dissipation of heat from the lion's body in hot climates. Consequently, if the American lion were indeed solitary, lacking the complex social, polygamous lifestyle of its Old World relatives, we would expect it not to possess a mane at all, or only an insignificant one, because the thermal advantage in not having one would outweigh the dominance-signaling advantage of having one (which would be far less important in a solitary, monogamous species). Yet if it did lack a (notable) mane, it clearly cannot present itself as a plausible candidate for North America's maned mystery cats.

Another important, undeniable fact highlighted by Grayson is that animals identical to North America's black panthers are also being reported in regions of the world where survival of prehistoric lions (and lionesses) is totally incongruous (such as Great Britain), as well as from regions where lions have never existed at any time (such as Australia). Compared to the escapee/release theory, the *P. (l.) atrox* explanation is clearly inadequate here.

In addition to Grayson's comments regarding the *P. (l.) atrox* concept, another equally formidable and significant objection to this exists. Namely, the blunt fact that, judging from reconstructions of the American lion in life based upon fossil evidence, it simply did not resemble the maned cats being seen there today. Whereas North America's maned mystery cats resemble modern lions in both overall size and physical appearance, the American lion was very different indeed.

To begin with, it was at least one quarter larger than the largest of modern lions, hence its earlier, alternative name, the great cat. Furthermore, relative to today's lions the American lion's face was shorter, with a broader nasal region, and its limbs were notably longer, so much so, in fact, that it is considered to have been a truly cursorial felid, far more so than any modern lion. There is also the afore-mentioned issue of its putative manelessness to keep very firmly in mind, and preserved fur samples attributed to *P. atrox* that have been found in Argentine Patagonian caves and distinguished from those of jaguars are reddish with spots, not tawny and unspotted as in modern-day lions (*Comptes Rendus Palevol*, November-December 2017).

American lion *Panthera atrox*, reconstruction (Sergiodlarosa/ Wikipedia)

Arguing in favor of post-Pleistocene survival of *P. (l.) atrox*, Coleman has noted that only one American lion fossil skeleton has been recovered for every 30 of its saber-tooth contemporary *Smilodon fatalis* from the famed tar pits of Rancho La Brea, near Los Angeles, California. This is a startling ratio, one that has been put forward by many paleontologists as evidence for the intellectual superiority of the American lion. Coupled with this is the fact that, relative to its body size, the brain of the American lion was larger than that of any Old World counterpart, either from the Pleistocene or from the present day. Thus, one might expect that *P. (l.) atrox* would stand a better chance of persisting undetected into modern North America than would *Smilodon*.

Yet in spite of the American lion's intellectual level, we cannot ignore the seemingly irreconcilable morphological and behavioral problems noted here. Consequently, although certainly a captivating concept, the theory that America's maned and black pantheresque mystery cats constitute a relict population of *P. (l.) atrox* is ultimately untenable, at least in my opinion, with the escapee/release theory reasserting itself as both the most plausible and the most parsimonious solution to these mystery felid forms.

As for those occasional episodes involving a maned lion-like and a black panther-like cat being seen together by various eyewitnesses, the most reasonable explanation is that in each of these exceedingly rare instances, they were a couple of escapees/releases (probably from a circus or private collection) consisting of a lion and a melanistic leopard that had originally been reared and maintained together prior to their escape/release. Rearing cubs of different Big Cat species together is by no means uncommon in captivity, especially in smaller, private zoos, and is sometimes even done purposefully to facilitate interspecific matings in the hope of obtaining exotic-looking hybrids to draw in visitors.

Lastly, but very tantalizingly, early Native American traditions in Maine tell of an extremely large, ferocious mystery cat termed the lunkasoose. This was reputedly distinct from the puma, and was said to possess a leonine mane...

Santer: North Carolina's Mystery Cat

The santer is a largely-forgotten feline anomaly that does not seem to slot readily into any of the categories of American mystery cat documented here. Hailing from North Carolina, this mystifying creature terrorized a number of local communities in the State during the last years of the 19th century, featured in many reports by the *Wilkesboro Chronicle* and the *Statesville Landmark* newspapers, and resurfaced in 1977 via a most interesting two-part article by Angelo Capparella III, published in the periodical *Shadows* and quoting from many of the original newspaper reports.

Initially treated humorously by the media, alarm generated locally by continued livestock killings and also by indications that more than one animal specimen was involved inspired a more serious attitude in later santer reports. Sadly, very little was recorded concerning its physical appearance, so no confident conclusion regarding its identity can be offered. The *Wilkesboro Chronicle* noted on 5 May 1897 that a lynx had escaped from a circus visiting the area the previous year, and inferred that this may be the santer.

On 9 June, it announced that the santer had been caught—and that it resembled a large "shepherd dog"—but the news also recalled rumors of a second animal on the loose. Certainly, all was not over because on 20 October 1897 Charles Smoot alleged that he had recently seen a santer, describing it as "striped from the end of its nose to the end of its tail." On 31 May 1899, a santer sighting was claimed by a Mrs. Smoak, who stated that it was grey and between a cat and a dog in size.

Due to the radical conflict in descriptions, it would seem that yet another cryptozoological composite had been "created" by reports of totally different animal types having been attributed erroneously to a single animal form. Capparella believed that although some sightings were probably of non-felids, the real santer was a genuine large cat form. Moreover, the santer's alleged boldness and lack of fear when in the presence of humans suggested that a puma identity was also unlikely, and he leaned instead towards the santer being a currently unknown felid. Yet the santer has failed to be reported since the 1890s (a mystery wild beast roaming around South Iredell in May 1934 was considered by locals to be the santer or an offspring of this beast, but no proof to support such claims was obtained).

Consequently, it would seem more likely to have been based upon a single animal of unexpected form (an exotic escapee/release once again?), whose exploits set the scene for over-enthusiastic identification thereafter of anything large as the culprit, which is reminiscent of the Surrey puma-fox situation and the Shing Mun tiger-Alsatian.

Tennessee's Multicolored Mystery Cat

This very baffling mystery cat was first documented cryptozoologically in my "Menagerie of Mystery" column in *Strange Magazine* (spring 1998). On 17 October 1997, Bryan Long posted an online note (in "The Cryptozoology Zone" of AOL's Parascope site) revealing that about a year earlier, a group of hunters in Tennessee had supposedly shot a very large cat of such unusual appearance that they never reported it for fear of going to jail for having killed the animal. Nevertheless, details of its appearance eventually filtered out, and according to Bryan, who saw a photograph of its carcass, it was cheetah-like in form but had a blood-red head and paws, a red line running from the back of its head to its tail (which was also red), and a golden-brown body patterned with black stripes and spots. Apparently, its carcass was skinned and the pelt hidden in a basement.

Curious to learn more, I contacted Bryan, and on 30 June 1998 I received a short email from him (Long, pers. comm.). In it, he revealed that the photo he had seen had belonged to a co-worker who had since sold it to a local university biology student (name unknown). Moreover, the cat had been shot near Tennessee's Jackson County, and some local farmers claimed to have seen similar cats in the past but not for years now. Identities such as jaguar and jaguarundi were subsequently proposed by various people, but neither species fits the animal's description.

The jaguar does not have a red head, paws, dorsal line, or tail. And although the jaguarundi does have a red color phase, the eyra, this is red all over with no fur patterning, and is certainly not cheetah-like in form. Could the red coloration simply have been dried blood? As I have not encountered any further reports of this or any other similar felid, Tennessee's cat of many colors remains an enigma.

Living Saber-Tooths in North America?

Based upon its current fossil record, North America's famous Pleistocene saber-tooth, *Smilodon fatalis,* became extinct 10,000-11,000 years ago. As noted in my book *Still In Search Of Prehistoric Survivors* (2016), however, there are a couple of very unexpected reports on file of supposedly living saber-tooths spied in the USA.

In December 2009, I learned from British cryptozoological researcher Richard Muirhead that in or around 1913, two such cats had allegedly been shot dead by the US Cavalry in Arizona. His source for this information was an American cryptozoologist, Andrew Ste. Marie, who in turn had heard about the incident from Joe Taylor, a museum curator. No-one apparently knew what had happened to the cats' carcasses afterwards, unfortunately, so there seems no way of confirming their taxonomic identity. I have since heard rumors of a photograph circulating online that depicts someone standing alongside a newly shot-dead saber-tooth, but this may well be a fake, digitally-created image. In any case, I haven't been able to locate it, so it remains just hearsay, at least for now.

Also in 2009, Richard Murhead mentioned that yet another of his American cryptozoological contacts, Jerry Padilla, claimed that one night in 1946 "a very close relative now deceased" saw a saber-tooth in northern New Mexico on an old, remote mountain road near the Philmont Scout Ranch in the Sangre de Cristo Mountains. The cat's color was that of a lion. With its only eyewitness no longer alive, however, yet again there is no direct way of investigating this cryptid further, except perhaps by interviewing local residents there in case others have seen such an animal. Having said that, I consider it highly unlikely that a carnivorous mammal as large, unmistakeable, and highly-specialized as a *Smilodon* saber-tooth species could survive in modern-day USA without its existence having long since been formally revealed.

The final selection of North American mystery cats to be considered here can only be described as a purring potpourri in which the unexpected and the unlikely are sometimes difficult to distinguish and disentangle from one another.

Blue Lynxes and Wampus Cats

Most people would be hard-pressed to believe in blue-furred lynxes, but in fact incontrovertible evidence for their existence is at hand. In 1938, Ernst Schwarz reported in the *Journal of Mammalogy* that the skin of just such a cat had been donated by Mrs. Charles D. Walcott to the United States National Museum (although very regrettably, having been obtained through the fur-trade, all that was known about this most significant skin's origin was that it had come from somewhere in Alaska).

Rather than exhibiting the familiar fawn color of normal lynxes, its pelage was bluish-grey all over. True, it did possess the usual scheme of pale inner limbs and belly, but whereas the tail-tip and ear-tufts of typical lynxes are black, those of this specimen were simply a deeper shade of blue-grey—indeed, the black coloration was totally absent. This demonstrates that it was not a chinchilla-phenotype mutant. Instead, as Schwarz stated, it clearly represented a dilute mutant (i.e. homozygous for the Full color gene's recessive dilute mutant allele, already proposed as a partial explanation for Fujian's blue tigers in Chapter 3, and responsible for the maltese shade in domestic cats).

Enquiries made to the fur trade confirmed earlier unsubstantiated eyewitness accounts because this mutant form occurs once or twice in every thousand lynx skins, according to various furriers. A drab blue lynx was also noted by Henry Poland in his book *Fur-Bearing Animals in Nature and Commerce* (1892), together with a yellow specimen for which felid geneticist Roy Robinson suggested in a 1976 *Genetica* paper that the recessive non-extension of black mutant allele of the Extension of black gene was responsible (it replaces black pigment with yellow or red). Poland's book listed examples of blue bobcats too, again evidently dilute mutants, plus red specimens that once more indicate the non-extension of black mutant allele.

Equally intriguing is the felid featured in a frustratingly vague newspaper report that appeared on 3 August 1823 in Kentucky's *Louisville Courier* and in Boston's *New England Farmer*. It referred to a large unidentified cat seen near Russelville, Kentucky, and described as:

> A tiger of brindle color with a most terrific front—his eyes were as the largest ever on any animal around these parts.

It was fired at from 50 yards, yet stood its ground for 12 shots before making off apparently unhurt and at full speed. An exotic escapee/release, a cryptozoological curio, or a muddled misidentification? Sadly it is very doubtful whether we shall ever uncover the answer.

Back in the early pioneering days of North America, a rich, if highly-imaginative, corpus of folklore sprang up among woodsmen and lumberjacks in particular, featuring all manner of bizarre beasts collectively dubbed "fearsome critters," and including a wide range of much-dreaded albeit decidedly dubious felids that unsuspecting, gullible strangers often believed were real. Perhaps the most infamous example was the wampus cat, which allegedly occurred in many different forms, each more terrible than the last. Take, for instance, the whistling wampus. This, supposedly, is a huge black beast inhabiting Waldron in Arkansas, deriving its name from its beguiling whistling cry (said to sound like "hoo-hoo"), and possessing a notable appetite for lumberjacks!

If the existence of an aquatic saber-tooth in tropical Africa seems improbable (see Chapter 4), then how much more so an amphibious panther in central Missouri? This unlikely beast is the subject of numerous lofty legends recalled by the local inhabitants (especially in the vicinity of Roark Creek's valley), who refer to it as the gallywampus, and assert that it swims like a colossal mink and occasionally attacks terrestrial livestock.

Nor to be overlooked (literally!) is the Oklahoma wowzer. According to traditional "fearsome critter" yarns from the backwoods, this mega-felid resembles a puma in coloration but is about 5-6 times bigger! One of these Brobdingnagian beasts allegedly destroyed an entire wagon train in early pioneer days, killing more than 30 oxen in a few minutes (it simply bit their heads off!), but not harming the people themselves.

Also sizeble is the glawackus, which is variously likened to a puma, leopard, or lion. One was supposedly spied in 1939 at Glastonbury, Connecticut. In November 1944 it reappeared, this time at Frizzleburg, Massachusetts, where it was seen fighting a bull. Strangely enough, its only known eyewitnesses seem to be newspaper journalists!

The splinter cat behaves very like a feline rhinoceroses because it likes nothing better than to charge headlong into trees, crashing its specially-reinforced brow against their trunks in a thunderous collision. The purpose of such belligerent behavior is to flush out the raccoons and wild bees upon which this animated battering ram feeds.

Cactus cat (William M. Rebsamen)

We should not forget the cactus cat either, reports of which supposedly originated from travelers passing through Tucson's cactus-bearing expanses during the 19th century. They described it as resembling a spiky-furred domestic cat but with the added attributes of a bifurcate tail and razor-sharp blades of bone on its forelimbs. It was said to use these blades to slash the base of large cacti in order to release the sap. Once the sap had fermented, the cactus cat would drink it heartily, then depart upon its inebriated way, screeching in drunken delight!

Needless (although sad) to say, there is very little justification for believing that wampus cats, wowzers, and cactus cats are anything other than terrors of the traditional tall-tale variety (several additional examples are documented in my book *Cats of Magic, Mythology, and Mystery*, 2012).

More recently, during the late 1990s, the Ozark howler bemused cryptozoologists for a short time. This bizarre creature, reputedly frequenting the more remote regions of Arkansas, Missouri, Oklahoma, and Texas, was likened to a black panther in overall form but readily distinguished by the pair of prominent horns that it bore upon its head; it earned its name from its weird cry, likened to a combination of a moose's bugle and a wolf's howl. Not surprisingly, this was soon exposed as a felid of the fraudulent kind by American cryptozoologist Loren Coleman and others.

Nevertheless, there is a tantalizing tie-in with one pseudo-cat that merits investigation. I learned from British folklore author Michael Goss that in British Columbia, Canada, totem poles produced by various Native American nations sometimes depict a felid that they adamantly class as a real but completely separate form from the puma, and which they refer to as the how-how. Many indigenous names for animals are onomatopoeic, derived from the sounds that they make. Is it just coincidence that "how-how" sounds remarkably similar to the hoo-hoo cry of the whistling Wampus cat?

And Finally…

No chapter on North American feline mysteries and marvels would be complete without a mention of the celebrated cherry-colored cat exhibited by that famed American showman Phineas T. Barnum. After having paid their money to see this wonderful animal, a few of its visitors later protested to Barnum that they had been both surprised and disappointed to discover that it was nothing more than an ordinary black cat.

Barnum, however, was quite unmoved by their remonstrance. For as he reminded them, he had promised them a cherry-colored cat and, after all, some cherries are black!

There's a lesson for cryptozoology in there somewhere...

Chapter 6
Mexico and Central/South America: Onzas and Jaguaretes

My own first impression of a living jaguar was not one of pure pleasure...For it is his instinct to remain always hidden from the eye, and it confuses and maddens him to be held in place by bars and gazed openly at.
— W.H. Hudson (on seeing a caged jaguar),
Harmsworth Natural History

Part 1—Mexico

When the Spaniards invaded Mexico during the 16th century, they named its large cat forms after species familiar to them back home in the Western world. Thus the puma *Puma concolor* was christened the leon, and the jaguar *Panthera onca* the tigre (names by which these two species are still known in many regions of Latin America today). According to traditional zoology, no other large-sized felid exists in Mexico.

Onza: Cheetah-Like or Puma-Like?

Conversely, modern-day inhabitants of Sinaloa, the province running 400 miles along the Sierras down Mexico's western coast, acknowledge the existence here of a third large cat form, referred to by them as the onza. This name is found in records dating back again to Mexico's first Spanish settlers. Moreover, it is derived from the Latin word "uncia," which in turn refers to the Old World cheetah. Does this mean, therefore, that the Spaniards in Mexico encountered a cat form that reminded them of a cheetah? Apparently it does, because the Sinaloans describe the onza as being superficially puma-like but with a more slender body and notably long limbs.

The most detailed onza narratives consist of *The Onza* (1961), a regrettably scarce book written by longstanding onza investigator Robert Marshall; a *BBC Wildlife* article and two *ISC Newsletter* reports written by former ISC Secretary J. Richard Greenwell during the mid-1980s; and Neil Carmony's more recent book *Onza! The Hunt For a Legendary Cat* (1995). These constitute the principal sources of information I utilized during the preparation of this onza history and should be consulted for further details.

What may be the earliest known reference to this felid (although the term "onza" was not actually used here) was made by Bernal Díaz del Castillo (c.1496-1584), under the command of the infamous conquistador Hernán Cortés. In a report some time after having visited the famous zoo of the Aztec king Montezuma, Castillo recorded seeing "tigers [jaguars] and lions [pumas] of two kinds, one of which resembled the wolf." In view of the wolf's long limbs and the corresponding characteristic of the onza, it is traditionally assumed that Castillo's "wolf-cat" (known to the Aztecs as the cuitlamiztli) was indeed the onza.

In any event, by the 18th century "onza" was well established as the official name for Mexico's third type of large cat. Consequently, the Jesuit missionary Father Ignaz Pfefferkorn—posted in 1757 to the remote, little-explored north Mexican province of Sonora—referred to "the animal which the Spaniards call onza," describing it as being far less timid than the puma but of similar shape except for its rather longer, thinner body. Comparable onza descriptions, but this time from Mexico's Baja California, were also given during the same period by two other Jesuits: French-born missionary Father Johann Jakob Baegert (working with the Guaricura Indians) and Mexican teacher-scholar Father Francisco Javier Claviego (aka Clavijero). Further onza reports and rumors were recorded during the 19th century, including the interesting belief of some Mexicans that the onza resulted from puma x jaguar matings.

In the early part of the 20th century, American zoologist George H. Parker discovered that the onza is a familiar but much-feared felid within the Mexican state of Durango. Its inhabitants informed him that the eyes of the onza were yellow and "look like balls of fire in the night," and that this mysterious cat was held to be the most dangerous animal in the Durango mountains. Nevertheless, with a distinct lack of physical evidence to examine, other zoologists paid little or no attention to such accounts, dismissing onzas as nothing more than misidentified or emaciated pumas.

In 1938, banker Joseph Shirk of Peru, Indiana, traveled to Sinaloa's San Ignacio district to hunt jaguars on La Silla Mountain and was accompanied by guides Dale and Clell Lee, renowned hunting experts from southeast Arizona. The higher altitudes of the San Ignacio district are relatively little-explored due to their extremely difficult terrain, and are characterized by a gradation from subtropical to semi-arid vegetation. According to reliable local inhabitants of Sinaloa, this remote locality is home not only to jaguars but also to onzas.

Following their arrival, the Lee brothers were unable to locate any jaguar signs. However, they had noted another form of cat here and therefore set their dogs off in pursuit of that one instead. Within a short time, a sizable specimen of this unidentified felid was treed and later shot dead, but upon examining it, the Lees were greatly surprised to discover that despite their very considerable hunting experience, they had before them the carcass of a cat form that they were completely unable to identify! Although it was reminiscent of a puma, its body was longer and much slimmer, and its limbs and ears were notably lengthier too. This animal was then measured and photographed, and the skull and skin were retained by Shirk.

Dale and Clell Lee with the Shirk onza, 1938 (International Society of Cryptozoology)

If only the remainder of the specimen had been preserved too—instead it was butchered and disposed of. For it transpired that when, after returning home to Arizona, the Lees formally announced their exciting find, they were shocked and alarmed to receive nothing but ridicule and disbelief from the scientific world. Lacking a complete specimen as evidence that could be thoroughly examined, their case for a new form of felid collapsed, and the Lees avoided further publicity on the subject. The skull of their specimen was apparently deposited in an unnamed American museum during the 1960s, and its origins are now unknown, whereas the onza itself remained entrenched as a feline persona non grata. A lithe, puma-like cat with freckle-like limb markings, identified locally as an onza, was trapped and shot some time prior to World War II in Mexico's Barranca de las Viboras ravine by American writer J. Frank Dobie, as he recorded in his book *Tongues of the Monte* (1935), but its skin was later destroyed by bugs.

These unhappy incidents, coupled with the all-consuming events of World War II, could quite conceivably have jettisoned the onza into scientific obscurity forever, had its history not attracted the attention of another Arizonan hunter, Robert Marshall. Rather than pursuing the onza with rifle bullets and eyepiece, however, Marshall chose instead to arm himself with paper, pen, and traveling boots, and set about documenting every onza report, rumor, and legend that he could uncover, and seeking out any onza relics that might exist. He eventually succeeded not only in amassing a very considerable stock of data concerning this mystery cat, but also in discovering an incomplete skull (its lower jaw was missing), reputedly derived from an onza originating from Los Frailes, Sinaloa. His researches climaxed in 1961 with the publication of his fascinating book *The Onza*, but science remained unwilling to acknowledge its subject's existence as a separate felid form.

The next significant step in the history of the onza—although it was not seen as such at the time—occurred in 1969, when Phil C. Orr described a new species of fossil Pleistocene puma from a perfectly-preserved skull, mandible (lower jaw), and scapula (shoulder blade), plus various vertebrae and ribs. Documenting it in a paper published by the *Bulletin of the Santa Barbara Museum of Natural History, Department of Geology*, he formally named this new cat *Felis trumani* (now *Miracinonyx trumani*—see later). Further finds in subsequent years enabled paleontologists to reconstruct a detailed picture of the cat's likely appearance in the living state, and they found that it bore a striking similarity to the Old World, modern-day cheetah (as did, although to a rather lesser degree due to its larger head and overall size, its Pleistocene antecedent—*Felis* [now *M.*] *studeri*, another fossil puma, formally described by Donald E. Savage in 1960 and nowadays often deemed conspecific with *M. trumani*).

Daniel Adams and co-workers from the University of California later suggested that *F. trumani* was a specialized puma whose evolution had paralleled that of the cheetah, occupying in the New World the latter felid's niche in the Old World as a cursorial hunting cat (*Science*, 11 March 1977).

By the end of the 1970s, conversely, after having analyzed in great detail the *F. trumani* fossil remains so far uncovered and comparing them with those of the Old World cheetahs, Adams had changed his mind (*Science*, 14 September 1979). As a result of his studies, he came to a startling conclusion, namely that pumas and cheetahs were very closely related, sharing a common origin, and that *F. trumani* and also

F. studeri were more than just pumas morphologically paralleling cheetahs—instead, they were genuine cheetahs themselves!

Adams based his belief upon numerous detailed and highly specific structural similarities between the American species (especially *F. trumani*) and the cheetahs of Africa and Asia, similarities that would not be expected if the New World cats were simply mimics rather than true relatives of the latter felids. Thus, Adams removed *F. trumani* and *F. studeri* from the genus *Felis* (still housing the normal puma at that time) and reassigned them instead to *Acinonyx*, the cheetah genus, so that they were then known respectively as *Acinonyx trumani* and *Acinonyx studeri*. Suddenly the cheetah was no longer an Old World speciality; it was now a native American too.

Adams's revelation meant that a remarkable situation now existed. Namely, that Pleistocene America possessed, in the form of *A. trumani* in particular, a bona fide cheetah whose appearance in life would have compared very favorably with that of a mysterious long-limbed cat of modern-day America whose native name, onza, derives from a word meaning cheetah.

Needless to say, this formidable coincidence did not remain undetected for long. German felid researcher Helmut Hemmer of the University of Mainz was aware of the onza through reading Robert Marshall's book and became intrigued by the close correspondence on a number of different morphological points between Truman's fossil cheetah and the alleged appearance of the onza. Following some researches on this subject during the early 1980s, Hemmer formulated the daring hypothesis that these two cat forms were indeed one and the same species, i.e. that *A. trumani* had in fact survived beyond the Pleistocene into the present day as the identity of the Mexican onza. He made public his bold proposal via a paper presented at the Third International Congress of Systematic and Evolutionary Biology, held at the University of Sussex, England, on 7 July 1985. Yet despite their apparent compatibility, this morphological marriage of *A. trumani* to the onza was not destined to last.

Hemmer's hypothesis brought the onza to the attention of J. Richard Greenwell, then the Secretary of the International Society of Cryptozoology, who became extremely interested in this cryptic cat and instigated his own personal onza investigations. He teamed up with puma specialist Troy Best from the University of New Mexico to uncover the long-lost Shirk onza skull (he had already learned that Robert Marshall still owned the Los Frailes skull). Sadly, their search was not successful. However, while seeking this one, another hitherto unrecorded onza skull was uncovered.

It had been obtained in 1938 from Sinaloa's La Silla Mountain—the same time and the same location as for the Shirk specimen. Little wonder then that when Greenwell and Best first received details of it, they assumed that this must be the lost Shirk skull. But eventual comparison of the newly-discovered specimen housed at Philadelphia's Academy of Natural Sciences with good photographs of the Shirk skull revealed that the two were in fact separate specimens. Further investigations revealed that this newly-identified onza skull had been obtained by R.R.M. Carpenter, another hunter who had been accompanied by the Lee brothers prior to World War II.

Richard Greenwell's researches continued fruitfully with his visit during summer 1985 to Sinaloa. For here he succeeded in locating a third onza skull, of a specimen shot during the mid-1970s by the late Jesus Vega (and once again in the very same location from which the previous specimens had been obtained). Greenwell visited Vega's son, Manuel Vega, and also fellow rancher Ricardo Urquijo, who now owned the Vega onza skull. Greenwell found it to be in perfect condition, and brought it back with him to Tucson on kind loan from Urquijo. Greenwell duly documented all of these significant finds in the *ISC Newsletter* for winter 1985.

Robert Marshall (left) and Ricardo Urquijo with the Vega onza skull (International Society of Cryptozoology)

With three skulls—Los Frailes (incomplete), Carpenter, and Vega, and all very similar to one another—now available for study, Hemmer at last had the much-needed onza material with which to put his hypothesis to the test. This he did by comprehensively comparing the morphology of these specimens with fossil skulls of *A. trumani*. And the result? Hemmer sadly concluded that although the onza skulls did display certain interesting similarities with the latter, some very notable differences also existed between the two forms, which clearly demonstrated that *A. trumani* and the onza were not one and the same species after all.

Even so, the possibility still existed that the onza was a close modern-day relative of the fossil species, transitional between this and the normal puma. But to determine whether or not this was correct, much more material than three skulls would be required. Nothing less, in fact, than a complete specimen, for full scientific examination, both anatomical and biochemical. Nevertheless, when seeking a creature as elusive as the onza, which inhabits terrain as notoriously remote and difficult as that of La Silla Mountain, such a requirement could quite conceivably take a lifetime or more to fulfill. In reality, thanks to a remarkable piece of good fortune, it took less than six months!

As documented fully by Greenwell in an extensive, revelatory *ISC Newsletter* report (spring 1986), on the evening of 1 January 1986 rancher Andres Rodriguez Murillo, from Sinaloa's San Ignacio district, was deer hunting in the vicinity of his house (in a valley behind Parrot Mountain) when he encountered a large, crouching cat that he assumed to be a jaguar. Fearing that it was about to attack him, he shot it. Examining the body, however, he found that it was not a jaguar after all and appeared to be different from a puma too. Remembering Greenwell's visit to this area a few months earlier, his interest in unusual cats, and his meeting with Manuel Vega, Rodriguez returned to his ranch with the carcass and contacted Vega forthwith. Vega arrived at once to examine the cat and identified it positively as an onza.

This initiated a rapid chain of communications, which ultimately resulted in the specimen being preserved in excellent condition by freezing at a commercial fishery company. Greenwell was swiftly informed of its discovery, and on 19 February a scientific team headed by Greenwell, Best, and blood tissues/electrophoresis specialist Ned Gentz arrived in the Mexican town of Mazatlán to dissect the specimen at the Regional Diagnostic Laboratory of Animal Pathology, an agency of Mexico's federal Ministry of Agriculture

The onza was a four-year-old female, of head-and-body length 113 centimeters (approx. 45 inches); tail length 73 centimeters (approx. 23 inches), which was very long relative to that of a female puma of comparable head-and-body length); and body weight 27 kg (59 pounds), which was notably less than that of a comparably sized female puma, even allowing for slight weight loss due to freezing. In overall appearance, it did indeed resemble a very long-limbed, slender puma, but in addition it exhibited certain features not displayed by normal pumas, e.g. noticeably long ears, and small horizontal stripes on the insides of its forelimbs.

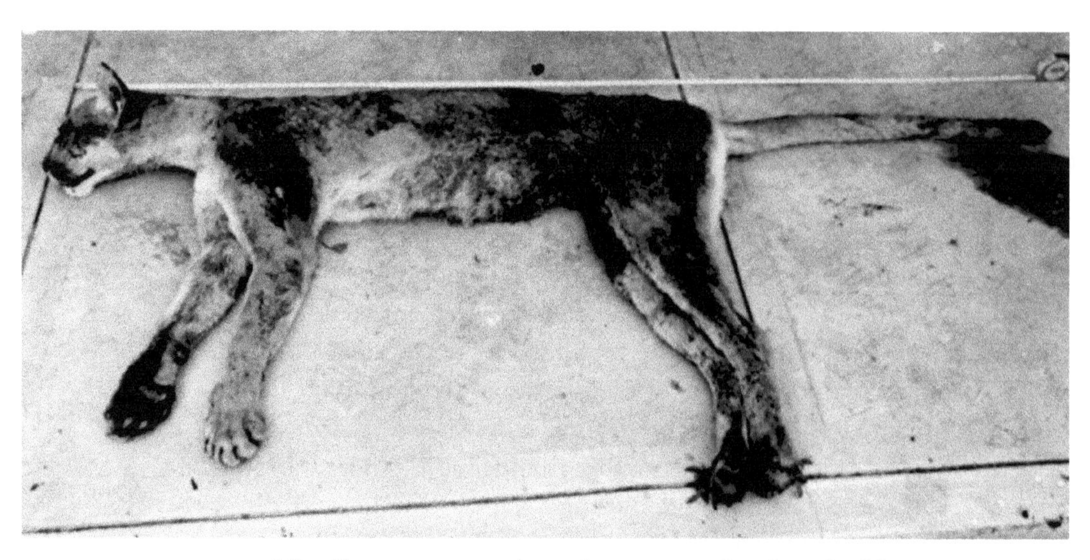

The Rodriguez onza (International Society of Cryptozoology)

Needless to say, with such an exceedingly valuable specimen, its external morphology was studiously examined prior to dissection. Only after it had been photographed from every imaginable angle, and every conceivable measurement had been recorded, was the first incision made. Furthermore, even the dissection itself was videotaped, to ensure that no vital detail had escaped documentation. The major organs were preserved, together with a front and a rear foot. Tissue samples from various organs were prepared for electrophoresis (a method of protein separation via the passage of an electric current), the result from which would then be compared with those obtained from other species in order to determine the taxonomic relationships of the onza. Hair and blood samples, again for interspecific comparison, were also taken, plus selected skeletal material (including the skull and a set of limb bones).

Among the most interesting discoveries resulting from the dissection and the first preliminary findings from the preserved material are the following. The body of the Rodriguez onza possessed adequate fatty tissue, thereby verifying that its gracility was natural, rather than emaciation brought about by prolonged starvation or metabolic disorders. In fact, it had been a very healthy specimen—even its parasite content was very limited—and had eaten a deer as its last meal (its stomach contained deer hooves). Moreover, its reproductive organs were normal, thereby discounting a hybrid identity. Its limb bones compared with those of adult male pumas; hence, if (as with pumas) the male is larger than the female in the onza, it is likely that male onzas are taller than male pumas.

During their joint onza paper presented at the July 1987 ISC members meeting at Edinburgh's Royal Museum of Scotland, Greenwell and Best announced that research upon the Rodriguez onza was still in progress. This research was channelled along two principal lines: osteological (examination and comparison of its skeletal tissues) and biochemical (genetic and protein analyses of various tissue samples from the onza and from other cat forms, involving Prof. Roy Mackal, a prominent biochemist and spare-time cryptozoologist from the University of Chicago).

No longer, therefore, was the determination of the onza's actual existence the prime objective, now the search was on for its identity. The following possibilities exist:

1) An abnormal (freak) puma.
2) An established puma morph.
3) A specialized puma subspecies.
4) A new species.

If it were found that both onzas and pumas can be born within a single litter, this would show that they belong to the same species, and that the onza, therefore, either is nothing more than an abnormal, freak puma or is an established puma morph, at least at present. (With its localized Mexican distribution and its morphological differences from normal pumas, one could perhaps speculate, as with the king cheetah, that it is evolving into a separate taxonomic form.)

In view of the fact that a puma subspecies already exists within the known Mexican distribution range of the onza, it would seem that a puma subspecies is an unlikely onza identity. However, if the lifestyle of the onza is such that it does not compete directly with the puma, this identity could be more tenable (i.e. implying sympatric speciation).

Even if it has originally descended from the puma, further researches into the anatomy, physiology, and ecology of the onza may determine that it is now sufficiently distinct from the puma to warrant classification as a separate species. Indeed, it is possible that with the Pleistocene extinction of the genuine American cheetahs *A. trumani* and *A. studeri*, the onza represents a sideline that evolved from the puma to fill the cursorial niche that the cheetahs left vacant. And the reason that the onza apparently exists only in the mountainous regions of northwestern Mexico rather than in the grassland regions of America with which one would normally associate a cursorial animal could be that it has been exterminated from such areas by virtue of their rapid colonization by humankind. And of course, if the onza does come into contact with the puma but does not interbreed, this is positive evidence in favor of its totally separate taxonomic status.

What is particularly intriguing is the fact that three novel cat forms—onza, Kellas cat, and fitoaty, represented by specimens or excellent, unequivocal photographic evidence—have been formally discovered since 1983, and in all three cases the most notable morphological feature is gracility. Could a fundamental genetic mechanism be at work, capable of yielding gracile mutants in felids, regardless of species or zoogeographical location? This would be as significant a discovery as a new cat species. Only time, and continued research, will tell.

It is very rewarding to learn that Richard Greenwell (who passed away in 2005) was strongly opposed to any deliberate shooting of further onzas. Comprehensive treatment of the material preserved from the Rodriguez specimen should reduce the need for such activity. Moreover, with respect to osteological material in particular, it is quite possible that other onza remains are currently housed unknowingly in museums around the world, masquerading as puma relics. A rigorous examination of this material—as was being pursued by Best during the 1980s and 1990s—may well result in additional onza remains.

Indeed, during the late 1980s I learned of one possible example while communicating with Ralph Molnar, then curator of the Queensland Museum, Australia. Molnar informed me that while he was researching for his Masters degree at the University of Texas during the 1950s, a large cat specimen—a road accident victim—had been found at the Texas-Mexico border, and had been brought back to the university. As far as he could recall, it had been reduced to skeleton form for preservation before the researchers had realized that it was not a puma—which, until then, they had automatically assumed that it was. Sadly, no further information was known to Molnar concerning this specimen, as by that time he had completed his course and had left the university.

If the researchers had thought that this mysterious cat were simply a puma, evidently it must have closely resembled one externally, with its differences only becoming apparent from skeletal examination. This is the very situation that has been rife throughout the onza's modern-day history. Could the latter specimen have been an onza? After I informed him of it, Richard Greenwell investigated this incident but nothing further was uncovered. Nevertheless, it is not inconceivable that another onza specimen waits to be unveiled.

Furthermore, around the time of the shooting of the Rodriguez specimen in January 1986, an onza was allegedly captured alive and held for several days in captivity at a ranch in northern Sonora, where it was supposedly photographed too. Tragically, however, when no-one showed any interest in it, its owner shot it and threw its body away. As for the photographs, these have yet to make any public appearance. And in early 1987, yet another onza was reportedly shot in Sinaloa, this time by a wealthy Mazatlán businessman, but true to form its remains were not preserved.

And this is as far as the onza's history had progressed by the time of this book's original edition in 1989, with, crucially, the findings of the research conducted by Best, Greenwell, and others (including feline molecular genetics expert Stephen O'Brien) upon the Rodriguez onza specimen still in progress and therefore unpublished. During 1998, however, a formal paper providing full details of the discoveries made in relation to that specimen was at last committed to print, in the ISC's own peer-reviewed scientific journal, *Cryptozoology*—which is when everything became extremely interesting and also extremely contentious, as I subsequently revealed in my book *The Encyclopaedia of New and Rediscovered Animals* (2012):

> Early biochemical tests had failed to uncover any characteristics differentiating the Rodriguez specimen from pumas, which would seem to suggest that onza and puma are very closely related. However, this conclusion stems from a fundamental assumption that, although widely accepted in the zoological (and especially the cryptozoological) community, has never been confirmed. Namely, that the onza specimen whose tissues provided these biochemical results, i.e. the Rodriguez specimen, really was an onza!
>
> Is it conceivable, however, that Manuel Vega had been mistaken, and that this animal had merely been a malformed or infirm puma that only outwardly resembled a genuine onza? After reflecting upon this disturbing possibility for some time, in January 1998 I heretically aired it within an onza article of mine [published by the British cat magazine *All About Cats*]—and now it seems that my suspicions may indeed have been justified. A few months after my article appeared, the much-delayed volume of the ISC's journal *Cryptozoology* covering the years 1993-1996 was finally published, and contained a report by a research team featuring Prof. Stephen O'Brien, an expert in feline molecular genetics. The team's report revealed that after conducting comparative protein and mitochondrial DNA analyses

> using tissue samples taken from the Rodriguez onza and from specimens of known North American cat species, the results obtained for the Rodriguez onza were found to be indistinguishable from those of North American pumas. Of course, this does not mean that all onzas are pumas, but how savagely ironic it would be for the most celebrated onza specimen not to be an onza after all.

Yet even if so, a solution to the riddle of the onza's identity may yet be in sight. On 15 April 1995, an alleged male onza was shot behind Parrot Mountain, this time by rancher Raul Jiminez Dominguez. Later that same day, after having been frozen, the onza's corpse was examined by two biologists from the National Autonomous University of Mexico, who took tissue samples away with them for electrophoretic analysis. The remainder of its carcass was preserved and dissected at Mazatlán for future study.

No further information has been released concerning this specimen, but let us hope that it will eventually provide a precise, unequivocal answer to the longstanding question of the onza's taxonomic identity. Until then, Mexico's feline enigma will remain a cryptozoological controversy, a far cry indeed from its popular yet sadly premature image as an erstwhile mystery cat whose reality is no longer in doubt.

One final point: subsequent to the publication in 1989 of this book's original edition, *A. trumani* and *A. studeri*, plus another closely related American species, *A. inexpectatus*, were rehoused within the genus *Miracinonyx*, as *M. trumani, M. studeri*, and *M. inexpectatus*. Moreover, genetic studies utilizing ancient DNA extracted from the fossil bones of the American cheetah *M. trumani* have revealed that today's puma and the American cheetah both evolved from an ancestral felid that entered North America from the Old World perhaps some eight million years ago, via the Bering land bridge formerly connecting Eurasia and North America, and that the morphological similarities between American cheetahs and Old World cheetahs is due not to close taxonomic affinity but rather to convergent evolution after all. In other words, despite its common name the American cheetah is not a real cheetah, more a puma masquerading as one.

Hunting Hyaenas and Man-Faced Felids

One further tantalizing matter to consider regarding the onza is whether it really is the same creature as the "wolf-cat" spied by Castillo in Montezuma's zoo. For although one would naturally assume this to be the case, it is by no means conclusively established. After all, despite its long limbs, the onza does not really resemble a wolf to any degree. Is it conceivable, therefore, that the Aztec "wolf-cat" was actually some other, still unknown beast? I would be inclined to dismiss this idea totally were it not for a certain intriguing fossil species.

Although more closely related to cats, hyaenas are quite dog-like in outward appearance. Moreover, whereas typically looked upon as Old World species, some hyaenas did exist at one time in the Americas too, which leads us into the most fascinating aspect of this subject.

In several different Mexican localities, skeletons have been found of a truly remarkable hyaena, which lived during the Pleistocene, i.e. a contemporary of *M. trumani*. However, this species, the hunting hyaena *Chasmaporthetes ossifragus*, shared much more than a geological time period and a geographical locality with the latter American felid.

For *C. ossifragus* was a cursorial dog-like hyaena possessing notably long, slender limbs, and a very gracile body. In short, its appearance in life would have been the very epitome of the descriptive term "wolf-cat." Is it possible that *C. ossifragus* did not die out during the Pleistocene but instead persisted at least to the time of the Aztecs in the more remote Mexican mountain lands? No reports of such a creature are known today, hence even if it had survived to such a recent date, it is surely extinct now.

I feel that it is less likely for Mexico's mountains to possess two mystery gracile carnivores than one in modern times, especially when they may have competed with one another to some extent, having adapted to occupy similar ecological niches. Nevertheless, it is certainly a thought-provoking coincidence that a creature fitting Castillo's description even more closely than the onza should have existed within this very same country at least as late as the Pleistocene.

As a final Aztec anomaly with possible pertinence to the onza, it should be noted that Castillo was not the only early Spanish visitor to describe a strange cat-like creature from Mexico. No less a personage than Christopher Columbus himself reported not only the puma and the jaguar but also a third great cat in his famous letter of 7 July 1503 to the King and Queen of Spain. In this he related:

> A cross-bowman slew a beast that resembled a large cat, but was much bigger and had a face like a man. He transfixed it with an arrow from the breast to the tail. Nevertheless it was so fierce that he had to cut off an arm and a leg. When a wild boar, which had

> been given to me as a present, caught sight of this beast its bristles stood on end and it fled with all speed...Although the great cat was stricken to death and the arrow still transfixed its body, it immediately attacked the wild boar, encircled its mouth with its tail and squeezed it vigorously. With the one arm it had left, it throttled the wild boar's throat as one strangles a foe.

The onza does allegedly have a shorter face than a puma. Conversely, the seemingly prehensile tail of Columbus's beast is difficult to reconcile with an onza (or indeed any other feline) identity. Perhaps, as suggested by Herbert Wendt in his book *Out of Noah's Ark* (1959), it was an unknown species of carnivore that has since become extinct. During the conquistador invasion of Mexico, the Aztecs' great libraries, containing numerous natural history tomes, were totally destroyed. The loss to zoology and cryptozoology of these irreplaceable sources of information may be beyond even the most grandiose of estimates.

Ruffed Cat

In any event, the onza may still share Mexico with one other mystery felid. To my knowledge, however, this second crypto-cat's existence is based solely upon a single account written by the American zoologist, veteran cryptozoologist, and popular writer Ivan T. Sanderson in the form of an article published in April 1973 by *Pursuit*, the official periodical of the Society for the Investigation of The Unexplained (SITU), which Sanderson founded. In his article, no name was allocated to the creature, but for reasons that will soon become evident, I shall refer to it here as the ruffed cat.

During 1940, traveling alone through Mexico on a mammalogical collecting trip, Sanderson arrived at an unnamed mountain settlement in the state of Nayarit's Sierra mountains. These are totally separated from the neighboring Sierra Madre Occidentale ranges and possess their own distinctive flora and fauna. At this settlement, a few locals spoke Spanish, and Sanderson let it be known that he was seeking specimens of a particular form of squirrel. The locals promised to obtain some for him, and in the meantime they brought along a number of skins of other mammals, in the hope of selling these to him too.

Among them was a large and very tough skin of a most unusual cat. The skin measured just over six feet from nose-tip to tail-base, with a further foot and a half constituting its relatively short tail. Of course, it is difficult to say how closely these measurements reflected this cat's actual length when alive because it would be virtually impossible to ascertain how much the skin had stretched or shrunk during drying. Intriguingly (in view of the onza), Sanderson noted: "The legs appeared to have been rather long compared to, say, a house-cat or a puma." The paws were still attached to the skin and were very large, well-furred, and splayed, with most of them still possessing their claws, which were bright yellow in color and very clearly retractile. The cat's face was short (again like the onza).

The cat's pelage was firm dorsally, soft ventrally, and predominantly brown in color. The head and shoulders lacked markings, but the flanks and upper limbs bore a series of wavy stripes in alternate light and dark shades of brown, whereas the lower limbs were very dark brown in color, almost approaching black. The ridge of the spine running along its back was also dark brown, and (as far as Sanderson could recall) so too was the tail.

By far the most outstanding feature of the skin, however, was that the hairs sited just behind its shoulders appeared to grow forward to yield a large ruff encircling the neck and covering the ears from above and behind.

Sanderson stated that he bought this skin and another smaller pelt of this same strange cat form that was in poorer condition but with sharper stripes. Together with other mammal skins that he bought, these felid pelts (which apparently cost a lot more than any of the others) were sewn up in several layers of sacking and were finally stored in the government jail of Belize, which Sanderson was using as a base.

Tragically, however, during a subsequent trip the jail was lashed by a severe hurricane and, being at sea-level, was completely flooded out. As a result, everything not in bottles was totally ruined, including the skins.

This is a most compelling but, sadly, rather intangible affair because there is so little in the way of hard facts that can be followed up. The area itself is unnamed; the skins are lost; and Ivan Sanderson, who travelled there alone and hence was the only named eyewitness to them, can no longer be questioned, as he passed away in 1973. The only hope is the possible existence still of the jail—perhaps, if it does survive, records regarding the skins may too, which could add some details to the account left behind by Sanderson.

What could these mystery cats have been? Lynxes have relatively long legs and short tail, large paws, and thick fur around the neck, which can look a little like a ruff, but no lynx possesses anything remotely as extensive as the ruff described by Sanderson for the Mexican ruffed cat. Moreover, as a trained zoologist and a mammal collector, Sanderson would surely have compared its skins specifically with lynxes if they had appeared comparable with such. Instead, the detailed description that he gave does not fit that of any known felid.

The only sequel that I know of to this sorry saga is that Sanderson claimed that he later saw another skin of this very distinctive cat form on sale at a tourist store in the big market at Colima, situated at the south end of Nayarit's mountain block. Regrettably, however, the price that its owner was asking was far beyond what Sanderson could afford.

It would seem that the only way to follow up any aspect relating to the Mexican ruffed cat is to do what Sanderson suggested at the end of his article: "...pay a visit to the market in Colima. They may still have unknown cat skins for sale." They may indeed.

Since this book's original edition in 1989, two additional snippets of information that may have possible relevance to the ruffed cat saga have come to my attention. The first of these was a brief but tantalizing paragraph in Harold T. Wilkins's book *Secret Cities of Old South America* (1952):

> At Atitlan, in Guatemala, is another monstrosity carved on a big boulder. It takes the form of a cat with a sort of ruff at his throat, which points to the east. On top of the boulder is a basin, but what purpose this cavity served, unless to hold a human victim's sacrificial blood, is unknown.

Guatemala is of course situated immediately to the south of Mexico and, indeed, only separated from it as an independent nation in its own right during the early 1800s. Consequently, it would not be implausible if the ruffed cat had once existed in Guatemala as well as in modern-day Mexico.

The second, related item is an email posted on the cz@onelist.com cryptozoology discussion group by CFZ founder Jonathan Downes on 1 June 1998:

> I think that there is still evidence for there being a new and undiscovered species of large felid in Mexico. Across Mexico City, for example, there are a number of pieces of stylised statuary which appear to show a puma like animal with a brushlike mane like a punk-rocker's Mohican! This animal seems to be a well known archetype within the Mexican culture.

Yet where is the mystifying ruffed cat today? I know of not a single modern-day report of a living cat fitting its instantly recognizable description. How ironic it would be if the skins encountered by Sanderson were from the very last representatives of a felid so distinctive and, seemingly, so greatly feared too, that in ages past humans had been sacrificed to a graven image of it, but which had subsequently slipped into extinction before the modern scientific world had even had time to acknowledge its existence. From deity to deceased, in just a few centuries?

Part II —South America

During the late Cretaceous Period (roughly 64 million years ago), South America became separated from North America, and after splitting off from Antarctica 35 million years ago it became—and remained—an island continent until the Pliocene's close. During more than 30 million years of complete insularity, South America's mammalian fauna thus evolved along totally independent lines from those of all other land masses, yielding many unique taxonomic groups.

During that time, no true felid existed here, but about 2 million years ago South America's independence ended when a land bridge formed, the isthmus of Panama (constituting present-day Central America), which connected it with North America. This enabled a great interchange of species to occur between the two continents, and among those mammals entering South America were the ancestors of the true felids existing here today. Of these, all but one species are Small Cats, including their largest modern-day representative, the puma.

Yet without any doubt, the feline embodiment of South America's verdant vista is its single species of Big Cat—the jaguar *Panthera onca*. Owner of a magnificently marked pelage, its elegance has tragically resulted in the jaguar being ruthlessly hunted into near or total extinction over much of its range. Formerly distributed throughout South America, it is nowadays confined largely to the Amazon basin, with a few highly endangered populations lingering principally in Ecuador, Peru, Bolivia, and Argentina.

The lengthiest fully-authenticated specimens on record were two that measured over nine feet in total length. At the other end of the scale, diminutive jaguars also exist, which in Honduras are referred to rather unflatteringly as rat-tigers.

Compared to the leopard, the jaguar occurs in a far smaller range of pelage varieties, but certain interesting, and sometimes very contentious, forms are on record.

Paraguayan Ghost Jaguars

As noted by Alan Rabinowitz in his book *Jaguar* (1987), in many parts of South America (but especially Paraguay) the native people fervently believe in the existence of pallid, ghost-like jaguars, occasionally glimpsed amidst the dense, shadowed jungle foliage. Needless to say, such tales were dismissed as myth and superstition by most Western travelers to these parts. However, in his authoritative two-volume tome *Apuntamientos para la Historia Natural de los Quadrúpedos de Paragüay y Rio de la Plata* (1802), the famous Spanish soldier and naturalist Félix de Azara described one jaguar skin shown to him as being so pale in color that the rosettes could only be discerned in certain lights. Might this have been a complete albino mutant? Similarly, a greyish-white skin, bearing only faint markings on its flanks and belly, was revealed to 19th-century Swiss naturalist Johann R. Rengger, during a wildlife expedition and lengthy sojourn in Paraguay (1818-1825). He was informed by the hunter concerned that its claws were white, again suggesting a complete albino.

Interestingly, on 19 January 2012, two white jaguar cubs with pale grey rosettes but normal green eyes were born to a typical wild-type father and a melanistic mother at the Aschersleben Zoo in Germany. The first white jaguars born in captivity as far as is known, they are most probably leucistic, as indicated by the pale, washed-out appearance of their coat but their normal eye color.

In contrast to these exceedingly rare (although clearly not mythical) white jaguars, melanistic individuals are much more common (and are the result of a dominant mutant alelle). Referred to in Latin America as "black tigers," they tend to be noticeably large, especially in Brazil's Mato Grosso. But according to various antiquarian zoology tomes and native Guyanan beliefs, yet another type of "black tiger" would seem to exist within this continent, one that allegedly is very different morphologically from the typical melanistic jaguar. Nevertheless, its precise identity has never been satisfactorily ascertained, but its name is the jaguarete.

Jaguarete

The name "jaguarete" (from the Guarani name "yaguarete," meaning "great beast") is used today in mainstream zoology merely as a synonym for the jaguar. Conversely, this was not always the case, and in cryptozoology it still refers to a much more mysterious cat.

Nowadays an all-but-forgotten felid, this mysterious melanistic cat was referred to as the couguar noire ("black puma") by the eminent 18th-century naturalist Georges-Louis Leclerc, Comte de Buffon in his 36-volume magnum opus *Histoire Naturelle* (1749-1804); and as the jaguarete (a less ambiguous name, which I therefore prefer and shall use hereafter) by his equally eminent naturalist contemporary Thomas Pennant in Pennant's *History of Quadrupeds, Vol. 1* (1781).

However, in the virtually verbatim version of Pennant's description that appeared in English naturalist and wood-engraver Thomas Bewick's tome *A General History of Quadrupeds* (1807), its writer S. Hodgson referred to it merely as the black tiger. Hodgson's choice of name would seem to imply that the jaguarete is truly nothing more than a straightforward melanistic jaguar. Yet neither the illustration that accompanied Pennant's description nor the woodcut (by Bewick) that accompanied Hodgson's account is compatible with such an identity. To quote Pennant:

Jaguarete in Pennant's tome, 1781 (public domain)

> Head, back, sides, fore part of the legs, and the tail, covered with short and very glossy hairs, of a dusky-color; sometimes spotted with black, but generally plain: upper lips white: at the corner of the mouth a black spot: long hairs above each eye, and long whiskers on the upper lip: lower lip, throat, belly, and the inside of the legs, whitish, or very pale ash-color: paws white: ears pointed. Grows to the size of a heifer of a year old: has vast strength in its limbs. Inhabits Brasil and Guiana [Guyana]: is a cruel and fierce beast; much dreaded by the Indians; but happily is a scarce species.

In addition, Hodgson noted that it frequents the seashore, and that it preys upon a variety of creatures, including lizards, fishes, and alligators (i.e. caimans—true alligators only occur in the USA and China), as well as devouring turtles' eggs and (rather curiously) the buds and leaves of the Indian fig.

Moreover, in 1778 German naturalist Johann Christian Daniel von Schreber had formally described the jaguarete in his multi-volume mammal encyclopedia *Die Säugetiere in Abbildungen nach der Natur mit Beschreibungen* (1774-1804), and had christened it *Felis discolor* ("two-colored cat"). Yet, paradoxically, the included color illustration of it merely showed a mid-/dark-brown cat resembling a normal puma.

Buffon's couguar noire (top) and Bewick's "black tiger" (bottom) (public domain)

What could the jaguarete be? On first sight, a black jaguar identity seems most likely, as in most specimens the rosettes can indeed be seen as cryptic markings against the coat's abnormally dark coloration. However, the black jaguar is dark dorsally and ventrally, just like the leopard's black panther morph and other melanistic felid individuals, thereby contrasting markedly with the near-white underparts, lower jaw, and paws of the jaguarete. Of course, it may be that the jaguarete is nothing more than an inaccurate description of a black jaguar, but arguing against this is the statement in a footnote by Pennant that two jaguaretes were actually shown in London during the 18th century, hence their appearance would have been familiar to naturalists of that time.

Another possibility exists, one that although on first sight seems rather radical is very compatible with the jaguarete's morphology. The mutant non-agouti allele is not the only abnormal form of the Agouti gene. A second one is a recessive form known as the black-and-tan mutant allele. In mammals homozygous for this allele, the dorsal pelage is black, but the underparts are light or cream in color (the nostrils and eyes, and often the inside or back of the earlobes too, may also be outlined with pale fur). A jaguar homozygous for the black-and-tan mutant allele would correspond very closely with the descriptions and depictions given by Pennant, Hodgson, and Bewick of the jaguarete.

An equally thought-provoking alternative to that suggestion is the possibility that the jaguarete constitutes a pseudo-melanistic form of jaguar, equivalent to the Asian and African pseudo-melanistic leopards described previously. Again, these had very dark upperparts and much paler underparts. Having said that, I have seen photos of two different pseudo-melanistic jaguars in captivity, and neither of them bears a resemblance to illustrations of the jaguarete.

A third (and probably the most plausible) solution to the mystery of the jaguarete is that like a good many other cryptids it was in reality a non-existent composite beast, erroneously created by combining together reports of wholly different animals. In the case of the jaguarete, those animals would seem to be normal South American melanistic jaguars (explaining the spots) and rare but nevertheless real South American melanistic pumas (explaining the two-tone color scheme—recall the description given in Chapter 5 of the black puma shot in Brazil by William Thomson in 1843).

Worth noting here is that only one clear photograph of an alleged black puma is known. Reproduced here, this photo depicts a dead specimen shot in 1959 by Miguel Ruiz Herrero in the province of Guanacaste along Costa Rica's north Pacific coast. Estimated to weigh 100-120 pounds, its carcass is seen alongside Ruiz's herdsman, but what happened to it afterwards is unknown. Note that its undersurface is very pale, almost white, in stark contrast to its very dark upper surface, thereby corresponding very well indeed with some of the images that I have seen of the jaguarete. Indeed, certain of these even include the puma's characteristic black facial bar, as seen, for instance, in the Pennant illustration.

Miguel Ruiz Herrero's black puma (public domain)

In addition, judging from the evidence of an extremely interesting 19th-century hand-colored engraving that I purchased some years ago, it would appear that a very similar puma to the one shot by Ruiz Herrero was actually exhibited alive for a while in England, at the London Zoo. (See Appendix 3 for full details and two versions of the engraving.)

Yana Puma

As if the jaguarete had not muddied—and muddled—the taxonomic waters sufficiently in relation to black pumas and black jaguars, South America may also be home to a further melanistic mystery cat, and one of quite prodigious size too, as I documented in another of my books, *The Beasts That Hide From Man* (2003). Known as the yana puma, it may even have been the inspiration for one of Sir Arthur Conan Doyle's most famous short stories, "The Story of the Brazilian Cat," first published by *Strand Magazine* in 1898, then republished in a 1923 collection of his horror stories, *Tales of Terror and Mystery*. Here is a condensed version of my book's account of the yana puma:

> Several years ago, I read a short story by Conan Doyle called "The Brazilian Cat," published in 1923, which featured a huge, ferocious, ebony-furred felid that had been captured at the headwaters of the Rio Negro in Brazil. According to the story: "Some people call it a black puma, but really it is not a puma at all." Yet there was no mention of cryptic rosettes, which a melanistic (all-black) jaguar ought to possess, and it was almost 11 feet in total length, thereby eliminating both puma and jaguar from consideration anyway. Hence I simply assumed that Doyle's feline enigma was fictitious, invented exclusively for his story. But following some later cryptozoological investigations of mine, I am no longer quite so sure.
>
> To begin with: in *Exploration Fawcett* (1953), the famous lost explorer Lt.-Col. Percy Fawcett briefly referred to a savage "black panther" inhabiting the borderland between Brazil and Bolivia that terrified the local Indians, and it is known that Fawcett and Conan Doyle met one another in London. So perhaps Fawcett spoke about this "black panther" and inspired Doyle to write his story. But even if so, it still does not unveil the identity of Fawcett's panther. Black pumas are notoriously rare—only a handful of specimens have been obtained from South and Central America (and none ever confirmed from North America). Conversely, black jaguars are much more common, and with their cryptic rosettes they are certainly reminiscent of (albeit less streamlined than) genuine black panthers, i.e. melanistic leopards. However, the mystery of Brazil's black panthers is far more abstruse than this...
>
> Yet as it seemed to be nothing more than a non-existent composite creature—"created" by early European naturalists unfortunately confusing reports of black jaguars with black pumas—the jaguarete eventually vanished from the wildlife books. Even so, its rejection by

> zoologists as a valid, distinct felid may be somewhat premature. This is because some reports claimed that the jaguarete was much larger than either the jaguar or the puma—a claim lending weight to the prospect that a third, far more mysterious black cat may also have played a part in this much-muddled felid's history.

In 1992, in the International Society of Cryptozoology's scientific journal, *Cryptozoology*, Peruvian zoologist Peter J. Hocking presented some previously unpublished anecdotal evidence from native reports for the existence amid Peru's remote tropical forests of four different types of mystery cat, all possibly new to science. A sequel paper by Hocking appeared in the *Cryptozoology* volume for 1993-1996 (but which was not actually published until 1998). The first of these feline cryptids is of great relevance to the subject of black pumas documented here and is known locally as the yana puma; it has been dubbed by Hocking "the giant black panther." Returning to my coverage in *The Beasts That Hide From Man*:

> In an article published by the journal *Cryptozoology* in 1992, Dr. Hocking revealed that this particular Peruvian mystery cat is said to be entirely black, lacking any form of cryptic markings, has large green eyes, and is at least twice as big as the jaguar. Moreover, the Quechua Indians term it the yana puma ("black mountain lion"). This account immediately recalls Conan Doyle's story of the immense Brazilian black cat. The yana puma is apparently confined to montane forest ranges only rarely visited by humans, at altitudes of around 1,600 to 5,000 feet. If met during the day, when resting, it is generally passive, but at night this mighty cat becomes an active, determined hunter that will track humans to their camps and has sometimes slaughtered an entire party while they slept, by lethally biting their heads.
>
> When discussing the yana puma with mammalogists, Hocking has frequently been informed that it is probably nothing more than a large melanistic jaguar. Yet as he pointed out in his article, such animals do not attain the size claimed for this mysterious felid (nor do melanistic pumas)—and the Indians are adamant that it really is quite enormous. Nevertheless, the yana puma could still merely be a product of native exaggeration, inspired by real black jaguars (or pumas) but distorted by superstition and fear.
>
> However, a uniformly black, unpatterned felid does not match either a black jaguar or a black puma—yet it does compare well with Conan Doyle's Brazilian cat. Moreover, as some jaguarete accounts spoke of a black cat that was notably larger than normal jaguars and pumas, perhaps the yana puma is not limited to Peru, but also occurs

> in Brazil. Is it conceivable, therefore, that Doyle (via Fawcett or some other explorer contact) had learned of the yana puma, and had based his story upon it? If so, it would be one of the few cases on file in which a bona fide mystery beast had entered the annals of modern-day fiction before it had even become known to the cryptozoological—let alone the zoological—community!

Particularly intriguing with regard to the yana puma is an illustration that I came upon while browsing through Sir William Jardine's tome *The Natural History of the Felinae* (1834). It was a magnificent watercolor drawing by James Hope Stewart of an alleged black puma from Paraguay, dubbed "*Felis nigra*," with big green eyes and an unpatterned coat.

Yet unlike other abnormally dark pumas on record, the Jardine individual was black all over, rather than black dorsally and paler ventrally. Consequently, its uniformly black, unpatterned pelage, together with its large green eyes, accord well with native descriptions of Peru's yana puma.

Indeed, could it be that the three so-called black puma specimens documented in this book (the Brazilian example shot by William Thomson, the Costa Rican example shot by Miguel Ruiz Herrero, and the London Zoo's 19th-century live example), all of which had pale underparts, were nothing more than exceptionally dark examples of the puma's grey morph rather than actual melanistic specimens? After all, normal leopards and jaguars have pale underparts too, but their melanistic versions are black all over. Surely, therefore, a genuine melanistic puma should be the same. If so, maybe *Felis nigra* and the yana puma are genuine melanistic pumas, but as yet we have no specimens to examine.

Felis nigra, as portrayed in Jardine's tome, 1834 (public domain)

Intriguingly, an alternative identity for the yana puma that has done the cryptozoological rounds is one that is not even feline but is instead ursine: the suggestion being that it is actually South America's dark-furred spectacled bear *Tremarctos ornatus*. To my mind, however, it seems highly unlikely that local hunters, with detailed knowledge of their homeland's fauna, would confuse a bear with a very large cat.

Onça-canguçú

Yet another black-furred South American mystery cat is a remarkable Brazilian felid known locally as the onça-canguçú. Its existence has lately been confirmed via the procuring of physical remains by Dutch zoologist Marc van Roosmalen, who has discovered numerous new and unclassified mammalian forms in Brazil during his researches there over the past three decades, but its taxonomic identity currently remains undetermined. I have documented all of Marc's discoveries in my book *The Encyclopaedia of New and Rediscovered Animals* (2012), so here is what I wrote about his mystery cat:

> **White-Throated Black Jaguar**
>
> Last, and most mysterious, of all, this unclassified big cat, which is known locally as the onça-canguçú ('bigger jaguar that goes in pairs'), resembles a very large black (melanistic) jaguar *Panthera onca*, but, uniquely, has a white throat and a tufted tail. Moreover, unlike normal melanistic jaguars, which, when viewed at certain angles, can be seen to be rosetted, the onça-canguçú is pitch-black with no coat patterning whatsoever. Marc has yet to see this creature personally, and also narrowly missed the opportunity to inspect one pelt—a hunter who had killed one of these cats threw its pelt away shortly before Marc arrived asking about this feline cryptid. Happily, he later obtained both a pelt and a skull, which should greatly assist in determining the onça-canguçú's zoological status.

Onça-canguçú painting (William M. Rebsamen)

As only its throat (as opposed to its entire ventral surface) is white, as it compares in size with a large black jaguar (hence it is evidently bigger and burlier than a puma), and as it has a tufted tail (an extraordinary feature possessed only by the lion among known felids), I do not personally consider the onça-canguçú to have any relevance to the question of whether black pumas exist in South America.

However, it will be most interesting to discover what DNA analyses on samples of the pelt owned by Marc reveal, and how closely the skull compares anatomically with those of jaguars and those of pumas.

Speckled Tigers

The second member of Hocking's quartet of Peruvian mystery cats is the so-called "speckled tiger," claimed by locals to be as big as a jaguar but with a proportionately larger head and a unique pelage consisting of a grey background covered with solid black speckles. There is no known species of South American cat alive today that fits this description, so what could this mottled mystery cat from the montane tropical forests of Peru's Pasco province be?

A jaguar with a freak coat pattern and coloration is the most reasonable explanation, but this solution poses problems. Also relevant to Paraguay's "ghost" jaguars documented previously, the pelage of a complete albino jaguar (i.e. homozygous for the complete albino mutant allele of the Full color gene) would have white background coloration and normal but white rosettes visible only in certain lights, like watered silk. Even a chinchilla-phenotype specimen (analogous or homologous to the white lions of Timbavati and/or the white tigers of Rewa) would have normal rosettes, probably grey or pale brown. So too would a leucistic specimen.

Genetically, the presence of solid black speckles reported for Peru's "speckled tiger" rather than well-formed rosettes is anomalous. The only comparable case is that of the speckle-coated servaline morph of the serval *Leptailurus serval*, plus a couple of servaline-like cheetahs with speckled coats that I have dubbed cheetalines (see Chapter 4).

One other controversial cat reported from South America that is somewhat reminiscent of Peru's speckled tiger is the cunarid din, mentioned by the Wapishana Indians of Guyana and Brazil to Stanley E. Brock. In *Hunting in the Wilderness* (1963), Brock described this strange cat as follows:

> The cunarid din is quite like the ticar din [normal jaguar], except that the ground colour is nearer white than orange or yellow. The Indians say that the white kind always attain a much larger size than the former, but this is doubtful as a fact. The spots are often finer on the fore quarters and spaced further apart, and there are noticeably fewer spots within the rosettes along the sides of the body, giving the skin a rather leopard-like appearance.

Moreover, during my earlier-mentioned browsing through Sir William Jardine's tome *The Natural History of the Felinae* (1834), I was startled to discover a color plate of a very odd-looking jaguar whose paler-than-normal coat lacked this species' familiar, clearly-defined rosettes and instead was patterned entirely with a heterogeneous array of solid black speckles and blotches.

Jardine's speckled jaguar from Paraguay, 1834 (public domain)

According to the plate's caption, this jaguar was a native of Paraguay. Consequently, always assuming of course that it had been depicted accurately, this suggests that speckled jaguars or jaguar-like cats have also occurred there in the past. Perhaps they still do so today.

Siemel's Mystery Cat

During the early part of the 20th century, renowned hunter Sacha Siemel shot a very odd-looking cat in Brazil's Mato Grosso whose identity has never been formally established. It was a heavily-built animal in appearance, with a fawn pelage that bore brown spots and also a dark stripe along the spine. Siemel referred to it as a sucuarana—the local name for such cats—but what could this enigmatic felid have been?

It was certainly no mere morph of some already known species, unlike, for example, rare red-striped ocelots on record (which according to felid geneticist Roy Robinson probably represent non-extension mutants of the normal ocelot). The most plausible suggestion is Siemel's own opinion, as recorded by Richard Perry in his book *The World of the Jaguar* (1970). Namely, that it was a puma x jaguar hybrid. If so, this would be of very great zoological significance. Almost all mammalian hybrids (including the leopons and ligers noted previously) are interspecific. However, if Siemel was correct, his felid would be an intergeneric Small Cat x Big Cat hybrid (the puma being a Small Cat belonging to the genus *Puma*, the jaguar a Big Cat belonging to the genus *Panthera*).

Moreover, virtually all intergeneric specimens of any kind have been bred in captivity as opposed to being wild-born. Successful hybridization between pumas and jaguars in captivity has been alleged; otherwise the only known intergeneric Small Cat x Big Cat hybrids are certain puma x leopard crossbreeds, again bred in captivity. As I documented extensively in *Cats of Magic, Mythology, and Mystery* (2012), probably the most famous of these were the several litters of puma x leopard hybrids bred in 1898 at German animal dealer Carl Hagenbeck's Tierpark (which moved premises to Hamburg's Stellingen quarter in 1907). One was a pumapard (male puma x leopardess hybrid) reared by a fox terrier bitch that was displayed at Hagenbeck's Tierpark during the 20th century's first decade. This specimen resembled a puma in overall form but was noticeably smaller than either of its progenitor species and was marked with pronounced rosettes and blotches. It also had a very long tail.

One of Hagenbeck's pumapards is preserved as a taxidermy specimen at Tring's Natural History Museum in Hertfordshire, England, which was originally the personal zoological museum of Lord Walter Rothschild. There is some confusion in various online accounts as to whether this specimen, small in size, is one and the same as the pumapard reared by a fox terrier. However, in *The Living Animals of the World*, a two-volume multi-contributor animal encyclopedia from 1901, a photograph of the terrier-reared pumapard was published with a caption stating that the animal was now dead: "...and may be seen stuffed in Mr. Rothschild's Museum at Tring"—which would seem to confirm that they are indeed one and the same individual. I was fortunate enough to be able to view this very unusual feline hybrid when visiting Tring in 2011.

A comparable cat, yet derived from the reciprocal cross (male leopard x female puma), thereby making it a lepuma, was purchased from Hagenbeck by the Berlin Zoo in 1898 and was said at the time by German zoologist Heck to resemble "a little grey puma with large brown rosettes." Documenting this same animal in 1968, German cat expert Helmut Hemmer described it as being fairly small with somewhat faded rosettes present upon a background pelage color resembling that of a puma. Puma x leopard hybrids obtained by artificial insemination are also on record.

Pumapard taxidermy specimen (Karl Shuker)

Judging from these individuals, one would expect a puma x jaguar hybrid to bear spots on a puma-colored background, which is precisely the appearance of Siemel's cat. Tragically, his specimen was not preserved, and the chances of a second puma x jaguar hybrid of wild-born origin being obtained are slim indeed. As noted by Perry, the fact that two large felids co-exist within South America suggests that they do not compete directly (and hence do not come into contact to any extent with one another). Consequently a mating between these would be extremely rare.

Having said that, cats resembling Siemel's shot specimen and similarly deemed by local hunters to be bona fide puma x jaguar hybrids can still be found along the Brazil's Araguaia River—at least according to professional hunter Lloyd Williams from Dalhart in Texas, that is, who referred to them as sussuaranas when making this bold claim in 1960.

Judging from its similar name, moreover, the sussuarana may well be akin to the soasoaranna. According to the celebrated explorer Sir Robert H. Schomburgk, writing in 1840, this was the name given by the Orinoco Indians to a mysterious puma-like cat of the Orinoco savannas that they readily distinguished from the normal puma (which they referred to as the wawula, and which existed both on the savannas and in the coastal area). The Indians claimed that the soasoaranna differed from the normal puma in that its head was small in proportion to its body, its body long, its forefeet very stout, and its tail more than half the length of its body, ending in a tuft of black hair.

Valero's Rock Jaguar and Red Jaguar

American correspondent Ted Leonard kindly brought to my attention some years ago a fascinating book that mentions two Brazilian mystery cats that were previously unknown to me. Written by Ettore Biocca, first published in English in 1970 (it was originally published in Italian), and based upon firsthand testimony related to him by its subject, the book is *Yanoáma: The Narrative of a White Girl Kidnapped by Amazonian Indians*. It recounts the remarkable true-life story of Helena Valero, who was abducted as an 11-year-old Italian girl by Yanoáma natives back in the 1930s and reared by them in the Amazonian jungle.

One of these crypto-felids was known locally as the rock jaguar and was briefly witnessed one day by Valero while in the company of some Yanoáma women and hunters. She described it as follows:

> It was morning that day and we had seen among the rocks, as if in a window, a jaguar's head. It was a kind of jaguar which I did not know: it wasn't one of those spotted ones or those red ones that they call kintanari. It was a brown jaguar and it had long hair on its head: it was the rock jaguar.

If this description is accurate and authentic, I suspect that it was not a jaguar of any kind but rather some other, unidentified large-sized cat, brown in color, with what seems to have been a mane. Intriguingly, that is not the only description on record of such a felid from South America. A maned mystery cat has also been reported from Ecuador.

But what of the equally anomalous kintanari or red jaguar that Valero alluded to? Unfortunately, that single brief mention quoted above is the only time that this strange creature is referred to anywhere in the book.

Just as there are freak all-black (melanistic) and all-white (albinistic) jaguar individuals on record, might there also be occasional all-red (erythristic) specimens? Certainly, erythristic individuals have been documented with certain other felid species, including the leopard, tiger, and jaguarundi. Alternatively, perhaps it was not a jaguar at all but instead some other large felid with reddish fur—a burly rufous puma, possibly?

Warracaba Tigers

The warracaba (or waracabra) tiger, as it is known to the native people of Guyana (formerly British Guiana), differs from the typical jaguar in an extremely significant way with respect to behavior. For whereas the recognized jaguar (whether spotted or black) is a solitary hunter, the seldom-sighted warracaba tiger allegedly hunts in packs, which in turn may contain dozens of individuals. Needless to say, any felid that hunted in this manner would be a very special kind of cat indeed.

Not surprisingly, therefore, the warracaba tiger has attracted considerable interest from travelers to Guyana. In an *Animal Kingdom* article from 1957, the eminent American naturalist and author William Bridges incorporated an impressive series of reports concerning this animal, dating back to the end of the 19th century, including the following selection.

In his book *Twenty-Five Years in British Guiana* (1898), Henry Kirke, a former Sheriff of Demerara, noted:

> There is a mysterious beast in the forest called by the native Indians the "waracabra tiger." All travellers in the forests of Guiana speak of this dreaded animal, but strange to say, none of them appear to have seen it. The Indians profess the greatest terror of it. It is said to hunt in packs (which tigers never do), and when its howls awake the echoes of the forest, the Indians at once take to their canoes and wood skins as the only safe refuge from its ravages.

Indeed, this was precisely the action taken by Indian attendants of British explorer C. Barrington Brown upon hearing (although not seeing) the approach of one such pack in an incident occurring at the edge of Guyana's Curiebrong River during the mid-1800s. On this occasion, a single boat was used as the means of escape, which Brown boarded too. Enquiring the nature of these evidently much-feared felids, Brown was informed by the Indians that they were small but exceedingly ferocious tigers; that they hunted in packs; and that they were not frightened by camp fires or anything except the barking of dogs. Upon crossing the river, however:

> ...a shrill scream rent the air from the opposite side of the river, not two hundred yards above our camp, and waking up echoes in the forest, died away as suddenly as it rose. This was answered by another cry, coming from the depths of the forest, the intervals being filled up by low growls and trumpeting sounds, which smote most disagreeably on the ear. Gradually the cries became fainter and fainter, as the band retired from our vicinity, till they utterly died away.

Brown remarked in his book *Canoe and Camp Life in British Guiana* (1876) that the cry of these beasts resembled that of the waracabra bird (better known as the grey-winged trumpeter *Psophia crepitans*, a predominantly glossy-black relative of the cranes, coots and bustards), hence the name "waracabra tiger." These latter mystery animals are called the y'agamisheri by the Accawoio Indians, who state that they vary in both size and color, and that as many as a hundred individuals can constitute a single pack. Little wonder that Brown's Indian companions were so desperate to depart. The prospect of meeting up with a hundred or so jaguars (even under-sized ones) all at once would surely daunt even the most courageous of human hunters!

In his book *Among the Indians of Guiana* (1883), author and explorer Sir Everard im Thurn alleged that he had actually encountered three warracaba tiger eyewitnesses but admitted that it was clear that the tale related by one of them was much exaggerated. Im Thurn also offered his own suggestion concerning these fabled felids, opining that these reports had taken their roots from the fact that puma families occasionally travel together.

During the early part of the 20th century, Lee S. Crandall, who went on to become the General Curator of New York's Bronx Zoo, spent time working in Guyana and encountered many reports of the warracaba tiger. Once again, however, he never met an Indian who affirmed unequivocally that he had not merely heard but had also actually seen any of these mysterious creatures.

This latter aspect is a frequent but notably perplexing component of warracaba tiger reports—the creatures are heard but never seen. Consequently, as a solution to the mystery of the warracaba tiger and especially to this notably strange facet of their case history, Crandall proposed the following elegant explanation.

Namely, that this beast was not a special form of jaguar at all. Instead, it was simply some animal species that hunted in packs at night, yet which voiced such terrifying sounds while doing so that no Indian had ever been brave enough to investigate the identity of these sounds' originators—as a result they never realized that this aurally abhorrent creature was in fact already known to them by sight during the daytime.

Crandall even named the species that he felt was responsible, an animal that is neither jaguar nor, in fact, any form of felid, but is one of South America's most unusual species of wild dog: namely, the bush-dog *Speothos* (formerly *Icticyon*) *venaticus*, a very curious, little-known canid not closely related to other species.

Measuring no more than three feet in total length and a mere foot in shoulder height, it is dark reddish-brown dorsally and virtually black ventrally (rather rare amongst non-melanistic mammals). The bush-dog's distribution extends from Panama and Colombia to northern Venezuela, Brazil, Guyana, Surinam, French Guiana, northernmost Ecuador, eastern Peru, northern Bolivia, and Paraguay. According to certain reports, it does hunt in packs (indeed, it may spend its entire life in packs), but in general behavior is exceedingly secretive.

Bush-dog
(Karl Shuker)

Worth noting was the impression by botanist Nicholas Guppy (who had spent much time in Guyana) that, whereas the older Indians still believe that packs of warracaba tigers exist in the more remote mountainous regions, the younger Indians seem more disposed to believing the Western identification of them as bush-dogs.

Certainly, as far as its distribution, hunting behavior, and general elusiveness are concerned, the bush-dog does compare favorably with the legendary warracaba tiger (and, as the latter is not normally seen, morphological comparisons are superfluous). Conversely, the famous hideous scream of the warracaba tiger contrasts sharply with the relatively feeble whine voiced by bush-dogs. Also, it is rather difficult to believe that the Guyanan Indians, frightened or not, could really confuse—visually and/or aurally, singly and/or in packs—a bush-dog with any form of jaguar. The mystery of the warracaba tiger may not be solved after all.

Mitla

The bush-dog has also been proposed as the identity of another mystery "felid." This latter crypto-cat is known as the mitla and was briefly referred to by the famous lost explorer Lieutenant-Colonel Percy H. Fawcett in his book *Exploration Fawcett* (1953), compiled by his son Brian. While describing various strange animals of the Madidi, in Bolivia, Percy noted: "In the forests were various beasts still unfamiliar to zoologists, such as the mitla, which I have seen twice, a black doglike cat about the size of a foxhound."

Ivan T. Sanderson claimed that during an animal collecting trip to South America, he had unsuccessfully attempted on several occasions to shoot one of these creatures. He also claimed that he had actually obtained a legless native skin of one, which he likened to that of a huge black serval with pricked ears and tiny lynx-like tail. Regrettably, however, he did not mention what happened to this cryptozoologically priceless skin afterwards. (Bearing in mind that he also suffered the earlier-mentioned loss of the Mexican ruffed cat pelts, the mitla pelt's apparent disappearance is doubly frustrating—or curious...?)

In September 1965, Jersey Zoo director Jeremy Mallinson set off on a one-man, two-month-long expedition to Bolivia in search of the mystifying mitla, but he failed to uncover the secret of its identity. In his book *Travels in Search of Endangered Species* (1989), Mallinson offered the following thoughts regarding its possible identity:

> By the time we paddled our way across the confluence of the Abuna with the Madeira...I recognised that I had not thrown any further light on the question of whether Colonel Fawcett's legendary animal had ever existed or not. Perhaps the mitla had been nothing more than a melanistic form of one of the several species of South American tiger cats or, as has been suggested, the black form of the jaguarundi which can grow to about the size of a foxhound and could, to a non-zoologist, appear to be half-dog, half-cat. Both Señor Carlos and Professor Gaston Bejarano had confirmed that the black form of the jaguarundi occasionally occurred in the north-eastern regions of Bolivia. However, I had learnt one important fact from my travels in this great integrated region of rivers and forests: that while these remoter areas of the Amazon basin still remain in existence, the forests could well harbour such animals as the mitla that are still strange to science, but it would only be by chance if their presence ever came to light.

Yet the jaguarundi *Puma yagouaroundi* is famously short-legged and long-bodied, so much so that it is sometimes dubbed the otter cat, but it is definitely not dog-like. Nor does it correspond with the black serval comparison offered up for it by Sanderson.

Another option that has been suggested is a strange, sometimes black-furred (sometimes cream-furred) relative of weasels called the tayra *Eira barbara*. Again however, the tayra is not overtly dog-like, and in any case it is already well-known to and well-recognized by the local Indians, who sometimes even keep specimens as pets.

In his book *Searching For Hidden Animals* (1980), veteran American cryptozoologist Roy P. Mackal suggested that the mitla could be the bush-dog, which as seen from the description of this animal, it is a somewhat more plausible possibility than those others noted here so far. Having said that, however, the bush-dog is reddish-brown in color, not black, as described for the mitla (although juvenile bush-dogs are indeed uniformly black). Moreover, it is relatively short-legged, not recalling the foxhound comparison given by Fawcett for the mitla.

In any case, if the mitla is truly a feline canid rather than a canine felid, in my view there is an even better, eminently more suitable candidate for it—one that I have nominated in various writings down through the years (having first proposed it in the original 1989 edition of this book), but which, oddly, had not previously been put forward in this capacity. The species I am alluding to is the small-eared dog *Atelocynus microtis*, also known colloquially as the zorro, and constituting the only member of its genus.

This is a very aberrant yet only sparingly-studied, notoriously-rare species known almost exclusively from various zoo specimens. Even the full extent of its exceedingly disparate distribution range is undetermined. It has been recorded with certainty from Bolivia, Brazil, Peru, Ecuador, Colombia, and Chile, but possibly from Venezuela and Panama too. Its head-and-body length ranges from just over two to three feet, plus another foot of tail, and it stands one to one-and-a-half foot high at the shoulder. Its pelage is black or dark brown dorsally, shading ventrally to a dull reddish-brown. But its most distinctive feature—and which is also the one that is most significant with respect to the mitla—is its gait. For the small-eared dog moves with a strikingly feline grace and lightness, totally unlike the less agile movements of other canids.

Small-eared dog in vintage photograph from 1913 (public domain)

Interestingly, until recently the small-eared dog had not been reported from Bolivia itself but was known to occur in some of Bolivia's neighboring countries. So, as it has never been well-studied in the wild, it would not have been too surprising if this elusive species were formally discovered in Bolivia too at some stage in the future, a prediction additionally supported by its compelling correspondence to descriptions of the mitla. And sure enough, a reliable observation of the small-eared dog in northern Bolivia at 14°25'47.9994"S, 63°13'47.9994"W by biologist R. Wallace was recorded during the early 2000s. This constitutes its species' most southerly record, in fact, and is in close proximity, moreover, to the Madidi where Fawcett reported the mitla—thereby enhancing the prospect that small-eared dog and mitla are one and the same creature.

Undeniably, its distinctly cat-like movements, combined with its dark color and overall size, yield a notable correspondence with the mysterious mitla. So perhaps Fawcett's dog-like cat will ultimately prove to be a cat-like dog.

At the same time, the history of the onza suggests another solution. Perhaps, by referring to the mitla as dog-like, he meant that it had long legs (remember the "wolf-cat" controversy concerning the onza?). If so, could the mitla be in any way related to the onza? The tantalizing fact is that onzas have been reported not only from Mexico but also from South America, a fascinating but little-known snippet of information revealed by J. Richard Greenwell and Troy Best in the onza paper they presented at the ISC's 1987 members meeting in Scotland. Perhaps Fawcett saw a dark color morph of the onza, but only future explorations of Bolivia's remote jungles (no easy matter even today) can provide the solution.

Pack-Hunting Mystery Cats of Peru, Venezuela, and Ecuador

Guyana's warracaba tiger is the most publicized pack-hunting mystery cat on record from South America, but it is not the only one. Examples have also been reported from Peru, Venezuela, and Ecuador.

As mentioned when discussing the yana puma and speckled tiger in this chapter, during the 1990s Peru-born zoologist Peter J. Hocking collected native reports concerning four mystifying cat forms allegedly existing in Peru but which are not known to science. The third one is the so-called "jungle wildcat," reported from montane forests in Peru's lower Urubamba River valley. Apparently, it is no larger than an average domestic cat, is patterned in a varied assortment of blotches, and has noticeably long fangs. Far more distinctive, however, is its apparent proclivity for hunting in packs, containing 10 or more individuals.

On 19 June 2019, Michael Merchant from Maine gave me the following interesting information via Facebook:

> When I was in Venezuela the Pemon Indians told me of two undescribed species of felines they were familiar with in the local jungle. One they said was huge, the size of an African lion and they were very fearful of, saying to see it was to die. The other was a smaller, cougar sized cat that travelled in packs, with the younger ones travelling in the trees, the adults tending more to the ground, hunting as a group.

The smaller of these two is especially intriguing. Not only is it allegedly a pack-hunting cat but also it is the only example of such a cat known to me in which the youngsters and adults are segregated into arboreal and terrestrial hunters—very strange indeed.

While visiting southern Ecuador's Morona-Santiago province in July 1999, Spanish cryptozoologist Angel Morant Forés learned of several mystery cats said to inhabit this country's Amazonian jungles. Upon his return home, he documented them in an online field report, entitled "An investigation into some unidentified Ecuadorian mammals," which he uploaded in autumn 1999 onto French cryptozoologist Michel Raynal's website, the *Virtual Institute of Cryptozoology*, and from where I downloaded a copy of it (fortunately, as it turned out, because, as so often happens in the ephemeral world of cyberspace, it now seems to have vanished). These very intriguing crypto-felids included two different alleged pack-hunting forms.

One of them is the tsere-yawá, which is also said by the native tribes to be semi-aquatic. Angel was informed that this three-foot-long felid hunted in packs of 8-to-10 individuals, and was brown in color, like the brown capuchin monkey whose local name, "tsere," it shares. In 1999, a young man named Christian Chumbi from Sauntza allegedly saw eight of these cats less than 50 feet away in the river Yukipa. Unfortunately, there are insufficient morphological details available to attempt any taxonomic identification of this mystery felid. Interestingly, the small-eared dog inhabits Ecuador and is known to be semi-aquatic—it even has partly-webbed feet. So might this reclusive canid species once again be in contention as the identity of a supposed mystery cat?

Alternatively, otters are social creatures, so could the tsere-yawá actually turn out to be lutrine rather than either feline or canine?

Indeed, one South American species, the marine otter *Lontra felina*, is so cat-like in outward mien (hence *felina*) that it is even referred to colloquially as the sea cat (it is predominantly coastal in distribution but will sometimes enter rivers in search of freshwater crustaceans). The other three species of South American otter currently known to science are the neotropical river otter *L. longicaudis*, the southern river otter *L. provocax*, and the aptly-named giant otter or saro *Pteronura brasiliensis*.

The second Ecuadorian feline pack-hunter is known as the jiukam-yawá. As Angel was only able to collect second-hand reports of it, not personal eyewitness accounts, however, he declined to document this cryptid in his field report.

With so little in terms of morphological details to analyze, the supposed pack-hunting felids of Peru, Venezuela, and Ecuador currently remain enigmatic to say the least. However, should any zoologist with cryptozoological interests be visiting any or all of these South American countries on official research business at some stage in the future, they should consider devoting some of their spare time there to the questioning of local inhabitants concerning these cryptids in the hope of obtaining additional details.

After all, when dealing with creatures as paradoxical as feline dogs and canine cats—not to mention a semi-aquatic cat—every snippet of information procured is a major bonus that may conceivably shed much-needed light upon these baffling beasts' taxonomic identities.

Tapir Tigers and Rainbow Tigers

The tsere-yawá and the jiukam-yawá were not the only mystery cats that Angel learned about during his visit to Ecuador. The locals also gave him details regarding:

1) A white-coated cat with solid black spots known as the shiashia-yawá, recalling the cunarid din of Guyana and Brazil, and Peru's speckled tiger, but smaller (said to be intermediate in size between a jaguar and an ocelot). Angel considers it possible that this felid is merely an albinistic jaguar, but as already discussed in relation to the speckled tiger, such an identity would not explain its solid black spots, which sound very different from the familiar rosettes of normal jaguars.

2) A huge dark-grey cat with massive paws called the tapir tiger or pamá-yawá, because it hunts Amazonian tapirs *Tapirus terrestris* and is said to be as big as a tapir itself. In 1969, Juan Bautista Rivadeneira, a Macas settler, observed one of these mega-cats at a distance of only 60

yards or so for around 10 minutes as it emerged from the Morona river and walked lazily across a sandy beach before disappearing from view. The tapir tiger is said to inhabit the Trans-Cutucú region, and also the area encircling the Sangay volcano.

3) A large maned mystery cat that several hunters reported to Angel, but none had personally seen. On 31 August 2001, however, he noted on the cz@yahoogroups.com cryptozoology discussion group that he had recently been contacted by an Ecuadorian woman living in Spain about this unnamed maned mystery cat, who claimed that she had once seen one, several years previously, and that it inhabited the mountains near Santo Domingo in northwestern Ecuador.

4) The water tiger or entzaeia-yawá, bushy-tailed and bigger than a jaguar (documented later in this chapter).

5) Finally, but most fascinating of all, the aptly-named rainbow tiger (aka rainbow jaguar) or tshenkutshen. According to the Shuar Indians in the Macas region, this mystery cat is reputed to be the size of a jaguar, and black in color, but ornately decorated with several stripes of different colors—black, white, red, and yellow—on its chest, "just like a rainbow," in the words of one native hunter interviewed by Angel. Said to inhabit the Trans-Cutucú region, Sierra de Cutucú, and the Sangay volcano area near Chiguaza, Ecuador's mystifying rainbow tiger is described by the Shuar as having monkey-like forepaws and being an exceptionally good tree-climber, leaping from tree-trunk to tree-trunk at great speed, and greatly feared as an extremely dangerous animal.

Rainbow tiger (tshenkutshen) representation, based upon eyewitness descriptions (Tim Morris)

A rainbow tiger may well have been killed in 1959 by Policarpio Rivadeneira, a Macas settler, while walking through the rainforest of Cerro Kilamo, a low mountain near the Abanico River. He had seen the creature leaping from tree to tree and, scared that it would attack him, shot it. When he examined it, he discovered that it was a jaguar-sized cat but instantly distinguishable from all cats that he had ever seen by virtue of the series of multicolored stripes running across its chest, as well as by a hump on its back, and also by its clawed but otherwise remarkably simian forepaws. Sadly, Rivadeneira seems not to have retained the creature's carcass, or even its pelt, so as yet there is no physical evidence available to verify this extraordinary felid's existence.

Personally, I find it difficult to believe that any felid would exhibit such a dramatic pelt. Conversely, its arboreal adeptness recalls the southeast Asian clouded leopards, so I have less problem accepting this aspect of the rainbow tiger.

Striped Tigers and Surviving Saber-Tooths?

Last but by no means least of Hocking's quartet of Peruvian mystery cats is in my opinion the most interesting example of all; he dubbed it the "striped tiger." This elusive inhabitant of hilly and lowland rainforests in Peru's Ucayali and Pasco provinces is described as being as large as a jaguar, with reddish fur bearing tiger-like stripes (but white in color rather than black)—a striking characteristic that delineates it instantly from all known New World felids.

Having said that, there are reports of unidentified striped cats spied elsewhere in South America. As related in his travelogue *The Cloud Forest* (1966), while traveling through Paraguay American naturalist/novelist Peter Mathiessen met a seaman named François Picquet, who mentioned seeing "...a rare striped cat not quite so large as a jaguar and very timid, which is possessed of two very large protruding teeth." According to Picquet, this mystifying felid inhabits the mountain jungles of Colombia and Ecuador, and led Mathiessen to wonder whether "...the saber-toothed tiger, like the cougar, had long ago established itself here in a smaller subspecies and had thus survived the Ice Age extinction of its North American ancestor."

Curiously, in the September 1998 issue of the French magazine *Science Illustrée*, there is an account of Picquet having observed a living saber-tooth in Paraguay in 1984. Two sightings of a saber-tooth by the same person in the same country but separated by two decades—or simply a muddle of facts by the media?

Might Mathiessen's speculation concerning Picquet's big-fanged striped mystery cat being a surviving saber-tooth have any merit? Whereas it has been suggested by some cryptozoologists that the New World lion *Panthera (leo) atrox* may still survive in parts of North America (Chapter 5), its seemingly less intelligent saber-tooth contemporary *Smilodon fatalis* is most assuredly extinct there. The principal reason for its demise is generally believed to be the disappearance of important prey species. If, however, as indicated by the discovery of sculptured *Smilodon* bones at Rancho La Brea, California, this mighty felid was contemporary with *Homo sapiens*, human persecution may also have contributed to its extinction. (Incidentally, back in 1907, California's *Smilodon* representative was classified as a separate species in its own right and dubbed *S. californicus*, but it is nowadays deemed to have been conspecific with *S. fatalis*.)

Are South America's striped mystery cats surviving *Smilodon* sabertooths? (Markus Bühler)

Certainly, overkill by humans has been implicated as a prime reason for the relatively rapid disappearance of a number of large mammals in the New World. Potentially inoffensive and sluggish giant herbivores such as the South American ground sloths and armadillo-like glyptodonts would undoubtedly have been easy prey. Conversely, even the least intelligent carnivores, by virtue of their inherent powers of concealment and stealth (assisted enormously by this continent's dense rainforests), would have constituted a far more difficult target. Moreover, South America possessed (and still possesses) considerably more species of native ungulate and other herbivores than North America, including many that were/are smaller than ground sloths and glyptodonts, thus yielding a plentiful and diverse source of prey for compatibly sized carnivores.

During the Pleistocene, *Smilodon fatalis* existed not just in North America but also in Peru. In addition, a 50% larger species, *Smilodon populator* (sometimes incorrectly termed *Smilodon neogaeus*, a nomen nudum) inhabited Brazil and Argentina, and was even bigger than the biggest modern-day lions. (Indeed, following the recent rediscovery hidden away in storage at Uruguay's National Museum of Natural History in Montevideo of a truly gargantuan *S. populator* skull, it is possible that this species weighed up to 1000 pounds!) With the extinction of South America's herbivorous megafauna, however, a smaller felid requiring less food would be selected in preference to much bigger forms. Consequently, although *S. fatalis* and *S. populator* were therefore doomed, if *Smilodon* had gradually reduced in size (an evolutionary trait exhibited by several large mammalian carnivores elsewhere during the Pleistocene, such as the lion, cheetah, tiger, jaguar, brown bear, and spotted hyaena), this smaller version could have subsisted upon smaller prey, which in turn may have enabled it to persist into the present day amid some of this continent's more secluded jungles. Plus, a notably shy disposition would confer a great advantage upon any creature whose principal enemy was humankind, by aiding concealment.

Is it truly possible, therefore, that amidst some of this still poorly-explored continent's more remote and inaccessible regions a last remnant of saber-toothed cats does indeed survive, avoiding the native Indians and remaining unknown to Western science? Quite apart from Picquet's testimony, a few other tantalizing reports exist that suggest such an exciting if startling suggestion may not be impossible.

In a *Transactions of the Ethnological Society* paper from 1863 concerning the stone cells of Chiriqui in Panama, C.C. Blake documented this pertinent snippet of information:

> At Timana, in New Granada [now Colombia], sculptured stones have been figured by Mr. Bollaert, representing a feline animal, the proportions of whose teeth slightly exceeded those of existing cats, and might possibly indicate a modified descendant of the extinct *Machairodus* [*Smilodon*] *neogaeus* of Brazil.

Of course, the sculptures may be idealized, rather than literal, representations of the animal that served as their subject. At the same time, various ancient artistic works are known that depict animals only scientifically revealed in modern times, as with the gerenuk, okapi, Grevy's zebra, and Roxellana's snow monkey, for example. Indeed, it is not unreasonable to suppose that clues to the existence of animals currently unknown to science still await discovery within other examples of early art ensconced in our world's myriad of museums and art galleries. So the stones of Timana should not be dismissed out of hand, especially as they are not the only pieces of evidence in favor of modern-day saber-tooth survival within South America.

In an article published by *The Field* (29 August 1942), Patrick Chalmers alluded very briefly to a cave painting discovered in South America that supposedly depicts a saber-tooth attacking an odd-looking animal that some paleontologists consider to be a representative of the genus *Macrauchenia* (basically camel-like in form, although possibly possessing a short nasal trunk, and belonging to the now-extinct taxonomic order of litoptern ungulates). This painting was also noted in Bruce Chatwin's book *In Patagonia* (1977); he had been informed by Patagonian polymath priest Father Palacios that it was approximately

10,000 years old, and was located at Lago Posadas in Patagonia.

Most archaeologists believe that humans first entered South America approximately 15,000 years ago. Consequently, if this painting depicts a genuine saber-tooth and macraucheniid, then we must assume that both of these species were still living at, and possibly even after, the Pleistocene's close (rather than having died out some time before, as paleontologists currently think), in turn making the prospect of modern-day saber-tooth survival rather less implausible.

On 12 September 2001, I received a very interesting email from Spanish biologist Gustavo Sanchez Romero, based in Tenerife, one of Spain's Canary Islands. It reads:

> The purpose of my mail is that one year and half ago I traveled to Venezuela (I have relatives there, since a lot of people from the Canaries migrated there back in the fifties, sixties, seventies and even eighties) including my parents.
>
> Well once there I took a trip to visit Salto el Angel (Angel Falls) the largest waterfall in the world, being almost 1000 m. high. It departures from the Auyan Tepui, a lofty flat top mountain (Tepui being the local name) only found in Venezuela. Once there I heard about Alexander Laime, the person you mention in your book (Prehistoric Survivors) who saw the prehistoric aquatic dinosaurs [sic—plesiosaur-like beasts] once bathing in a lagoon. Also the guide in the zone told me about "El tigre dantero" meaning the "Danta eating tiger". Danta is local name for tapir, and he told me that it was the size of a cow, and the surprising characteristic about it is that it was supposed to have huge fangs (canines), just like the prehistoric saber tooth tiger does have! I thought about it a little and then I have read in some books about similar descriptions from Paraguay and Ecuador. I thought that maybe you would like to hear this little story, so I hope it is useful to you!

It certainly is, because although the guide's claim contained little in the way of morphological details other than the creature's body size and huge canines, it nonetheless extends considerably the geographical distribution of reports appertaining to mega-fanged mystery cats in South America. Nor is this my only information regarding the tigre dantero.

A second piece came from a correspondent who prefers to be identified publicly merely by the user name "Bradypus Tamias." On 9 June 2019, I received an email from him containing these details:

I was recently looking for possible newer information about the unnamed "saber-toothed cats" reported from Ecuador and Colombia, and tried using some Spanish-language keywords in my search. This led me to a video, entitled "EL WAIRARIMA: Monstruos del Amazonas Parte 2 |Criptozoologia|Terror," which revealed two interesting pieces of information.

The first is that these cats do actually have a name, at least in Venezuela: "tigre dantero" (dancing tiger) or "wairarima". According to ethnological books and blog posts, this is the name used in Venezuela, and the cat it's applied to is apparently larger than a jaguar, but the Spanish-speaking online cryptozoological community seem to have adopted the term to refer to those nameless, sub-jaguar-sized fanged cats reported from the montane forests. You might already have known about these names, but I've never seen them mentioned in English-language sources before, nor had I read that the cryptids themselves were also reported from Venezuela.

On 11 June 2019, Bradypus emailed some additional details:

I feel I have to say that the name wairarima/tigre dantero may not always refer to the fanged cats. Descriptions in (modern) ethnological books about the Pemon Indians, and an old description from William Beebe, make it sound very much like the Ecuadorean pama-yawa [aka tapir tiger, the huge black mystery cat documented by me earlier in this present chapter] (or just an oversized melanistic jaguar); big, dark grey, and semi-aquatic, with no mention of fangs or a short tail. Also, it seems that "tigre dantero" might actually mean "tapir tiger," just like pama-yawa, instead of "dancing tiger," as I translated it in my last email. On the other hand, a 1991 Venezuelan sighting described the wairarima as light brown, with large fangs, a short tail, and stocky front legs. This is obviously very different (and very much like a saber-tooth) but it was explicitly identified by the eyewitness, a Pemon Indian, as a wairarima. So I'm not quite sure what's going on with it.

This latter translation of "tigre dantero" ties in precisely with the version provided in 2001 by Gustavo Sanchez Romero (and also, as Bradypus Tamias noted, with the tapir tiger of Ecuador), but "wairarima" is new to me. Nevertheless, the existence of such terms as well as the video clearly indicates that this Venezuelan mystery cat is indeed well known among the Hispanic cryptozoological community, even if not—until now—among its English-speaking equivalent.

The most extraordinary report known to me that concerns an alleged surviving South American saber-tooth is one that was kindly brought to my attention by mystery cat investigator Phil Bennett during the 1980s. According to one of his own correspondents, a large felid weighing 160 pounds was shot in Paraguay sometime during 1975 and was referred to locally as a "mutant jaguar." However, a zoologist by the name of Juan Acavar is alleged to have examined its corpse, discovered that it possessed fangs measuring 12 inches long, and suggested that it could actually be a *Smilodon*. Nevertheless, to avoid disturbing the local people, the authorities apparently decided to stay with the mutant jaguar identity.

If this remarkable account is valid, it would obviously rank as one of the greatest zoological discoveries of modern times, but sadly all my attempts to trace Juan Acavar, the whereabouts of the specimen itself, and any corroborating information to ascertain whether this incident really did happen in the first place have so far met with failure. Similarly, neither Phil Bennett nor his correspondent succeeded in obtaining further details. Consequently, I clearly cannot vouch for either its authenticity or (assuming that it is authentic) its accuracy. Indeed, one readily apparent distortion in this account must surely be the fang (upper canine) measurement given. The longest upper canines so far recorded from any known felid type are those of *Barbourofelis fricki*, a Nebraskan nimravid from the Pliocene, yet even these formidable fangs "only" attained a length of approximately eight inches.

Yet due to its tremendous significance, if this incident did occur, it seemed wisest to include it here—just in case. Also, by bringing it to public attention, it is possible that the full facts (if any) will ultimately be uncovered.

An even more dramatic identity for South America's big-fanged striped mystery cats than a surviving species of saber-tooth was proffered by veteran cryptozoologist Bernard Heuvelmans. In his annotated checklist of apparently unknown animals (*Cryptozoology*, 1986), he suggested that Picquet's mystery cat of Colombia and Ecuador was "much more likely" to be a surviving *Thylacosmilus*. Indigenous to South America and a member of the extinct sparassodont taxonomic order of mammals, *Thylacosmilus atrox* was a jaguar-sized but marsupial-related counterpart to *Smilodon* and other genuine saber-tooths, paralleling these latter creatures to a remarkable extent morphologically, especially in relation to its huge canines, due to convergent evolution.

However, *Thylacosmilus* became extinct long before the genuine saber-tooths, dying out during the Pliocene epoch shortly after the formation of the isthmus of Panama had connected South America to North America, because this intercontinental land-bridge enabled more advanced mammalian predators from North America such as *Smilodon* to invade South America and out-compete *Thylacosmilus*. For this reason alone, therefore, I consider it far less likely that Picquet's cat was a surviving *Thylacosmilus* rather than a surviving *Smilodon* or an unknown form of modern-day true cat.

In more recent times, the saga of Peru's striped tiger and also its afore-mentioned speckled tiger took a dramatic step forward. In 1994, Hocking succeeded in obtaining a skull of each of these mysterious forms (both skulls are of females), and I was later sent some color photographs of them by Angel Morant Forés. Unfortunately, the photos only depict the skulls face-on, placed alongside a jaguar skull for comparison, but the striped tiger's skull is visibly narrower than either of the others. Unexpectedly, its canine teeth are no longer than those of the jaguar (but it is possible that only the canines of the male protrude in the manner described by Picquet). In contrast, those of the speckled tiger are extremely robust, more like tusks than fangs—indeed, the entire skull is far sturdier than those of the striped tiger and jaguar.

At that time, the skulls themselves were due to be closely examined by a selection of American felid specialists, at least one of whom had already announced that the striped tiger's skull did not correspond with any species known to science. However, nothing was published regarding their findings (always assuming that such examinations did indeed take place, which is unclear).

In March 2014, conversely, the results of a morphometric analysis conducted on Hocking's speckled tiger skull and striped mystery cat skull was published in the journal *PeerJ*. Featuring Hocking and British paleontologist Darren Naish, the research team concluded that both skulls fell within the range of cranial and mandibular (lower jawbone) variation assignable to the jaguar. Here is their paper's abstract:

> We sought to resolve the identity of the skulls using morphometrics. DNA could not be retrieved since both had been boiled as part of the defleshing process. We took 36 cranial and 13 mandibular measurements and added them to a database incorporating nearly 300 specimens of over 30 felid species. Linear discriminant analysis resolved both specimens as part of *Panthera onca* with high probabilities for cranial and mandibular datasets. Furthermore, the specimens exhibit characters typical of jaguars. If the descriptions of their patterning and pigmentation are accurate, we assume that both individuals were aberrant.

Based upon these two skulls, therefore, it would seem that neither the speckled tiger (dubbed an "anomalous jaguar" in the paper) nor the striped tiger (dubbed in it the "Peruvian tiger") constitutes a new species. I hardly need point out, however, that conclusions drawn from a sample size of just one for each of these two cat forms can scarcely be said to be definitive, especially when not only do their decidedly non-jaguarine pelage patterns remain unexplained, but also there is no absolute confirmation that these two skulls actually did originate from a speckled tiger and a striped tiger (the pelt that originally accompanied the striped tiger skull was sold to person(s) unknown before the skull came into Hocking's possession). It should also be pointed out that both skulls were at the very edge of the jaguar's known morphometric variation. Nevertheless, it is good to know that some serious science has finally been applied in relation to these mystery felids and its findings made public. If only this could happen more often with more cryptids.

Water Tigers

In *On the Track of Unknown Animals* (1958), Heuvelmans painstakingly demonstrated the totally separate nature of two South American mystery beasts, which hitherto had been lumped together as a result of confusion and contradiction propagated by a number of scientific authorities at the end of the 19th century. One of the mystery beasts concerned—a cow-sized but harmless herbivorous animal known to the Tehuelche Indians and others of Patagonia as the ellengassen—could be a surviving ground sloth. The other unknown animal, the iemisch, is very different, as will now be revealed.

Iemisch: Aquatic Cat or Giant Otter?

A puma-sized amphibious carnivore, it is known to the Tehuelche as the iemisch, translated as "water tiger" (and thereby corresponding to its modern Spanish name of "tigre de agua"). Judging from the descriptions given by native Indians and also by Western observers of the iemisch, morphologically it compares most closely with a giant-sized otter. Moreover, such an identity has been put forward for it by both Heuvelmans and Mackal, who consider that it may even exceed in size the world's largest known otter species, the saro or giant otter *Pteronura brasiliensis*. Measuring five feet in head-and-body length plus a further three feet of tail, this magnificent animal is itself puma-sized but, sadly, appears to be extremely rare (confined to areas of Brazil, Guyana,

Uruguay, Venezuela, and northern Argentina).

However, there is one notable problem when attempting to assign an ultra-large otter identity to the iemisch. Namely, the latter beast's notorious ferocity, which inspires terror in its human neighbors. Although I agree that on morphological grounds the existence of a truly gigantic otter in parts of South America could solve the mystery of the iemisch, I find it very difficult to believe that any otter, regardless of its size, could engender such horror. An amphibious felid perhaps, but surely not an otter? Other researchers have similar doubts; after all, even the enormous saro is a notably shy creature, except when defending its territory against rivals from its own species.

Yaquaru: The Real Water Tiger?

Consequently, Heuvelmans and German anthropologist Robert Lehmann-Nitsche (who spent 30 years working professionally in Argentina) considered that such stories have arisen through confusion between the otter and traditional tales and memories of the jaguar (now extinct in Patagonia but formerly present there and known to the Araucan Indians as the nahuel). Heuvelmans cited the following accounts in support of this suggestion.

The Austrian Jesuit missionary Martin Dobrizhoffer, in his famed work *Historia de Abiponibus* (1783)—a history of a race of people known as the Abipones inhabiting Argentina's extreme north—mentioned the yaguaro:

> In the deepest waters there usually hides an animal larger than any hunting-dog, called tigre de aqua by the Spaniards and yaguaro by the Guaranis. It has a woolly hide, a long and tapering tail, and powerful claws. Horses and mules swimming across these rivers are dragged to the bottom. Soon afterwards one sees the intestines of the animal, disembowelled by the tiger, floating on the surface.

In his own book, *A Description of Patagonia, and the Adjoining Parts of South America* (1774), fellow Jesuit Thomas Falkner mentioned a brief personal sighting of a water tiger leaping into the River Parana. Although Falkner provided no morphological details of this animal, he did supply a detailed water tiger description given to him by the local Indians to whom this is a familiar beast:

> It is called yaquaru, or yaquaruigh, which (in the language of that

> country) signifies, the water tiger. It is described by the Indians to be as big as an ass; of the figure the size of a large, over-grown river-wolf or otter; with sharp talons, and strong tusks; thick and short legs; long, shaggy hair; with a long, tapering tail. The Spaniards describe it somewhat differently; as having a long head, a sharp nose, like that of a wolf, and stiff, erect ears...It is very destructive to the cattle which pass the Parana; for great herds of them pass every year; and it generally happens that this beast seizes some of them. When it has once laid hold of its prey, it is seen no more; and the lungs and entrails soon appear floating upon the water. It lives in the greatest depths, especially in the whirlpools made by the concurrence of two streams, and sleeps in the deep caverns that are in the banks.

Heuvelmans noted that the large size, aquatic predilection, propensity for hauling horses and mules beneath the water, and fearsome claws of this beast are all compatible with the jaguar, which is true. However, there are also various profound difficulties to be overcome in order to identify it categorically with the jaguar.

One of these difficulties concerns the water tiger's alleged inhabitation of the lower depths and whirlpools of rivers. The jaguar is a good swimmer and pursues its prey into rivers without hesitation. Nevertheless, it is not sufficiently adapted for an amphibious existence to be able to dwell in the underwater locales noted for the water tiger. Equally, even allowing for the considerable flexibility and distortion associated with orally-preserved legends, it is still rather hard to believe that in local myths the primarily terrestrial jaguar could be converted into a fully-fledged aquatic carnivore.

Additionally, if the jaguar is involved in the water tiger history, I am very surprised that no mention is made of its extremely striking rosetted coat. Instead, the feature concerning the water tiger's pelage that is mentioned is its long shaggy hair—definitely not a jaguar characteristic. Similarly, although the jaguar does indeed possess rather formidable canine teeth, it would be a gross exaggeration to describe them as being tusk-like.

In short, the yaquaru does not correspond very closely at all with the jaguar. Moreover, not only do the hirsute coat and prominent teeth of the yaquaru distinguish it from the jaguar, they also clearly separate it from any form of otter.

Within *In Patagonia*, Bruce Chatwin listed five different Patagonian mystery beasts, and categorized the yaquaru (or yaquaro) as a separate creature from the iemisch (or jemische). While discussing the yaquaru, he also referred to a section in traveller George Chaworth Musters's memoir *At Home With the Patagonians* (1871) that mentioned the refusal by Musters's Tehuelche guide to cross the River Senguer for fear of encountering "yellow quadrupeds larger than a puma."

What can the yaquaru be? And how does it fit in with the already much-muddled mystery of the water tiger?

Judging from the evident dissimilarity of the yaquaru to both the iemisch and the jaguar, it would appear that an amalgamation of creatures has occurred, but one that has not previously been recognized as such. Rather than the iemisch (surely some form of giant otter) having been combined and confused directly with the jaguar (as proposed by Heuvelmans and Lehmann-Nitsche), I suggest that an intermediate link between these two animals is present, which has previously remained unidentified, yet which is the true water tiger. The identity of this hitherto-hidden link is that mysterious mammal referred to as the yaquaru.

From the above-documented descriptions, this cryptid can be seen to be a large and voracious carnivorous mammal that possesses a shaggy but non-patterned pelage, plus very large, tusk-like teeth (probably canines); and which, although less adapted for an aquatic life than the saro (unlike the latter, it can move well on land), is far more so than the jaguar and thus can exist even in turbulent and deep stretches of water.

Thus the yaquaru may be a South American equivalent of the amphibious saber-tooth cats allegedly existing in Africa (see Chapter 4). In short, yet another example of evolutionary convergence of form resulting from two different species occupying the same ecological niche on their respective continents, the Old World Pleistocene saber-tooths in Africa and the New World Pleistocene saber-tooths in South America.

One South American saber-tooth line may have remained terrestrial (explaining Picquet's striped mystery cat, the Timana sculptured stones, etc), whereas another became secondarily aquatic (yielding the yaquaru). The yaquaru's ferocity has been assimilated into reports of its fellow aquatic carnivore the iemisch; and by being a sizable aquatic cat, the yaquaru has infiltrated the Indians' ancient memories of the now-extinct Patagonian jaguar, thereby even acquiring a similar name, despite its very different morphology and habitat. Suddenly the tangled tale is tangled no longer—three totally separate mammalian forms are at last well and truly separated.

Further support for the possibility of an aquatic saber-tooth existing within modern-day South America is a letter written by a Guyanan missionary to his superiors in France during the 18th century. In this letter, recorded in Lieutenant-Colonel René-Antoine Ricatte's book *De l'Ile du Diable aux Tumuc-Humac* (1979), the missionary mentioned a mysterious animal called the aypa. Most excitingly, this latter beast was described as being an aquatic creature whose head and neck greatly resembled those of a tiger in shape, size, and fur, but possessed jaws containing exceedingly large teeth. Its body, anomalously, was allegedly covered in scales, but as Heuvelmans has suggested with the African dingonek, this may well be an optical illusion (Chapter 4).

To bring this history of the water tiger to a close, three more South American mystery beasts require inclusion here: the Ecuadorian entzaeia-yawá, the Guyanan maipolina, and the Chilean chongonga.

Entzaeia-Yawá, Ecuador's Very Own Water Tiger

As noted previously in this chapter, during his July 1999 visit to Ecuador Spanish cryptozoologist Angel Morant Forés learned from the locals of several purportedly unknown species of cat. All but one of these have already been discussed, so here is the final one: the entzaeia-yawá or water tiger.

The following account of this aquatic cryptid is excerpted from Angel's (formerly) online field report:

> According to the Shuar people the rivers of the Morona-Santiago province are the abode of a water-dwelling felid they call entzaeia-yawá. Unfortunately, almost all the accounts I gathered concerning this animal were rather vague. It appears that water tigers show a wide range of colour morphs (black, white, brown and reddish). They are said to be nocturnal animals as big or somewhat bigger than a jaguar and with a bushy tail. Entzaeia-yawá is regarded as a most dangerous creature and attacks on humans are not rare.
>
> Carlos Pichama told me how a water-tiger had killed his cousin's wife in the course of a fishing trip to the Mangusas river (not far from Suantza). After setting up a camp on the river shore, his cousin went on a hunting excursion to the rainforest leaving his wife alone. A couple of hours later, when he came back to the camp she had disappeared. Following her footprints he arrived to a sandy beach at a place where the river formed a natural lake and cried out for her, but to no avail. Upon further examination, he located the spoor of a water tiger which seemed to have been stalking his wife. Back in Suantza he told the story to her wife's parents who concluded that a water tiger had

> dragged her into the water. Next day, he and his brothers returned to the spot where the woman had been killed by the water tiger. The group of men exploded several charges of dynamite in the lake and saw the corpse of a long-haired reddish colored animal of big size come at the surface. All my informants described the paws of the water tiger as being like a duck's (In fact, when I showed them some drawings of animal spoor, they would point to otter tracks as those which most resembled the entzaeia-yawá's). When I showed the track of a bear's hind paw to a native hunter he claimed it was very much like that of a water tiger because it had a flat palm.

Based upon evident morphological and behavioral differences, spoor similarities notwithstanding, in his report Angel discounted the possibility that the water tiger was one and the same as the giant otter (saro). He also noted that none of his informants selected a picture of this species when shown a variety of animal pictures in an attempt to identify the entzaeia-yawá. The fact that they likened its feet to a duck's simply indicates that its feet were webbed, as one would expect with an aquatic mammal.

Maipolina, Mystery Marauder of Maripasoula

On 21 October 1962, a boy fell into the Maroni River at Maripasoula, Guyana. When his dead body was ultimately brought to the surface by local inhabitants and police, however, they immediately perceived that it had been partially devoured by some animal, and a doctor called to examine it confirmed this in his detailed report for the death certificate. The locals were adamant that the responsible party was a popoké, a most mysterious creature also referred to as the water-mother (by the Creoles) and the maipolina (by the Indians), but very greatly feared by everyone in the area. It is said to inhabit caves and hollows in the riverbank and lies in wait for its prey in whirlpools and beneath the water surface. It will attack adult humans and even their boats, and it was held responsible for six other violent deaths within just a few years at the time of the boy's death in the region of Maripasoula, Benzdorp, and Wacapeu.

Few people claim to have actually seen the maipolina, but Lieutenant-Colonel Ricatte (mentioned previously regarding the aypa) was able to interview one of those rare eyewitnesses, an Indian named Amaipeti, son of Touanke, the Big Chief of the Roucouyennes. In his book, Ricatte reported that one evening Amaipeti had seen the animal about 550 yards from Touanke's village (roughly 18 miles upstream of

Maripasoula). It was lying on a rock and was estimated by Amaipeti to measure just over nine feet long and just over three feet wide. Its feet were clawed, resembling the formidably-taloned hind paws of a three-toed anteater *Tamandua* sp., according to Amaipeti. Its ears were drooping, and he described its eyes as resembling those of a tapir (large, round, dark brown?). The creature's fur was short, the chest was whitish, and a stripe of a similar color approximately four to six inches wide ran down from the head along the animal's back, whereas the remainder of its pelage was fawn in color. Its tail appeared to resemble that of a cow. Most significant of all, however, were its teeth; according to the description that Amaipeti gave to Ricatte, they were comparable with those of a walrus!

This last-mentioned feature is immediately indicative of a saber-tooth. Its tail is also intriguing; if it resembled a cow's, then we must assume that it was long and terminated in a tuft of hair, comparable in fact with a lion's tail. It certainly does not sound like that of any otter. Consequently, a taxonomic alliance with the iemisch would seem to be ruled out on this score alone, as well as taking into account its extraordinary dentition and drooping ears.

Maipolina, based upon eyewitness descriptions (Markus Bühler)

A pelage paler ventrally than dorsally (countershading) is common in aquatic animals, assisting them to remain unseen from both above and below when submerged, hence this color combination would be expected in an aquatic saber-tooth. But if the maipolina is indeed such a felid, how can its short fur be reconciled with the woolly pelage of the yaquaru? Perhaps they constitute separate species, a long-haired form in Patagonia and a short-haired version in Guyana. Alternatively, they may constitute a single species, but one that is sexually dimorphic (i.e. the male and female differ notably from one another morphologically, as with the lion and lioness). Thus the yaquaru might be the male saber-tooth, hirsute like the lion, whereas the maipolina is the female, like the lioness lacking this hairiness.

In 2007, British cryptozoological field investigator Richard Freeman led a Centre for Fortean Zoology (CFZ) expedition to Guyana. While there, the team collected a series of detailed eyewitness accounts and descriptions of a mystifying aquatic animal known to the locals as the water tiger but which is seemingly different from the maipolina. According to this new data, Guyana's water tiger is spotted, bearing black markings resembling those of a jaguar on a white background, but has a striped head like a tiger. A pelt of one such beast, matching this description and roughly 10 feet in length including a long tail, which had been shot, was seen back in the 1970s by Joseph, one of the team's local guides. However, the water tiger can also apparently exhibit a range of other colors. If only that pelt had been preserved!

Intriguingly, a local man named Elmo from a township called Point Ranch claimed that the water tiger hunts in packs. Each pack is reputedly led by an alpha individual that Elmo termed "the master," which organizes the hunting carried out by younger members of the pack.

Chile's Chongonga

Last, but definitely not least, is a very remarkable aquatic felid from Chile that has a truly unique morphology. The following details were contained in an email sent to me on 17 June 1998 by British paleontologist Darren Naish, who was at that time a postgraduate student in Portsmouth University's paleontology department:

> One of my colleagues here at the department, Stig Walsh, is working on a Chilean bonebed, and consequently has spent a fair deal of time out there. Today he told me that he heard stories while there of an aquatic (river dwelling) cat, apparently with flippers, called the Chongonga. We have a sketch and an authoritative source regarding this beast...Ever heard of it? I may have mentioned previously the phocid [seal lacking external ears] that was recently seen in Chile—it turned out to be an otariid [eared seal].

Reading about a flippered cat made me think irresistibly of an otariid, i.e. seals with external ears. They include the fur seals and the sea lions, some of which can appear somewhat feline in superficial appearance. Having said that, if a feline lineage did adapt to an exclusively aquatic existence, its species are likely to evolve webbed feet and, eventually, flippers anyway.

Sadly, everything on file concerning South America's varied water tigers at the present time consists merely of anecdotal evidence or total speculation—at least until a good photograph if not an actual specimen can be obtained for unequivocal identification. Nevertheless, there are tantalizing reasons for wondering if some form of aquatic saber-tooth may indeed exist in various regions of this vast continent, apparently alongside a truly enormous otter and a ground sloth of bovine build! Any zoologist willing to devote the time (and finance) to search for such creatures may well obtain a herculean haul of major discoveries (not to mention a sliver of scientific immortality) for his or her labors.

A Last Mystery

Finally, in 1972, *Pursuit*, the periodical founded by American zoologist and cryptozoologist Ivan T. Sanderson, reported that a few years earlier a person from El Salvador in Central America had claimed that he occasionally hunted Bengal tigers ("tigres Bengalis") across the border in Honduras, explaining that they were descendants of some tigers that had escaped several years earlier from a circus and had since multiplied.

Now I realize that in *The Lost World*, Sir Arthur Conan Doyle's ferocious if fictitious explorer Professor Challenger said of future zoological discoveries to be made in tropical America that "nothin' would surprise," but whatever would he have made of Bengal tigers?!

Addendum: The Night Jaguar

According to an article in Chad Arment's *BioFortean Review* series, Colima in Mexico reputedly harbors an intriguing but little-reported mystery cat known variously as the night jaguar, carraguar, or renegrón. It is said to resemble a very large all-black jaguar but with coarser fur and is exceptionally ferocious and fearless. I am only aware of two reports, one extremely brief, the other more extensive, but both of them over a century old. Is this cryptic Mexican cat still being reported today?

Chapter 7
Australasia: Queensland Tigers and Emmaville Panthers

There in the bracken was the ominous spoor mark.
— M. Harris, "The Tantanoola Tiger"

Present-day higher mammals (i.e. excluding the egg-laying monotremes) are split into two basic taxonomic groups, variously termed infra-classes or clades: 1) the metatherians or marsupials (named after the characteristic marsupium or pouch possessed by most species); and 2) all other mammals, the eutherians or true mammals. The marsupials and eutherians constitute two totally separate lines of mammalian evolution, but wherever the two groups have met, the marsupials have generally been out-competed by the eutherians.

In Australia, however, the two groups did not encounter one another to any great extent until relatively recently (via human introduction of various eutherians to the continent). This is because by the time Australia had become totally isolated from all other land masses (about 100 million years ago, during the Paleocene) except for Antarctica (30 million years ago), marsupials had entered this island continent but eutherians had not. Unlike South America (also once an island continent), Australia never again became connected with any other land mass.

Consequently, with the exception of a few that were filled by monotremes (plus, very much later, by some small eutherian rodents and bats that found their own way here, prior to humankind's arrival), those ecological niches compatible for mammalian occupancy were taken up exclusively by marsupials, a feat achieved by adaptive radiation. That is, the ancestral marsupial forms diversified via evolution into a vast array of different morphological types, each adapted for a separate niche. Moreover, they became so specialized within their niches that many marsupial species actually came to resemble their eutherian equivalents outside Australia, a process called convergent or parallel evolution.

Marsupial counterparts of eutherians include the thylacine or Tasmanian wolf (paralleling the familiar eutherian wolf and other large canids), carnivorous marsupial "mice" (paralleling shrews), dasyures (mustelids and viverrids), marsupial mole (eutherian moles), wombats (marmots), phalangers (squirrels, there are even gliding versions called flying phalangers that closely correspond to those eutherian gliders the flying squirrels), hare wallabies (rabbits and hares), and kangaroos (deer and antelopes).

In view of such comprehensive morphological comparability of marsupial forms with eutherians, it is very surprising that no marsupial equivalent to any of the world's large eutherian felids (leopards, tigers, etc) exists in Australia today (especially as there is an abundance of suitable prey). No marsupial felid exists *officially*, that is...

Queensland Tiger (Yarri)

Nancy O'Brien of Cairns, Queensland, was someone with good reason for doubting this zoological tenet, judging from her own experience with a strange animal, which she recalled to *Wildlife in Australia* in June 1969:

> It was perched on my casement window top, and growling and snarling, and raking the air towards me with its right paw. Its eyes were wide open and a glittering green. So I sat up in bed and shook my walking stick at it and it leaped down. Being bright moonlight I saw the length of its body and that its tail was as long as its body and the stripes on it, from the small of its back to the butt of its tail. I immediately thought of it as a small half grown tiger.

In her accompanying drawing, the stripes of her unexpected visitor were reminiscent of a thylacine's, but its head was unmistakably feline.

Nancy O'Brien was by no means alone in having reported such a creature; dozens more describing large striped cat-like beasts have been recorded over many years from several northwestern regions of Australia. Moreover, Queensland has been a particularly frequent source of these reports, so much so in fact that the most commonly-used English names for this unclassified creature are Queensland tiger or Queensland tiger-cat, even with sightings reported outside this state. It is also referred to in cryptozoological circles by its native aboriginal name, the yarri—as recorded for it by explorer Carl Lumholtz in his book *Among Cannibals, An Account of Four Years Travel in Australia and of Camp Life With the Aborigines of Queensland* (1889).

One issue that has been a common source of confusion regarding the Queensland tiger can be noted at this point to prevent any ambiguity arising hereafter. A known species of mammal exists that is also commonly called the tiger cat and which also inhabits Queensland. However, this latter creature (referred to scientifically as *Dasyurus maculatus*, and belonging to the dasyure family of marsupials) is a much smaller beast than the unidentified Queensland tiger dealt with in this book, and is spotted, not striped, thereby having no bearing upon the reports noted here.

Following a familiar trend with mystery creatures, the Queensland tiger, although unknown to science, is anything but unknown to local inhabitants. For example, an Aboriginal painting from northern Australia was identified by anthropologist John Clegg in 1978 that could possibly portray a striped cat-like beast. And according to Bernard Heuvelmans such a creature has been known to the native Queensland Aboriginals from the earliest times. This is true for these regions' Western inhabitants too.

As far back as 1705, a report from the Dutch East India Company at Batavia, Java, mentioned a tiger existing in Australia, and on 7 November 1871 the Queensland tiger finally came to the attention of science when a letter from Brinsley G. Sheridan (Police Magistrate of Cardwell, Rockingham Bay, Queensland) was published in the Zoological Society of London's *Proceedings*. In his letter, Sheridan described how, while his 13-year-old son was strolling along a path close to the Bay's shore, his pet terrier took up a scent from some scrub near the beach and followed it, barking furiously, and pursued by the son (who was well-versed in bush lore despite his youthful age). About half a mile from where the scent was first taken up, the youngster observed an animal lying in the long grass, which he described as follows:

> As big as a native Dog [dingo]; its face was round like that of a Cat, it had a long tail, and its body was striped from the ribs under the belly with yellow and black. My dog flew at it, but it could throw him. When they were together I fired my pistol at its head; the blood came. The animal then ran up a leaning tree, and the Dog barked at it. It then got savage and rushed down the tree at the Dog and then at me. I got frightened and came home.

After listening to this account from his son, Sheridan made enquiries and learned that such a creature had also been seen in the neighborhood by other people, including Reginald Uhr, Police Magistrate at St George. Furthermore, Sheridan noted:

> Tracks of a sort of Tiger have been seen in Dalrymple's Gap by people camping there...the country is so sparsely populated, and the jungles (or, as we call them here, 'scrubs') so dense and so little known, that I have no doubt that animals of this kind exist in considerable numbers...

This compelling communication initiated a follow-up letter to the Society by Walter T. Scott from Cardwell's Vale of Herbert, dated 4 December 1871 (but not published until 5 March 1872). In it, Scott reported that Alfred Hull, Licensed Surveyor, had lately been working with a party of five men on the Murray and Mackay rivers, north of Cardwell. While in their tents one night between 8:00-9:00 p.m., they heard a loud roar nearby and emerged with guns at the ready. Although unable to spot the creature responsible, they did note an unusual track, precisely formed in the soft ground. Scott also recalled that in 1864, a bullock-driver known to him had mentioned seeing a "tiger" but was not believed at the time, as, in Scott's words, "he was a notorious liar." Now Scott wondered if he could have been telling the truth after all.

A second letter by Scott was published by the Society on 5 November 1872, and contained another tiger incident:

> Mr. Robert Johnstone, an officer of the Native Police, being in the scrub on the coast-range west of Cardwell with some of his troopers, had seen a large animal in a tree about forty feet from the ground, which on being approached sprang off to another tree about ten feet off, grasped it and descended tail first. The animal was said to have been larger than a pointer-dog, of a fawn-colour, with markings of deeper shade. Its head was quite round, and showed no visible ears; its tail was long and thick.

Many more sightings followed, and not just in Queensland. Some "tigers" were even shot, but with disappointing results. The notorious "Tantanoola tiger" was identified as a calf, a second from Gippsland proved to be a feral pig, a third from Riverina was "an escaped wild dog" (dingo?), and so on, as revealed by Gilbert Whitley in an *Australian Museum Magazine* article of 1 March 1940. Happily, however, some convincing reports were also still emerging, especially from Queensland.

Take, for example, the following selection of excerpts from eyewitness reports collected by Queensland tiger researcher Janeice Plunkett and documented in 1970 by Peter Makeig in a *North Queensland Naturalist* article. In 1910, a Kuranda inhabitant (to whom such creatures were evidently nothing special!) stated:

> Most of the tiger cats which I have killed were about four feet long and of fawn colour, with black stripes running across the body, which was fairly long, unlike an ordinary cat.

A fellow Kurandan noted in a separate incident some years later that:

> The creature...had a round face and four exposed "tiger teeth"...the other salient point in my opinion was the fact that big savage pig dogs were terrified of it.

A Tiaro dweller shot a supposed Queensland tiger in 1915 and recalled that such creatures were:

> ...slightly taller and heavier built than a domestic cat, with large head and strong shoulders. Also striped rings around the body. This specimen had a young one on each teat, approximately ten in all.

And according to a report from Bellenden Range in 1925:

> ...[an animal] about as large as a medium-sized dog rushed out and climbed a nearby tree.

Of these, the Tiaro specimen is rather anomalous. Whereas it sported the typical ring-like stripes, in comparison with most Queensland tigers reported, it seemed very small for an adult. Yet it is not the only diminutive specimen on record.

For example, moving from Queensland to Western Australia, Graham Soule noted in his book *The Mystery Monsters* (1965) that George Sumner of Port Hedland allegedly shot one such animal in 1905 near Katanning, southwest of Perth. It had a cat-like head, short-furred body striped with grey and black, and according to Sumner, it "was not a domestic cat gone wild."

Back in Queensland, a similarly small, striped creature was spied in or around 1900 by J. McGeehan, who fully documented it in a *North Queensland Naturalist* article in 1938. He had been walking through scrub behind his two dogs at about 4:00 p.m. on the Atherton Tableland when he heard a sound up ahead, which suggested that his dogs were attacking some animal, something that was making a sound rather like that of a possum "but harsher and more deeply intoned." Upon his arrival on the scene, he saw a creature that:

> ...partly resembled a large domestic cat, excepting for the body, which was rather light in build. The most striking part of its appearance was the well defined hoops of colour which encircled its body. These hoops or bands appeared to be about 2.5 inches in width, and the

> colours were white and dun alternating in perfectly marked circles.

As far as McGeehan could remember, this singular form of striping did not extend beyond the torso, its head, limbs, and tail being dun-colored. He described its head as being shaped more like that of a Pomeranian dog than a cat's but with smaller pricked ears. Its eyes were dark, and its fur seemed finer and shorter than a domestic cat's. He estimated the animal's height at 12 inches and its neck-to-tail-base length at 14 inches. Of particular interest were its teeth:

> I noticed that, when the mouth was opened, the top and bottom jaws, at the front, contained long fine fangs, but...cannot now recollect whether the number on each jaw was two or four...the fangs were in sets of two and about a quarter of an inch apart.

In other words, these fangs apparently were not canine teeth, like normal fangs, but incisors—bear this in mind for future reference. McGeehan could not specifically recall the tail's appearance but felt that it matched that of a large domestic cat. He learned in 1937 that similar animals had been seen in the scrubs of Babinda. Sadly he did not preserve the skin.

Another potentially significant episode on Atherton Tableland also ended negatively, as documented by Heuvelmans in his classic 1958 book. After having personally seen a strange and sizable striped creature near the Tully River's source, naturalist George Sharp later heard that a similar beast had recently been shot by an Atherton farmer. Naturally Sharp immediately made his way to the locality to examine the skin, which measured about 5 feet from nose-tip to tail-tip, but is no longer in existence.

A supposed Queensland tiger was actually trapped alive by a Mr. Endres of Mundubbera. It had a very short head and neck, and was striped, although the stripes were not complete circles. It was about 1.5 feet tall, and the length of a large domestic cat, which is reminiscent of the McGeehan and Tiaro beasts. Once again, however, its pelt was not preserved.

A particularly memorable encounter with a Queensland tiger was featured in the Brisbane Courier and later documented by Gilbert Whitley. While riding between Munna Creek and Tiaro, G. de Tournoeur and his companion P.B. Scougall spied:

> ...a large animal of the cat tribe, standing about twenty yards away, astride of a very dead calf, glaring defiance at us, and emitting...a growling whine.

As far as they could perceive through the torrential rain and approaching darkness, this formidable beast:

> ...was nearly the size of a mastiff, of a dirty fawn colour, with a whitish belly, and broad blackish stripes. The head was round, with rather prominent lynx-like [tufted?] ears, but unlike that feline there were a tail reaching to the ground and large pads. We threw a couple of stones at him, which only made him crouch low, with ears laid flat, and emit a raspy snarl, vividly reminiscent of the African leopard's nocturnal "wood-sawing'"cry. Beating an angry tattoo on the grass with his tail, he looked so ugly and ready for a spring that we felt a bit "windy"...

Queensland tiger, reconstruction based upon eyewitness descriptions (William M. Rebsamen)

Nevertheless, when the men rushed towards this creature and cracked their stock whips, it bounded away.

The famous Australian writer Ion Idriess claimed to have seen fearsome animals of this type on at least two different occasions on the York Peninsula. He noted that it was:

> ...as high as a hefty, medium-sized dog. His body is lithe and sleek and beautifully striped in black and grey. His pads are armed with lance-like claws of great tearing strength. His ears are sharp and pricked, and his head is shaped like that of a tiger.

In a rather graphic account, Idriess told how he encountered one such creature killing a kangaroo; and in another account he recalled seeing a second individual that this time was dead, lying alongside Idriess's famously-feared staghound. Although initially impressive contributions to the Queensland tiger data file, Idriess's detailed reports lost significance for quite some years when Queensland tiger investigators realized that those reports were in fact almost identical to the ostensibly fictional accounts that had been included much earlier by fellow writer D.H. Lawrence in his celebrated novel *Kangaroo* (published in 1923).

Recently, however, Australian cryptozoologist Malcolm Smith discovered that the Idriess reports were first published a year before Lawrence's novel, in a long-overlooked newspaper article (*Rockhampton Morning Bulletin*, 8 June 1922) under the pen-name Gouger, and that a brief reference to that article had even been included by Lawrence in his novel's Queensland tiger account (yet had somehow gone entirely unnoticed by all other cryptozoologists). So it was Lawrence who had been inspired by Idriess's reports and not the other way round after all, thereby rehabilitating Idriess very deservedly as a reliable Queensland tiger eyewitness after he had unfairly spent so long in the cryptozoological wilderness.

Unlike many crypto-cats, the Queensland tiger has frequently received favorable interest and attention from science. Indeed, in 1926 it was actually included in a formal tome, *The Wild Animals of Australasia* by Albert S. le Souef and Henry Burrell, who termed it the striped marsupial cat of York Peninsula and described it as follows:

> Hair short, rather coarse. General colour fawn or grey, with broad black stripes on flanks, not meeting over the back. Head like that of a cat; nose more produced. Ears sharp, pricked. Tail well haired, inclined to be tufted at end. Feet large, claws long, sharp. Total length about five feet, height at shoulder eighteen inches.

Reports concerning the Queensland tiger appeared in a scientific paper by G.H.H. Tate. And Ellis Troughton, former Curator of Mammals at the Australian Museum, included it in his definitive work *Furred Animals of Australia* (8th edition, 1965), containing various accounts I've recalled in this chapter. Summing up his section on this animal, Troughton commented:

> Although there is some divergence concerning the size of the animal and the disposition of the stripes, there seems some possibility that a large striped marsupial-cat haunts the tangled forests of North Queensland.

Even the eminent British zoologist Maurice Burton, noted for his skepticism concerning the Surrey puma and Loch Ness monster, looked favorably on the possible existence of the Queensland tiger. Although this is most encouraging, one statement in an *Oryx* article that he wrote on the subject in 1960 is certainly not—looking back, it now appears to constitute an unhappily accurate prophecy:

> It is possible that nothing more may be heard of the alleged "tiger-cat", that it may go down, or has gone down, before human settlement, or in competition with the wild dog and it is now extinct.

For it is all too true that notably fewer reports of the Queensland tiger have emerged during the past 60-70 years. One individual was described in a report from Mount Molloy in 1953, collected by Janeice Plunkett:

> The head was a good deal larger than an old tomcat, with teeth a lot like the extinct saber toothed tiger (not size but shape).

Note for further reference the remark about prominent teeth.

Well away from the centre of Queensland tiger country is the Otway Range. Yet in the early 1960s, while driving along the road from Beech Forest to Gellibrand, a man (name unknown) observed in this area a tiger-like animal with a feline head, and dark stripes that traveled towards the creature's rear. As far as the eyewitness could tell, its coat's background color was dark fawn.

During this same period, a two-foot-high beast with a heavy head, slim hind limbs, small hind paws, and marked with irregular black-and-white stripes on its body and its long blunt-ended tail, was spied in the vehicle headlights of a man driving through Emmaville, Victoria. This location is now synonymous with another mystery felid, as revealed later.

In 1968, a strange creature sighted at Queensland's Mount Bartle Frere was described as possessing a head which:

> ...appeared round and broad, its nose shorter and broader than a dog's. Some of its teeth appeared to protrude out and upward like tusks.

Note the markedly prominent teeth again. This report is yet another one collected by Janeice Plunkett, who became a well-known name during the late 1960s in the Queensland tiger saga. A secretary from Sydney, Plunkett had developed a keen interest in this Antipodean anomaly—so much so that by 1969 she had amassed a hefty dossier of eyewitness accounts and had begun a personal campaign to track down a specimen and uncover its precise scientific identity. Her cause was aided by a series of reports publicizing her work in Brisbane's *Courier-Mail* newspaper (2, 31 August, 25 October 1970).

In September 1970, Plunkett led a three-month search team into the rainforest areas near Queensland's Cairns, Mackay, and Maryborough regions, visiting such locations as the Atherton Tableland, New England Tableland, Glen Innes, and Toowoomba. A piece of evidence in which she was particularly interested was the unusual four-toed print reported almost a century earlier by Alfred Hull; it became her expedition's logo. Yet, sadly, despite her determination and enthusiasm, the Queensland tiger eluded all of her attempts to find it.

During the 1972/1973 winter, an Alsatian-sized creature with tusk-like upper fangs, dirty grey coat, and long tail, was reported from the Daintree River's Cape Tribulation side. Moreover, a long-tailed beast with a striped two-foot-long body, 1.5-foot tail, triangular head, and feline gait was spied by three witnesses near Cairns's Herberton-Ravenshoe road junction one evening in January 1973 at about 10:15 p.m. (and also apparently in much the same area on 23 December 1972). Both of these sightings were reported by the *Cairns Post* (17, 18 January 1973).

By this time, however, various other mystery carnivores Down Under had begun to attract cryptozoological attention too, so the Queensland tiger became increasingly overshadowed and overlooked. Reports still emerge intermittently, but hard evidence in support of its continued existence is scarce indeed. In August 1982, a feline beast of leopard-like size and gait, and with a heavily striped tail, was briefly seen about 110 yards away by Norwegian research scientist Per Seglen while driving between Nambung National Park (near Perth) and Badgingarra (*West Australian*, 24 August 1982).

In 1983, a supposed Queensland tiger was killed at Craignish. Photographs of it that appeared in the *Maryborough Chronicle* on 2 February 1983 depicted an animal with a long tail, lengthy fur, and large canines. Based upon its dentition, it was identified by Janeice Plunkett as a dog, an identity supported by thylacine expert Eric Guiler. A few additional recent(ish) reports can be found in my 2012 cat book and in my 2016 prehistoric survivors book, as well as in *Out of the Shadows: Mystery Animals of Australia* (1994) by Tony Healy and Paul Cropper, in Malcolm Smith's book *Bunyips & Bigfoots: In Search of Australia's Mystery Animals* (1996), and in various books and articles by Rex Gilroy.

Assuming that the Queensland tiger does still survive, what could it be? Eutherian tiger escapees/releases from captivity offer no explanation because its appearance bears very little similarity at all to *Panthera tigris*; even their striping patterns are different.

Conversely, in view of the overwhelming predominance of marsupials amongst native Australian mammals, it is hardly surprising that this taxonomic group of pouched creatures provides both of the more plausible candidates for the Queensland tiger's identity.

1) Thylacine (Tasmanian Wolf)

The first contender is a creature that is itself currently playing hide-and-seek with cryptozoologists—the thylacine (aka Tasmanian wolf or Tasmanian tiger) *Thylacinus cynocephalus*, a strikingly dog-like marsupial of roughly collie size, with dorsal transverse stripes upon its posterior back region and tail base. The thylacine's history is a singularly sad one. During the 19th century, it carried a bounty on its head as a supposed sheep-killer and was shot on sight throughout Tasmania. By the beginning of the 20th century, this once-common species, the largest modern-day carnivorous marsupial, was already rare. A few specimens were exhibited in zoos, but no captive breeding program was attempted. In 1936 the last confirmed specimen, an adult male named Benjamin, died in Hobart Zoo, with the wild population already utterly decimated. Ironically, just a few months before Benjamin died, the thylacine was finally made an officially protected species, but this was much too late—in reality it appeared to be extinct. Expeditions were launched to rediscover it, and these have continued right up to the present day. Several promising sightings (albeit greatly outnumbered by all manner of unconvincing ones) suggest that the thylacine may yet be holding on to existence by the skin of its polyprotodont teeth, but even if so it is without doubt one of the most gravely endangered of all mammalian species.

Thylacine (public domain)

On mainland Australia, its demise came very much earlier, the most recent fossils date at 2200 + 96 years BP (Before the Present-Day). So whereas the chances of a living thylacine being confirmed on Tasmania seem increasingly slim as the years roll on by, they would appear at first glance to be virtually non-existent on the island continent itself. Yet since the 1960s in particular, many controversial reports have been recorded that describe creatures resembling thylacines being sighted in several different mainland locations (but notably in the western and southeastern states), as documented by Eric Guiler in his definitive treatises *Thylacine: The Tragedy of the Tasmanian Tiger* (1985) and (co-authored with Philippe Godard) *Tasmanian Tiger: A Lesson To Be Learnt* (1998), as well as by numerous media reports.

Even more puzzling is the Ozenkadnook tiger of Victoria. A photograph of this creature taken in the 1960s by Rilla Martin of Melbourne shows a dog-headed beast whose foreparts seem to bear white stripes upon a black pelage. In view of its white rump and tail and the comparable whiteness of the surrounding foliage too, however, it is probable that these most unusual markings are merely shadow effects rather than actual stripes. Nevertheless, the identity of the animal itself is still a mystery. On 24 March 2017, columnist Peter Hoysted alleged in *The Australian* that the photo was a hoax, featuring a cardboard cut-out, but as he could only offer anecdotal evidence to support his claim, this "explanation" is, by definition, unconfirmed, and therefore unconvincing.

Another canine conundrum is the yokyn, the Aboriginal name for a strange dog-like beast of variable (sometimes brindled) color, stocky muscular build, and with long claws. It is familiar to Australian farmers and settlers, but not to scientists (*Fate*, May 1977). Could it be a mainland thylacine? Alternatively, as suggested by Ralph Molnar, erstwhile Curator of the Queensland Museum, it might be a dog x dingo hybrid (Molnar, pers. comm.).

And how should we classify the waldagi? In a *Mankind* paper from December 1973, Erich Kolig reported three canine terms from the Wolmadjeri language in the Kimberleys: "moran," a dingo; "gunjar," a domestic dog; and "waldagi," a sinister dog-like beast. Unlike many Australian mystery beasts referred to by the Aboriginals, the waldagi is definitely not a spirit form. So what is it?

Evidently, there is a pressing need for detailed investigation of these varied dog-headed mystery carnivores Down Under, in order to identify categorically the creature(s) involved. It is equally evident that to rule out entirely the possibility that thylacines exist in Australia would be rash. For it seems most unlikely that all reports of mysterious dog-headed beasts here simply involve misidentified dingos, domestic dogs, or hybrids of these—which in turn brings us to an all-important question: Is the Queensland tiger itself a mainland form of thylacine?

Queensland tiger reconstructed with Ozenkadnook tiger's markings (top) and Ozenkadnook tiger photograph (bottom) (Markus Bühler / public domain)

Although superficially promising—after all, the thylacine is (or was) a large striped carnivore native to Australia—there are several major problems with this identity. First and foremost, all known thylacines (Tasmanian and fossil mainland specimens, plus some remarkable mummified individuals preserved at Western Australia's Thylacine Hole) have unmistakably dog-like heads (hence its specific name, *cynocephalus*). Conversely, the Queensland tiger's head has always been described as being unquestionably cat-like. Can such a striking discrepancy be overcome?

While a molecular biology student at Rhode Island's Brown University during the late 1980s, Victor Albert was especially interested in the Queensland tiger, presenting a paper on this cryptid at the July 1987 ISC members meeting in Scotland, where he suggested that, although improbable, it would not be impossible for this mystery beast actually to be a short-faced thylacine form. Ever since the mainland and Tasmanian thylacine populations parted from each other by Tasmania's own separation from the mainland about 12,000 years ago, the two groups could have followed very different evolutionary pathways, so that today their descendants may be very dissimilar morphologically from one another. Although possible in theory, the mummified mainland thylacines and contemporary fossils demonstrate that as recently as about 2,200 years BP no such change had taken place, and it is not very likely that anything so drastic has occurred since then.

Another fundamental problem to be faced when attempting to equate the Queensland tiger with a thylacine identity concerns tree-climbing. For whereas the Queensland tiger displays feline arboreal agility, the thylacine is stoically terrestrial like its eutherian canid counterparts.

Coat coloration is another area of conflict. The striking black-and-white hoop-like stripes totally encircling the tail and the underparts of much of the Queensland tiger's body certainly do not correspond with the shorter black stripes present dorsally upon the thylacine's brownish coat, which never extend downwards as far as the underparts and do not encircle the tail at all.

Then there is the question of spoor. In an *Annals and Magazine of Natural History* paper from 1873, pioneering Australian zoologist Gerard Krefft had suggested that the strange four-toed print reported by Hull was really that of a thylacine or even just a domestic dog. Yet in the spoor of these latter beasts, the toes are spaced well apart from one another, whereas those in the spoor noted by Hull were aligned very closely together. Intriguingly, as I learned from Malcolm Smith, an engraving of another print very similar to this in terms of toe spacing, but showing five rather than four toe impressions, can be seen at the Carnarvon Gorge, in a display of Aboriginal rock art that features numerous engravings of other animals' footprints too. Yet unlike all of those, this particular example has never been identified with any known creature, native or introduced, living today in Australia. Nor indeed has the Hull paw print, as painstakingly analyzed by Smith in a *Journal of Cryptozoology* paper from 2012.

The Queensland tiger's exceedingly prominent tusk-like teeth and curved claws, so frequently reported, plus its leopard-like growl and its tail's very great length, are further features that serve well to differentiate it readily from the thylacine. And as a final but highly significant point, we must take into account the fact that most reports of cat-headed striped beasts originate from Queensland, whereas dog-headed striped animals are most commonly reported elsewhere, as revealed by examining the geographic spread of mainland thylacine reports documented by Guiler and by Smith.

If the two striped creatures are one and the same species, with the differences in morphology and behavior due merely to inaccurate reporting of sightings, we would expect a uniform distribution of both animal types. Consequently, the logical (if unexpected) conclusion that must be drawn from this is that if eyewitness reports are to be believed, Australia may house not one but two types of striped carnivorous mammal unrecognized by science—surviving thylacines and whatever the Queensland tiger is.

Supporting this solution to the situation on mainland Australia is the apparent existence of an exactly comparable condition on New Guinea too. According to le Souef and Burrell, writing in their book *The Wild Animals of Australasia* about areas outside Queensland where creatures resembling the Queensland tiger have been reported:

> We have had a striped carnivorous animal described from Northwest Australia, and Lord Rothschild states, from native reports, that a similar animal exists also in New Guinea.

At the same time, a distinctly dog-like form of unknown identity (but apparently separate from New Guinea's famous singing dogs *Canis familiaris hallstromi,* related to the dingo) exists on Papua New Guinea's Mount Giluwe (in the eastern half of New Guinea). This mountain just happens to be sited close to areas in which fossil thylacines have been found. Moreover, a thylacine-like beast known locally as the dobsegna has also been reported from New Guinea's western, Indonesian half, which was formerly known as Irian Jaya.

All in all, it would seem that whereas the thylacine is a plausible contender for the identity of the dog-headed striped beasts of western and southeastern Australia, we must look elsewhere for the solution to the mystery of the cat-headed tiger of Queensland—and thereby avoid (as pointed out by Victor Albert too) the danger of lumping together two clearly separate forms of Australian mystery carnivore.

2) Thylacoleonid (Marsupial Lion)

This second contender is the one most vehemently supported by Albert, myself, and the great majority of cryptozoologists. It is a creature that was itself a profound mystery for many years, yet was first described from fossils as long ago as 1859, and by one of the world's most renowned paleontologists, Richard Owen. In his *Philosophical Transactions of the Royal Society* paper, he formally christened it *Thylacoleo carnifex*—"flesh-eating pouched lion"—and pronounced that it had functioned as a bona fide marsupial felid. Since then, other related fossil species have been disinterred and described, yielding an entire thylacoleonid lineage, of which *T. carnifex* was not only its most formidable but also for a long time its most mysterious member. And all because of its teeth.

In eutherian felids, the canines are enlarged in size and the incisors very greatly reduced. In *T. carnifex*, however, the exact reverse was true. For whereas its upper canines were very small like eutherian felid incisors (its lower canines were absent altogether), its first pair of upper and lower incisors were huge, uniquely tusk-like structures that protruded from its jaws almost like a parrot's beak and were very reminiscent of eutherian felids' canine teeth.

Nevertheless, virtually every paleontological contemporary of Owen argued stridently that these immense incisors could not actually function as canines. However, these were not the only dental features that had suggested to Owen a carnivorous existence for *T. carnifex*. Its last pair of upper premolars and lower premolars possessed sharp blades (carnassials), a diagnostic feature of carnivorous mammals. Moreover, its eyes were set forward on its head, suggesting that it had binocular vision, again a notable characteristic of predatory species.

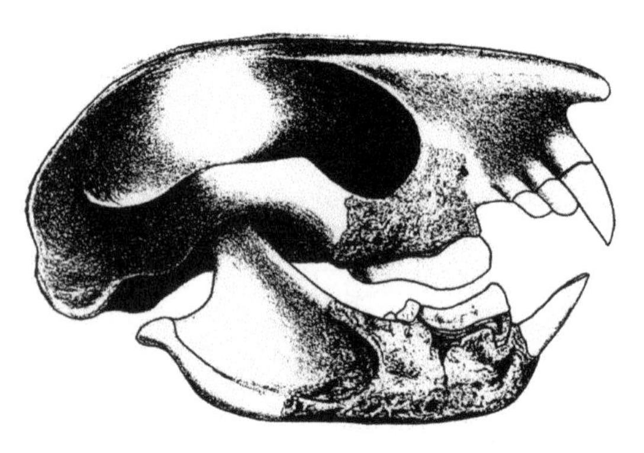

Thylacoleo carnifex skull, diagram (public domain)

Yet Owen's critics continued to deny flatly any possibility that *T. carnifex* was a meat eater. The reason for this surprising skepticism rested with this species' ancestry. For the remainder of its anatomy had revealed unequivocally that its closest relatives were not the carnivorous marsupials already known (dasyures, thylacine). In stark contrast, they were the leaf-eating phalangers, fruit-eating possums, and browsing macropods. Whoever had heard of a flesh-eating phalanger? It seemed as incongruous as a carnivorous koala!

Consequently, every conceivable attempt was made to reconcile this bizarre beast's extraordinary dentition with (at best) a conformist vegetarian diet or (at worst) anything that did not include meat. Suggestions were varied and exotic, ranging from foliage and crocodile eggs to cucumbers and melons. None, however, provided a comprehensive explanation for its total dental complement. Indeed, this dietary dilemma was conclusively resolved only in modern times, courtesy of the scanning electron microscope (SEM).

Scientists discovered that by examining their surfaces via SEM techniques, teeth used for the chewing of flesh and those used for the chewing of plant material could be readily differentiated from one another by the pattern of scratches they bore. Here at last was a definitive means of uncovering the cryptic lifestyle of *T. carnifex*. Its carnassials were duly examined and were shown to have been used for shearing, which in turn supported the belief that its mighty incisors had assumed the role of canine teeth.

After more than century, Owen's belief had been vindicated. Despite having descended from vegetarians, *T. carnifex* was indeed a carnivore, and therefore a veritable wolf in sheep's clothing (or, to be zoologically precise, a lion in phalanger finery!).

Some near-complete, fossilized skeletons of this species have now been uncovered, so that we have a good idea of its appearance in life. Namely, a leopard-sized, felid-shaped animal with a short, round head and powerful jaws housing those tusk-like, protruding upper and lower incisors (indeed, it is now believed to have possessed the strongest bite for its size of any known mammal). Despite having fairly long limbs, *T. carnifex* is now believed to have exhibited marked arboreal tendencies too, each forepaw possessing a pseudo-opposable thumb with a broad nail, which would have served it well in grasping branches (and probably even prey too), whereas its other toes each bore a formidable claw. Moreover, it is very probable that the pelage of *T. carnifex* was brindled or striped for camouflage purposes, like those of many eutherian felids, and it may have remained hidden up a tree, ready to ambush an approaching prey animal (most probably a kangaroo), dropping down upon the latter victim when unsuspectingly passing by underneath.

Up alongside the above reconstruction of *T. carnifex*, compare the following description of the Queensland tiger, compiled from the eyewitness accounts given here. Namely, a creature at least equal in size to a small leopard, with a feline build, longish limbs, and a distinctly catlike head whose jaws possess notably large, prominent teeth (actually described as tusk-like by some eyewitnesses). It also exhibits formidable claws, arboreal prowess, and a striped pelage.

It scarcely need be said that these two descriptions match very closely—in fact they could be two different eyewitness accounts of the same animal. Hardly surprising, then, that a surviving thylacoleonid (either *T. carnifex* itself or a modern-day descendant) is the favorite identity of most cryptozoologists for the Queensland tiger, a possibility that has gained significant credibility since the revelation that *T. carnifex* was truly carnivorous.

Remains of *T. carnifex* have been discovered at several different sites across Australia, including Darling Downs in Queensland. Most specimens are about 30,000 years old. However, carbon-dating techniques carried out in the 1950s on a specimen unearthed at Lake Menindee, New South Wales, and reported by B. Daily in an *Australian Museum Magazine* article from 1960, indicated that it died as recently as 6670-6470 BC. Although such a date, if accurate, would be a great boon cryptozoologically, it is more likely to be an underestimate, as dating techniques used at that time were not as precise as those utilized today. All the same, certain Tasmanian specimens have been dated as no more than 10,000 years old.

Thylacoleo model decorated with Queensland tiger markings (Rebecca Lang)

Yet if the Queensland tiger is a thylacoleonid, where are the missing fossils required to bridge the 10-millennium gap between itself and the Tasmanian specimens? Perhaps thylacoleonids did exist during this period, but in areas where fossil preservation is poor and where, in any event, paleontological investigations are difficult to carry out, in areas such as rainforests, for example. Certainly the diverse fauna typical of such habitats would supply more than enough prey to support a number of these carnivores. They would not even be in competition with the thylacine (either in the past or, if it still exists on the mainland, in the present). For the thylacine possessed a very different dietary preference (medium/small prey) from that of *T. carnifex* (large prey).

Victor Albert noted that the locality containing Mount Bartle Frere in northern Queensland not only includes the largest (and among the least explored) rainforest expanse in Australia but also has remained virtually unchanged for 30,000 years. So if any thylacoleonids once existed here, then it is possible that they are still doing so. Consequently, as Albert also pointed out, it is particularly thought-provoking that this very area just happens to be one of the principal centers for Queensland tiger sightings. Similarly, other major locations for such reports also have adjacent rainforests, e.g. Mackay, Maryborough, Cairns.

It only remains to say that the morphological and behavioral correspondence between *T. carnifex* and the Queensland tiger is sufficiently striking for only the most narrow-minded and pessimistic of zoologists to deny out of hand any possibility that a thylacoleonid still exists today.

One final point. What were the various "mini" Queensland tigers reported earlier? Although some may have been bona fide cubs, the Tiaro specimen was clearly an adult individual (and therefore not a Queensland tiger). According to Albert, a freak striped variety of the known (and normally spotted) tiger cat *Dasyurus maculatus* has been recorded at least once. It is possible, therefore, that a creature such as this is responsible for the "mini" tiger sightings, as the animals involved were all of dasyure size and build. Alternatively, they may have been nothing more exciting than feral domestic cats with unusual markings that led their observers to wonder whether they were something special.

I can think of no more apposite a conclusion to my Queensland tiger coverage than the following quote from a review of this book's 1989 edition that was penned by Colin Groves (from Canberra's Australian National University), one of the world's leading mammalogists, who also had a longstanding interest in cryptozoology. His review was published by the ISC's journal *Cryptozoology* in 1992:

> The Queensland Tiger has always seemed to me to need further scrutiny: those reported fangs are suggestive of a specialized carnivorous marsupial, such as *Thylacoleo*, as indeed Shuker points out. But as the Tablelands and Daintree Rainforests get tamed and explored, yield up their last mammalogical secrets, and get invaded by feral cats, we may lose the chance to find out.
>
> When the last rainforests have been logged, overrun by tourists, or dissected into tiny fragments, perhaps we will find a *Thylacoleo* skull on the mantelpiece of an old-timer, who will recount how his dogs killed the creature when he was a boy; and he will smile his enigmatic smile, and we will never know.

Let us pray that those prophetic lines never come to pass, that this most fascinating cryptid is not only real but also still in existence, and will one day be formally discovered, rescuing it from the sad, ignominious fate of slipping into extinction without ever having been recognized by science.

Australian Mystery Pumas and Black Panthers

As a land famed for its unique marsupial fauna and virtual absence of eutherian mammals, plus its separation from all other land masses for at least 30 million years, Australia must surely rank as one of the least likely places on Earth to house large feline creatures markedly reminiscent of Britain's Surrey puma and Exmoor beast, or North America's bona fide pumas and mystery black panthers. Nevertheless, reports of such beasts have indeed been documented from this island continent and have received an increasing degree of media attention in recent years. Furthermore, these felids are so similar to their UK and USA counterparts that if locality names were omitted from the representative selection of Australian puma/black panther incidents, it would be quite impossible to determine from which of these three totally separate regions the reports had originated!

The earliest of Australia's mystery black panthers to attract widespread notice hailed from New South Wales, with the town of Emmaville as the epicenter of events and encounters. Inevitably referred to thereafter in NSW as the Emmaville panther (even though, as will be seen, some sightings have been made far beyond this locality), it began receiving media coverage in 1956 and went on to establish itself for quite a few years as guaranteed headline material. From a sighting that he made of such an animal in October 1959, Sydney businessman Wallace E. Lewis described it as being jet black in color and slightly larger in size than a greyhound, with a long tail and feline head—a morphology reiterated in the many other eyewitness reports on record.

Although guaranteed to attract attention, the Emmaville panther's zoogeographically-unexpected appearance was not a primary reason for its fame or, more correctly, its infamy. This was due to a much more serious and much less palatable matter—the concomitancy of its appearances with the wholesale slaughter of sheep and other farm livestock. A direct relationship, or just coincidence?

To give one such example: between 1956 and 1957, 340 of Clive Berry's sheep were killed on his 4600-acre property at Pretty Gully, west of Uralla, by some carnivorous animal; more were killed in 1958. He dismissed the possibility that the culprit was either a dingo or domestic dog, emphazising that the manner in which his sheep had been killed bore the hallmarks of a felid rather than a canid attacker. He spent a considerable time attempting to capture their killer but met with no success.

The husband of Mrs. S. Godley, one of Berry's neighbors, had also lost sheep to a mystery attacker with feline characteristics, so Mrs. Godley made a plaster cast of the latter's spoor and sent it to Sir Edward Hallstrom (then Chairman of the Taronga Park Trust) for formal identification. Sir Edward announced that although it did not match that of either dog or dingo (thereby corroborating Berry's view), it did correspond with that of a panther or tiger. This sparked off what was to become a deep interest of his own in this mystery creature—so much so that he offered a reward of £500 to anyone capturing it alive or dead if it indeed proved to be a panther or tiger, and double that amount if it actually turned out to be a marsupial felid (*Australian Outdoors and Fishing*, April 1977). Neither reward was ever claimed, but sightings and alleged panther killings of sheep continued in New South Wales throughout the 1960s.

In 1966, some miles north of Nowra, a black panther-like creature was attacked in one of Samuel Knight's paddocks by his two dogs; these were both badly hurt in the ensuing battle and fled, leaving their pantheresque adversary to escape into the surrounding bush. Interestingly, only a few months earlier, a circus truck supposedly crashed while traveling between Nowra and Moss Vale, resulting in the escape of a black panther. If true, this could explain the origin of Knight's unexpected visitor, but in fact the whole circus incident seems to have been based upon unsubstantiated local rumor rather than fact. Anyway, it could not explain the welter of pre-1966 panther reports from this State.

By 1969, the Emmaville panther's supposed crimes had become even more heinous, from sheep-killing to tourist-scaring. Consequently 50 expert rifle-shooters were set upon its trail, but once again to no avail. Moreover, reports of panthers on the prowl were now emerging from other states too (*Sunday Express*, 2 March 1969).

In September 1972, farmer George Moir of Kulja, Western Australia, was shocked not only to discover several of his piglets dead with their hearts torn out and their throats ripped open but also to witness during this same period his panic-stricken sheep being rounded up, sheepdog-style, by two totally unfamiliar animals. As he drew nearer, he observed that they were uniformly black in color, stood two feet high at the shoulder, and possessed a long slender body with a tail of equal length. He realized that they were not dogs, appearing instead to be panthers that ran with a canter-like gait, the two forefeet coming down alternately, about eight inches apart. They made their exit by leaping over two fences (one of these beasts failed to clear the first fence, falling to the ground but then, according to Moir, climbing over it like a cat), and were chased by Moir and fauna warden Don Noble, But they were forced to give up the hunt after five miles, as the cats showed no sign of easing the furious pace of their flight.

Moir subsequently learned that his neighbors Alan and Ron Johnson had lost 14 of their own pigs to this (or another) pair of panthers, which they had seen on several occasions, as well as having heard them give voice to blood-curdling cries. Such sounds were already familiar to inhabitants of Emmaville and other NSW dwellers, such as farmer Ted Bell of Black Mountain, who recalled hearing such cries during the period when a mysterious large-sized felid had been reported in that area. Moir's unwelcome callers, or other creatures of the same type, paid a subsequent visit to his farm in November and were sighted by other inhabitants of the area, so that they ultimately became known as the Kulja panthers (*Perth Sunday Times*, 12, 18 November 1972).

News of these animals reached the State's Agriculture Protection Board, but it declined to intervene in the situation, its opinion being that they were not panthers at all, but were merely black kangaroo dogs. One such dog was shot, but panther sightings continued.

During the mid-to-late 1970s, reports of black panthers intensified dramatically. In a summer 1980 *Fortean Times* article, Australian cryptozoologist Paul Cropper examined a series of reports from New South Wales, whereas the Western Australian contingent attracted the attention of David O'Reilly, then a reporter for *The Australian*. Victoria had its share of panthers too; moreover, it was during this period and from this state that a further fact regarding Australia's out-of-place felids became widely known at last, namely that there was not one but two such forms of cat. In addition to accounts of jet-black pantheresque beasts, reports were also coming in concerning tawny-colored puma-like cats. In truth, these latter felids had been sighted for many years, but until now they had failed to attract the degree of attention received by the more exotic, eye-catching black panther type.

Demonstrating the clear dichotomy of form but evident overlap of distribution relative to these two felids is the following pair of sightings made during 1977. In April, while hunting near Horsham (about 200 miles northwest of Melbourne, Victoria), Neal Bothe, Ivan Bothe, and John Spencer reported seeing a creature that they described as resembling a puma, observed at a distance of only 33 yards (*Sydney Sun*, 28 April 1977). And in November, a jet-black beast was sighted in the Wimmera outback (about 223 miles northwest of Melbourne), again at close range, i.e. less than 55 yards away, by Wimmera Lands Department officers Bert Bray and Alan Knight. According to their detailed description of the animal, it was about 40 inches long with a lengthy cat-like tail, and stood about two feet tall. Its head was cat-like, with a very small nose and ears, its shoulders and rump were slender, and its coat was sleek (*Sydney Sun*, 25 November 1977).

The presence in Western Australia of these two forms was highlighted by the writings of David O'Reilly, including two major reports that appeared in *The Australian* on 5 and 6 June 1978. In these, he boldly contrasted the orthodox scientific view (that such animals were simply feral domestic cats or dogs) with those put forward by local inhabitants. With first-hand experience, including the all-too-available evidence of slaughtered livestock, these people were united in believing that Western Australia's mystery cats were genuine pumas and black panthers that had escaped (accidentally or via deliberate release) from captivity and were now thriving in a naturalized state in the Australian bush, preying upon wild creatures such as kangaroos and making forays onto bush-bordering farms in search of sheep and other stock.

Furthermore, with sightings emerging as far afield within the state as Warburton, Healesville, Montrose, Neerim, and Mt. Evelyn in the past few years (and many of them too precise and well-supported for an urban folklore explanation), it seemed evident that a fair number of these animals existed—little wonder then why the farming communities were so alarmed.

By 1981, David O'Reilly's personal investigations in the Perth area of Western Australia had brought in so much material concerning its out-of-place felids that he wrote it up in a book entitled *Savage Shadow: The Search For the Australian Cougar*. In 2011 it was republished with a foreword by Australian cryptozoologists Mike Williams and Rebecca Lang, whose own book, *Australian Big Cats*, had appeared in 2010. Moreover, 2012 saw the publication of *Snarls From the Tea-tree: Big Cat Folklore*, by David Waldron and Simon Townsend. For many years, however, *Savage Shadow* had been the only book devoted to these Australian anomalies, containing numerous reports and eyewitness accounts documented chronologically, and supplemented by data on physical evidence (claw marks, hair samples identified as feline, and animal kills—some represented by photographs—bearing the unequivocal signs of a felid attacker).

Sadly, however, what would have been the book's highlight proved to be as elusive ultimately as its subject. In *Savage Shadow*, O'Reilly recalled that a slide photograph supposedly depicting a wild Australian black panther had come his way during his investigations. Taken by a Barry Morris, allegedly near Carnarvon, when blown up it revealed a cat-like beast with jet-black pelage and a tail that curled high in a semi-circle above its body. Unfortunately, the $1000 price that Morris asked for the slide was more than O'Reilly's newspaper bosses (and also those of all other newspapers to which Morris later applied) were willing to pay. Hence the photo apparently has never been published.

Do pumas roam the Australian outback? (public domain)

During the 1980s, much attention was redirected to Victoria, with sightings of tawny cats and jet-black cats being recorded regularly each year, with the former type becoming increasingly evident. As before, individuals of both color forms were being seen in the same locations. For example, at Talbot, near Maryborough, a great tawny male cat was sighted in June 1985 by John Higgins and was seemingly spied by others too in the vicinity of the forest behind Talbot; whereas at nearby Daisy Hill a pair of the black felids was heard and seen on a number of occasions. Far more perplexing, however, was another revelation: that these two color forms appear to belong to the same species.

For according to a detailed *Maryborough Advertiser* report of 15 July 1987, a tawny-colored female living in the forest near Majorca had a half-grown black cub. Moreover, in the mid-1970s, after having been flushed by staff from the vineyard at Chateau Remy, a huge black cat returned later that day with its mate, a ginger-colored individual.

Most significant of all, the mystery cats, regardless of taxonomic identity, finally achieved a degree of official status in the eyes of governmental officials, inasmuch as during the 1980s the Victorian Government actually added pumas to the list of predators that can attack farm stock—something that seemed to have been happening with alarming frequency in 1987. So much so in fact that sophisticated American animal snares were used by local researchers in a bid to catch alive one of these elusive felids active in the St. Arnaud and

Maryborough regions. They were also made available to officers from the Department of Conservation Forests and Lands (staff from this Department actually made their own sighting in July 1987 of a large-sized felid, news of which was passed on immediately to senior executives at Melbourne).

As further proof of the more serious approach to the subject taken by Australian officials, provision was made at the Melbourne Zoo for the accommodation of any mystery cat caught alive, and a dossier of material collected on apparent puma predatory activity in central Victoria was to be forwarded for evaluation to the United States Department of Agriculture. They were also being sought by the Rare Fauna Group in the Grampians with high-tech sound and camera equipment.

It seems likely that if this policy of formal (and informal) investigation and determination to secure a specimen of these animals continues, the mystery will eventually be resolved. During the 31 years since this book's 1989 edition, many additional searches and plans have indeed been made, and sightings have continued to be reported, but as in the U.K., continental Europe, and (regarding mystery black panthers) North America, no major revelations have taken place—with one notable exception, as will be detailed below.

Meanwhile, the all-important question of what these cryptids actually are remains to be answered. Here are the major contenders:

1) Feral Domestic Cat?

As already noted in previous chapters, feral domestics do not normally exceed tame domestics in size (although in Australia they can weigh as much as twice that of their tame counterparts), and they certainly never attain a sufficient size to be capable of killing adult sheep and kangaroos in the meticulously clean manner typical of mystery "big" cat kills Down Under. Conversely, animals often seem larger than they really are when observed from a distance, hence some puma/black panther sightings made a long way off could conceivably have been of feral domestics.

Supporting this suggestion is an incident recorded in the *Sydney Daily Mirror* on 15 November 1977. A large cat—described in the report as puma-like with protruding yellow eyes, a two-foot-long tail, and fang-like teeth—had recently been shot dead by Tom Sega after stalking Tom and his father, Roman Sega, through the thick undergrowth on Cambewarra Mountain, near Nowra, NSW. Once dead, the creature was skinned by the Segas, who carried home its pelt but regrettably left its carcass behind on the mountainside. Paul Cropper was most intrigued by this incident and visited the Sega family in 1979 to see the skin for himself. But upon examining it and even allowing for two years of shrinkage and other inevitable distortions, he was unable to convince himself that it had come from anything other than a large feral domestic cat.

Worth noting, incidentally, is that although it was referred to as puma-like in the report, Paul Cropper discovered that in color its pelt was black (not tawny). If his identification of this specimen was correct, it serves well to show that even people familiar with a given area and its fauna can sometimes misidentify an animal encountered there if not seen clearly. Thus the concept of a feral domestic cat being mistaken for a bigger felid is no longer so difficult to accept, unless, of course, it is offered as the explanation for every mystery "big" cat sighting. After all, there is a limit even to misidentification.

Then again, what if some "black panthers" Down Under really are feral domestic cats but specimens much larger than traditionally accepted by science? As revealed by British paleontologist Darren Naish in posts for 4 March 2007 and 13 February 2012 on his *Tetrapod Zoology* blog, some remarkable videos have been taken in quite recent years of what appear to be exceptionally large, black, Australian feral domestics—individuals readily distinguished morphologically from black panthers, yet dramatically bigger than any typical black domestic cat, with shoulder heights of around two feet and total lengths in excess of four feet.

Perhaps the most famous example of such an animal, however, was not merely filmed but was actually shot dead in Gippsland, Victoria, by hunter Kurt Engel in 2005. Claims concerning its total length varied from 5.25 feet to over 6.5 feet, but no measurements could be independently confirmed as Engel discarded most of the carcass after photographing it (he did retain the tail). Unfortunately, the photos incorporate a degree of forced perspective, making the cat look bigger than it actually was; but even allowing for this effect, it is evident that the animal was unusually large if it were truly a domestic cat. To obtain a definite identification, Rebecca Lang and Mike Williams sent tissue samples from its preserved tail for DNA analysis at Monash University, and the results obtained from these tests revealed that this remarkable creature's identity was indeed *Felis catus* (*Melbourne Herald Sun*, 27 November 2005).

So could it be that a novel strain of giant black feral domestic cat is truly emerging Down Under? If so, more specimens need to be obtained and genetically assessed, not only to confirm their taxonomic identity but also to help uncover clues as to how and why such extraordinary feline evolution is taking place.

One of the genes responsible for black coat coloration in felids is pleiotropic, also increasing overall body size in these same animals. So could this dual genetic effect be engendering extra-large black feral domestics Down Under?

If such a phenomenon is genuinely occurring, it may not be limited to Australia either, because seemingly identical creatures are also being reported in such disparate regions of the world as North America, Britain, New Zealand, continental Europe, even the Hawaiian Islands. And the only common zoological denominator linking all of these regions is that they are all home to feral domestic cats. Once again, moreover, many of these far-flung black mystery felids being reported and loosely labelled in the media as black panthers or even black pumas simply don't look like genuine panthers or pumas. They are not big enough, and their body proportions are wrong, comparing far closer with those of domestic cats, but domestic cats of much bigger stature than normal. If rogue genetics are creating a strain of extra-large black feral domestic cats, however, this would instantly reconcile these morphological discrepancies.

2) Misidentified Canids?

Let's now consider the possibility that dogs and dingos (of which many black specimens have been recorded) are being taken for large-sized cats. Although this is very likely to be correct in some instances, all too often mystery cat eyewitnesses have expressly stated that the animals that they have seen did not resemble canids. Equally, paw prints, sometimes discovered after such sightings, have not always matched those of canids.

Moreover, there is even some disquieting evidence that officials may be falsely identifying cat-derived evidence as having originated from dogs, with one such claimed instance being highlighted in a *Hawkesbury Gazette* article for 17 February 2005. For quite some time, Mike Williams has been investigating reports of mystery felids across the Hawkesbury Range in New South Wales, and on several occasions scat samples reputed to originate from these felids have been obtained and sent for analysis to various scientific laboratories. Every time, however, they have been resolutely identified as canine, not feline.

Accordingly, Mike decided to put the accuracy (or otherwise) of such testing to a real test. He obtained (with permission and assistance) some scats produced by a black panther held in captivity at Bullens' Animal World in Warragamba, then sent them off as usual for identification (having deliberately refrained from stating what animal they were from) to two different laboratories—one of which, moreover, is a center preferred by the State Government in its own investigations of such material. And when the results came back—yes, you've guessed it—both laboratories had identified the scats as canine. In addition, fur samples obtained from the same black panther, which Mike sent off to a third laboratory, were identified there as domestic cat.

Not surprisingly, faced with such an extraordinary outcome, Mike now believes that the only hope of obtaining undeniable evidence for the existence of big cats in Australia is DNA analysis, concluding: "I can understand the fur result. Fur can be subjective...But with the scats, they had a stench of cat urine and gigantic fur balls with bones. For whatever reason, the experts used by the State Government are incapable of seeing the difference between felid and canid."

3) Native Eutherian Carnivore?

This is an immediate non-starter. With no fossil evidence to support the existence of eutherians in Australia prior to its permanent isolation 30 million years ago from Antarctica (and very much earlier than that from all other land masses), any identity involving a native eutherian carnivore can be written off as exceedingly unlikely. Even the so-called native dog of Australia—the eutherian dingo—was introduced by humans; it did not evolve here.

4) Unknown Marsupial Carnivore?

In view of Australia's marsupial dynasty, it is more reasonable to suppose, as with the Queensland tiger, that these so-called pumas and black panthers could be unknown marsupial carnivores. At the same time, whereas the Queensland tiger is markedly different in fundamental morphology from any other striped feline creature known in the world today, the Australian pumas and black panthers appear identical in size, shape, relative body proportions, and color to those also reported from Britain, continental Europe, and North America (as well as known pumas and black panthers from their native lands), suggesting a common explanation for many (if not all) of these animals, regardless of locality.

Having said that, there is at least one claimed sighting on file that, if genuine, threatens to disrupt this mainstream explanation. It was documented by veteran Australian cryptozoological investigator Rex Gilroy in his book *Mysterious Australia* (1995):

> Craig Black, a young fossicker [mineral prospector], was digging in a creek in Ben Lomond National Park one day in 1961 when he realised he was being watched by a large black "panther" further up the creek on the opposite bank. The animal emerged, then dashed across the shallow creek. It was apparently a female.
>
> "I am positive I saw that it was carrying a pouched cub," he said later to a ranger.

Needless to say, if this creature were indeed carrying a cub in a pouch, not only was it a marsupial but also it adds a further, and exceedingly dramatic, new dimension to the already complex enigma of black feline mystery beasts on record from Australia.

5) Escapee/Released Pumas and Black Panthers?

As previously discussed, an extremely plausible common origin does exist: escapees/releases from captivity. Indeed, at least three escapee non-native cat specimens are known to have been shot dead in Australia—a puma that escaped from a circus and lived wild for a time before being shot at St. Arnaud in Victoria, 1924; a second puma killed at Woodend, Victoria, 1960s and subsequently stuffed; and the so-called Broken Hill lioness (a bona fide lioness) in New South Wales, 1985. Still to be resolved, however, is the taxonomy of those numerous cats allegedly seen but not procured.

If two different species are involved, pumas and melanistic leopards would seem to be the most likely identities. Conversely, if only a single, dimorphic species is responsible (i.e. one having two distinct color morphs), inevitably we must look to the puma, certainly in relation to reports of pairs of Australian mystery cats that consist of one individual of each color. In short, we must accept that naturalized black panthers do exist in Australia and/or that pumas do too and have given rise either to a melanistic morph (a notably radical concept) or, more conservatively, to a very dark brown or grey morph.

Many books mentioning Australian mystery cats comment that reports date back to the 1940s, which ties in nicely with the well-voiced rumors that during World War II, American airmen stationed in various parts of Australia (notably the Western Australian coast) owned several puma and/or black panther mascots that they subsequently released into the wild, and which in turn have bred among themselves to yield a notable number of individuals living here today.

For many years, this very smooth answers-for-everything solution was widely accepted by believers in Australia's feline anomalies—until the summer of 1987 when the *Maryborough Advertiser* declared that this solution seemed to be little more than a cover-up. For its investigations of old reports had revealed that tawny felids and black felids were being hunted in the ranges on both sides of the highway around Victoria's Taradale and Kyneton as long ago as the 1880s!

Eldridge, an Asian black panther and squadron mascot of America's 13th Tactical Fighter Squadron, "Panther Pack": Are Australia's mystery pumas and black panthers escapee/released US Air Force mascots? (public domain)

Moreover, a lioness-like creature (puma?) was sighted in the hill country of Gunyah, near Trafalgar, by three men in November 1933; puma-like cats were being reported from Bendigo properties before World War II; and a beast described in the *Advertiser* as a black puma was actually killed at Kyneton in 1942—all thereby pre-dating the arrival of American airmen.

No longer does the airmen origin for such animals stand. We must also look to the more orthodox sources of captive animals—zoos, private collections, homes—from which escapes can occur. Also to be considered is the possibility that some such cats have been deliberately released by people who smuggled them into the country and then decided to lose the evidence. Yet even this solution has problems.

The importation and quarantine laws in Australia are very strict. Additionally, even once inside Australia, livestock moving between states (as in circuses or other traveling shows, for example) has to be recorded on entry into a state and checked out when it leaves, making the possibility of an escapee avoiding official notice a very slim one (Malcolm Smith, pers. comm.). Nevertheless, despite all of the improbabilities and regardless of their identities, mystery cats do exist in Australia. Perhaps the ongoing investigations taking place across this continent by professional and amateur researchers alike will soon obtain the elusive solution—and preferably of the type that can be accommodated alive and well at Melbourne, Adelaide, or Taronga Zoo.

Mystery Lions of the Blue Mountains

Staying with the subject of out-of-place non-native felids, one or more African lions *Panthera leo* on the loose is the conservative explanation generally offered by naturalists when faced with the enigma of the Blue Mountains' maned mystery beasts. West of Sydney, New South Wales, the Blue Mountains have long been associated with rumors and reports of huge cat-like beasts of ferocious temperament. Interestingly, they are not confined to modern reports but were well known to the aboriginals who once inhabited this range. They called them warrigals ("rock dogs"), a name sometimes applied nowadays to the dingo, but it is clear from their descriptions that their warrigals were something very different from a dingo.

Veteran Australian cryptozoologist Rex Gilroy has made a detailed study of the Blue Mountain lions. According to his accounts of these animals included in one of his articles (*Nexus*, June-July 1992) and in his cryptozoology book *Out of the Dreamtime* (2006), the aboriginals described them as six-to-seven-feet long, around three feet high, with a large cat-like head, big shearing teeth that protruded from their jaws, brown fur (sometimes light, sometimes dark; sexual or age differences?), and a long shaggy mane. Testifying to the continuing presence of these animals here is an accurate portrait of the terrifying leonine beast that approached three young shooters in the Mulgoa district south of Penrith, close to the Blue Mountains' eastern escarpment, one day in 1977; it fled into nearby scrub only when the alarmed trio fired at it. A similar creature had been reported from this same region in 1972, where it had allegedly been killing sheep.

Back in April 1945, a bushwalking party clambering down Mount Solitary's Korrowal Buttress made good use of their binoculars to watch four warrigals moving across Cedar Valley. And in 1988, some campers near Hampton, west of Katoomba, saw one for themselves; this area had been experiencing some severe cases of cattle mutilation, a feature that crops up time and again when charting sightings of warrigals.

Based upon the longstanding history of these animals, I find it difficult to believe that they could be escapee lions or suchlike. An undiscovered native species would seem to be a more tenable explanation, which is echoed by Gilroy, who proposes that the warrigal is a surviving species of marsupial lion. As the warrigal differs markedly in appearance from the Queensland tiger, however, we can only assume that if Gilroy's hypothesis is correct, and if the eyewitness accounts of such animals are accurate, then there are two separate species of marsupial lion currently prowling various portions of Australia's wildernesses. This is a remarkable concept, but nonetheless it would appear to be the only one that offers a satisfactory conclusion to this extraordinary saga.

New Guinea Mystery Cats

There are no scientifically-verified native marsupial or eutherian felids living in New Guinea. Having said that, I mentioned previously that, based upon native testimony, Lord Walter Rothschild had stated that a striped mystery beast resembling the Queensland tiger seemingly exists in New Guinea. Moreover, that is not the only feline cryptid reported from this vast but only partially explored island mini-continent. I know of at least two others, very different from one another morphologically, but equally memorable, albeit once again for very different reasons.

I documented the first of these in my 2012 cat book. On 11 November 2011, Malcolm Smith's internet blog, *Malcolm's Musings*, contained a fascinating post in which he reported that one of his neighbors, Esther Ingram, who had been born to missionary parents in New Guinea, once observed an apparent mystery cat at close range there when revisiting this great island as an adult. Her sighting occurred one evening during December 1999/January 2000 at a distance of only 20 yards, when the creature emerged from some jungle and crossed the road ahead while Esther and her father were being driven by her foster-brother in the Eastern Highlands province of Papua New Guinea (PNG). According to Esther's description as recorded by Smith, the creature was:

> ...very solidly built, and the head-body length was about five feet. Both Esther and her father were amazed at how huge it was...Esther, in particular, made an attempt to study as many details as possible. (Remember, it was very close.) The basic colour was white, with ginger "trimmings" on the tail and ears. Pale gingery, vertical stripes, not terribly well delineated, appeared on the sides, but they did not extend to the back, or dorsal surface, which was completely pale. She specifically noted that the forepaws were cat-like, rather than (say) hoofed like a goat's. She didn't get a glance at the rear paws. The tail was ginger and very long, hanging to the ground. I enquired about bushiness etc, to establish a comparison with a dog's. She said it was a bit coarser or fluffier than the body, but not much. On the body itself, the fur was smooth. The head was broad, short, flattish, and

> definitely cat-like. It did not protrude like a dog's. The ears were ginger, mottled with white, and hung down. They were not as long as a spaniel's, but they were definitely long and rounded, and gave every indication of being naturally floppy. It was this feature which amazed both of them (and me as well, as it doesn't sound anything like a cat's). Esther also thought she saw whiskers.

Very intrigued by Esther's account, Smith contacted Australian mammalogist Tim Flannery, an expert on New Guinea fauna, and asked his opinion as to what she may have seen. Flannery deemed it likely that her mystery beast had been a tree kangaroo, but Esther, born and raised in New Guinea and regularly returning there for visits as an adult, was very familiar with the appearance of such animals and did not agree with this identification of the creature that she had seen. Could it have been the New Guinea version of the Queensland tiger as claimed by Rothschild?

With no known large-sized mammalian predators other than imported canine forms, such as the nowadays exceedingly rare New Guinea singing dog, it would not be inconceivable, ecologically speaking, for an elusive feline cryptid to thrive here, plentifully supplied with wallabies, tree kangaroos, possums, rodents, bird life, and other potential prey species. How ironic it would be if the Queensland tiger, or something very like it, was ultimately discovered not in Australia but instead in its more mysterious northern neighbor, New Guinea.

The second of this island's mystery cats is more famous but for all the wrong reasons. Supposedly referred to locally as the moolah, it is just one of several very remarkable beasts that were said to exist here by Captain J.A. Lawson in his notorious book *Wanderings in the Interior of New Guinea* (1875). According to Lawson, he had landed here in 1871, and among his truly extraordinary alleged discoveries were the world's highest mountain (which he dubbed Mount Hercules, far taller than Everest yet could be climbed in just a day!), very large monkey-like ape-men, enormous herds of deer and buffalo numbering in their thousands, the world's tallest tree, flightless birds resembling ostriches or emus, and, his pièce de resistance, the moolah, a specimen of which he supposedly shot and which was ostensibly one and the same as India's Bengal tiger! Here is Lawson's description of his freshly-killed moolah:

> This animal was formed exactly like the Indian tiger, nor was it inferior in size; but it was a much handsomer creature. It was marked

> with black and chestnut stripes, on a white, or nearly white, ground. Its length from the nose to the root of the tail was seven feet three inches.

On fundamental zoogeographical grounds alone, however, Lawson's claims regarding the existence in New Guinea of the moolah and the other creatures listed above were arrant nonsense. Yet, incredibly, for some years afterwards they were widely accepted as gospel, until continued explorations of New Guinea finally confirmed the entire content of Lawson's book to be fictional—indeed, a veritable Munchausenesque satire on the whole concept of Victorian exploration rather than anything remotely factual. As for the true identity of the mysterious Captain himself, this has never been conclusively established. However, the most popular suggestion is that Lawson was actually Robert H. Armit (1844-?), a lieutenant in the Royal Navy with experience as an assistant surveyor in Australian waters, and later Honorary Secretary of the New Guinea Colonising Association.

New Zealand and Hawaiian Mystery Cats

Moolahs notwithstanding, not even the possibility of bona fide unidentified cats prowling Australia and New Guinea could be any more remote than the likelihood of mystery felids inhabiting New Zealand. With the possible exception of the scientifically unrecognized waitoreke (a mystifying otter-like beast now probably extinct), New Zealand's only native mammals are two bat species. All other furry fauna currently existing here has been introduced in modern times by humans.

Consequently it is extremely surprising to learn that reports have been intermittently documented from this dual-island country of large-sized, unmistakably feline beasts that elude all attempts at capture and even identification. The major outbreak to date of encounters with such animals occurred in 1977, as illustrated by the following selection.

On 8 July 1977, the *New Zealand Herald* reported that Graham Stevens, a Three Kings security officer, had been checking a building on his beat one evening in the Auckland suburb of Mangere, when he saw a large lion with big yellow eyes at very close range, just in front of his van. Upon notification, the Otahuhu police searched the area but found no sign of the creature, and the manager of the circus that was currently in town maintained that none of his lions had escaped.

Ten days later, New Zealand IYA Radio announced that a tiger had been seen in Kaiapoi, and during the last week of July various newspapers also carried news of this animal. It transpired that the sighting had been reported by a Kaiapoi woman who allegedly observed the animal in her garden at 4:00 a.m. At first, the police discounted the report as a hoax, but then spoor and droppings were discovered in the nearby sandhills, close to Pines Beach. As a result, a full-scale search was set in motion, involving a helicopter and officers armed with guns, headed by Police Inspector W.J. Perring, who was quoted by the *Sydney Sun-Herald* on 24 July as saying that he was "reasonably satisfied" that a large-sized cat, possibly a tiger, had indeed been in the area. Nevertheless, the search was unsuccessful and was eventually called off, amidst rumors that the culprit was a tiger kept as a pet by a local man, which presumably had escaped and was later secretly recaptured. None of this, however, has been verified, and like so many mystery cat incidents, the animal was simply never heard of again—nor for that matter was the Mangere lion.

Could they have merely been misidentified dogs? Quite possibly, although in the lion instance the animal was seen at very close range, making such an error rather more unlikely. No description of the Kaiapoi tiger's spoor or droppings was released, so we cannot use these to form any opinion regarding their originator's identity.

In 2003, following several sightings of a dark-brown puma-like cat in the agricultural area of Olinda on the Hawaiian island of Maui, officials set some traps in the hope of snaring the elusive felid (*Honolulu Star-Bulletin,* 17 June 2003). Yet, as almost inevitably when dealing with crypto-cats, nothing was caught.

As discussed in 1980 by B.M. Fitzgerald in a *Carnivore Genetics Newsletter* article, feral domestic cats, descended from escapee/released pets, are very common in New Zealand and are rightly blamed for the demise of a number of their endemic bird species. Even so, they could hardly be responsible for either of the large felids discussed here. The currently-held maxims of paleogeography and zoogeography dictate that New Zealand became isolated from other land masses long before either eutherian or even marsupial mammals had chance to reach it. Consequently, it would seem impossible for this country to be harboring an undiscovered native species of large feline carnivore.

Thus we must fall back once again onto the escapee/release theory for a plausible solution. All the same, it would be a lot more reassuring if more than just a very occasional supposed escapee/release could actually be caught.

Lewis Carroll devotees will fondly recall the White Queen informing a bewildered Alice in *Through the Looking Glass*:

> Why, sometimes I've believed in as many as six impossible things before breakfast.

A cryptozoologist in the making?

Chapter 8
Final Thoughts: Conclusions and Conservation

> *How often have I said to you that when you have eliminated the impossible, whatever remains, however improbable, must be the truth.*
> — Sir Arthur Conan Doyle, *The Sign of Four*

After having eliminated the impossible relative to mystery cats—viz. that all reports of all such creatures result from dimly-viewed dogs, manic mendacity, drunken delusion, or mass hallucination—we are left with an initially improbable but ultimately inevitable conclusion. Namely that mystery cats of wide diversity and worldwide distribution do indeed exist. Having discussed their likely identities individually in the previous chapters, when viewed collectively, three fundamental categories of modern-day mystery cats are present.

1) *Felids constituting species or subspecies currently undescribed by science (or at least unknown to science in the living state):*
These possibly include the mngwa, machairodontid-like terrestrial and aquatic forms from Africa and South America, Queensland tiger, Iriomote yamamaya, Ecuadorian rainbow tiger, and spotted lions.

2) *Felids constituting hitherto-unrecorded morphs or other non-taxonomic forms of known species:*
These probably include the black tigers and Fujian blue tigers, woolly cheetah, ndalawo, fitoaky, and quite possibly the jaguarete, damasia, and Transcaucasian black cat.

3) *Felids constituting known species occurring in unexpected, non-native localities.*
These probably include Britain's lynx-like cats; North America's maned leonine creatures; the puma-like felids of Britain, continental Europe, and Australia; and the black pantheresque beasts of Britain, continental Europe, at least some in North America, and Australia.

In addition, there are certain felids, such as Mexico's ruffed cat and onza, that seem fated to straddle two categories until genetic studies can determine unequivocally their correct classification.

The official discovery of any of the mystery cats documented in this book will be of great scientific worth, but some will be especially significant. For example, although the accidental/deliberate introduction by humans of certain small-sized known felids into foreign localities (e.g. feral domestic cats worldwide, jaguarundis into Florida) is already officially acknowledged, the definite discovery of pumas living and breeding wild in Britain or black panthers in Australia, for example, would demonstrate that much larger known felids can and do survive in out-of-place locales.

Moreover, such a radical revelation would require urgent attention not only from zoologists but also from respective governmental authorities. Until now, the only comparable situations concerning large mammalian carnivores involve packs of feral domestic dogs, which in Italy and the US, for example, have often occupied the ecological niche left vacant by these countries' wolves since their widespread eradication by humans.

Equally noteworthy would be the official discovery of blue and/or black tigers, adding very considerably to our knowledge of pelage-related mutant forms exhibited by felids. Similarly, it is possible that future genetic research conducted on onza material will reveal that this is actually an abnormally gracile mutant of the puma, created by a hitherto-unrecorded genetic mechanism.

With respect to unknown taxonomic forms, the prizes on offer are exceedingly great. Felids such as the spotted lions and Iriomote yamamaya may prove to be new subspecies; whereas the mngwa could even be a new species. Moreover, some may actually offer up even more to science than new species. If, for example, the Central African Republic's tigre de montagne proves to be a bona fide saber-tooth, this will resurrect an entire taxonomic (sub)family of felids, thus needing to be added to the list of animals known to exist today. Equally, if the Queensland tiger is found to be a surviving thylacoleonid, an entire resurrected family of Australian marsupials will need to be included in that list. Most dramatic of all, if the large-fanged striped cat of Colombia and Ecuador is discovered one day and actually turns out to be a living *Thylacosmilus*, an entire taxonomic order of South American carnivorous marsupials hitherto believed to have become extinct at least two million years ago will be raised from the dead! Little wonder that mystery cats are nowadays receiving much-warranted attention throughout the world by cryptozoological researchers as well as by mainstream zoologists.

Even so, their scientific significance is not the only reason why every effort should be made to discover and document these animals with all speed. Another reason—equally important but far more urgent—also exists, namely their extreme vulnerability. Down through untold centuries, felids have been persecuted relentlessly and mercilessly by humans, for three principal reasons.

Firstly, they are carnivores, and thereby a threat to livestock and also (at least in the eyes of humans, even if not in reality) to humans themselves. Secondly, by being fierce, the larger species in particular have constituted prized targets for trophy hunters. Thirdly, due to many felids possessing highly attractive coats, they have been greatly sought after by the fur trade. The collective result of these depredations can be ascertained readily by reading the latest IUCN Red List of Threatened Animals, which classes several full species of felid as Vulnerable, plus various additional species and subspecies as either Rare or Endangered.

If this can happen to known felid species, how much worse is the likely fate of those felids not recognized by science? After all, one can readily comprehend that an animal cannot receive official protection under conservation laws if it does not officially exist to begin with.

One ray of hope for the future of mystery cats lies in the fact that if they are indeed real, they have eluded scientific discovery right up to the end of the second decade of the 21st century, thereby demonstrating that they are very adroit (at least so far) at avoiding direct annihilation by humans—again for three reasons.

Firstly, they are very elusive. Secondly, in the case of many tropical forms, the native people inhabiting their localities have often refrained from killing them out of fear, or the imposition of taboos. Thirdly, "civilized" humans, whose armory of guns and scant regard for "primitive" customs could certainly pose a threat, is generally prevented from doing so by the frequently inhospitable and/or inaccessible nature of these cats' domains. However, humans can still decimate such creatures indirectly by destroying their habitat, and, therefore, the delicate ecological web of existence into which, like all life forms, they are inextricably interwoven. Most devastating of all for mystery cats is the continued destruction of the world's rainforests because these are home to the vast majority of the taxonomically significant mystery cats covered in this book.

Happily, the rainforests' known cats are receiving the earnest attention of notable international organizations such as the International Union for the Conservation of Nature and Natural Resources (IUCN), and the World Wide Fund For Nature (WWF). But what can be done for the tropics' mystery cats? This is where cryptozoology can come into its own. For although the discovery of a new species as a result of cryptozoological researches is in itself a great achievement, cryptozoology is uniquely able to accomplish an even greater one—the prevention of an animal species dying out before its actual existence is formally acknowledged. By painstakingly researching and pursuing an animal form dismissed by traditional science as folklore or fancy, cryptozoology can not only bring about its discovery but also, in so doing, make it eligible for conservation consideration, thereby saving it from continued obscurity and ultimate extinction within habitats already threatened with destruction.

Due to ignorance and arrogant disregard for travelers' reports and native peoples' narratives, the tragic disappearance of creatures before they actually received official recognition (let alone protection) is certainly not without precedent. This is well demonstrated, for example, by the extinction of Madagascar's elephant birds and giant lemurs, plus the enigmatic sea mink *Neovison macrodon* of New England and Canada. Although known to the local people and reported by travelers, these species were largely ignored by scientists, until, through habitat annihilation and excessive hunting by humans, they died out.

Many other species have also disappeared before having been formally recognized to our knowledge, but how many times has this occurred without our knowledge. Quite possibly a whole wealth of fascinating and irreplaceable species have been lost in this way, unknown and unrecorded, due simply to humans' continued stubbornness and reluctance to take notice.

Consequently, it is imperative that cryptozoology succeeds in its pursuit of mystery cats and mystery animals of every other kind. To date, no country provides funds for scientifically based cryptozoological research. Hence it is especially to be hoped that a major cryptozoological discovery will be made in the near future, not only for that particular species' survival, but also for the undoubted incentive that it will provide to granting bodies for the provision thereafter of cryptozoological research funds. This in turn could ultimately lead to the protection of other, currently unknown species too.

Although cryptozoology seems to be the only hope for engendering protection for mystery cats, if pursued too zealously it may actually have the very opposite effect. This was recognized by former ISC Secretary J. Richard Greenwell, whose statement that one onza specimen was enough for research purposes and that no more should be killed cannot be commended highly enough. Similarly, the Bottriells are to be praised for their sterling efforts to attract positive scientific interest in the king cheetah and secure its protection in the wild.

It is time for cryptozoology to come of age. It continues to gain interest and respect, but it must now accept a major responsibility too. No longer should it be content merely to seek out new animals, it must also ensure that its discoveries are conserved and perpetuated, otherwise its goals will be meaningless, its ideals empty.

Perhaps my documentation of mystery cat lore in this book may help, however slightly, to bring about the formal discovery of one such felid, whose "official" status can thereafter work to acquire its protection and its habitat's conservation. Certainly I could not request a greater boon of my own existence than to assist in the accomplishment of such an achievement. To quote a late, much-missed friend, who was both a keen cryptozoologist and a passionate conservationist: "Take only memories, Leave only footprints, Kill only time." (Trevor Beer, *Poachers Days*)

Wise words indeed.

Appendix 1
King Cheetahs in Asia?

This Appendix contains the complete text and both illustrations from my paper investigating evidence for the former existence of king cheetahs in Asia, as published in 2013 in the *Journal of Cryptozoology*, which is the world's only peer-reviewed scientific journal currently in print that is devoted to mystery animals. Reproduced with permission, the paper's full bibliographical reference is:

SHUKER, Karl P.N. A historical depiction of a king cheetah in Asia? *Journal of Cryptozoology* Vol 2, pp. 31-39 (Dec 2013).

A Historical Depiction of a King Cheetah in Asia

Abstract
A work of 16th-century Mughal art is examined that contains a representation of an anomalous striped cat not previously examined in the literature but which recalls the striped king cheetah morph, hitherto known only from Africa.

Introduction
The logo of the *Journal of Cryptozoology*, the king cheetah is a rare morph of the cheetah *Acinonyx jubatus*, from which it differs visually by virtue of its pelage's ornate patterning of stripes, blotches, and swirls, contrasting markedly with the familiar polka-dot spotting of typical, wild-type cheetahs. Although long known to native hunters in southern Africa and traditionally referred to by them as the nsuifisi or leopard-hyaena (Hichens, 1937), its existence remained unconfirmed by science until 1926. This was when a specimen was trapped south-east of Salisbury (now Harare) in what is today Zimbabwe, and its skin presented to Salisbury's Queen Victoria Memorial Library and Museum (Cooper, 1926). When the skin was subsequently loaned to the British Museum (Natural History), the museum's felid expert, Reginald Pocock, considered it to represent a new species of cheetah, which, when formally describing it, he named *Acinonyx rex*, the king cheetah (Pocock, 1927).

In subsequent years, a number of other 'king' skins were obtained, and some were clearly intermediate in appearance between kings and normal spotted cheetahs. As a result, the king cheetah was no longer deemed by Pocock and other zoologists to be a valid species, but merely a rare morph of *A. jubatus* (Pocock, 1939). This demoted status was confirmed in 1981, when a litter of cubs born to a pair of normal spotted cheetahs at the de Wildt Cheetah Breeding and Research Centre of Pretoria's National Zoological Gardens contained a king cheetah (Shuker, 1989). Just a few days later, a king cheetah cub was also present in the litter of cubs born to the sister of the first litter's mother (Shuker, 1989).

During the mid-1980s, king cheetah researchers Rudi J. Van Aarde and Ann Van Dyck revealed that the king cheetah's phenotype was due to a single recessive mutant allele, one that was quite probably homologous to the allele responsible for the blotched tabby phenotype of the domestic cat *Felis catus* that sometimes occurs in litters produced by mackerel-patterned tabby cats (Van Aarde and Van Dyck, 1986). This suggestion was recently verified when both the cheetah's king phenotype and the domestic cat's blotched tabby phenotype were shown to be caused by the same single mutation in a gene known as Transmembrane aminopeptidase Q or Taqpep (Kaelin et al., 2012).

Many morphs can arise spontaneously throughout a given species' entire zoogeographical distribution. In the cheetah, conversely, the king morph has exhibited a very localised distribution, seemingly limited to a triangular portion of southern Africa south of the Zambezi enclosing eastern and southern Zimbabwe, northern South Africa, and eastern Botswana (Bottriell, 1987). Only one possible exception is on record—a king cheetah skin obtained from a specimen reputedly shot during 1988 by a poacher in the Singou Total Fauna Reserve within the West African country of Burkina Faso (Frame, 1992). However, the specimen's provenance has never been formally confirmed, so it may conceivably have been killed elsewhere and its skin later brought to Burkina Faso.

In stark contrast, the cheetah's total range in historical times not only spanned much of Africa (south, central, east, west, and north) but also the Middle East and southwestern Asia, particularly India - where the Mughal (Mogul) emperors of the 16th-19th Centuries frequently used it as a hunting beast. Akbar the Great (1542-1605) allegedly owned a thousand cheetahs. Nowadays, however, Asian cheetahs are only known to survive in Iran and possibly also Pakistan, with probably less than a hundred specimens alive in total; the last known Indian specimens, three in total, were all shot in 1947 (Shuker, 2012). In addition, there are a few highly elusive small-spotted cheetahs reported from Egypt's Qattara region, whose exact taxonomic identity remains uncertain (Ammann, 1993).

Yet despite the cheetah's formerly extensive distribution, not a single example of a king cheetah has ever been documented from anywhere outside Africa. Consequently, the author was very interested when a historical Indian work of art depicting a mystifying cheetah-like cat with stripes killed during a royal hunt in India was recently brought to his attention, and decided to research this potentially significant case.

Methods

In October 2010, the author was contacted via email by Raheel Mughal, a longstanding cryptozoological correspondent, who mentioned that his Mughal ancestors in India used to house king cheetahs along with normal cheetahs in specially-constructed kennels and were used for hunting game. When the author informed him that king cheetahs have never been recorded from Asia, Mughal disagreed, stating that he had seen paintings of the Mughal emperors on hunting excursions being led by normal spotted cheetahs and also featuring striped king cheetahs.

When the author asked him where he had seen such paintings, Mughal replied:

> [It was in] 1995, I was 11 years old, this was at Dubai airport, on route from Pakistan. I asked the shop keeper and he told me that it was a depiction of the Mughal emperors and their cheetahs, I was particularly captivated by the cheetah at the centre of the image, it had as I described earlier stripes, swirls and was clearly different from the other cheetahs in the painting. Last year, I found a similar though different image on none other than Wikipedia, they have since taken it down (and I can't remember the name of the exact page where I saw it). I've just tried googling it, but sadly to no avail.

Although the author has spent time at Dubai International Airport on a number of occasions during various overseas travels, and has seen all manner of artwork on sale there, he has never seen anything depicting a king cheetah-like cat. However, as it is an extremely big airport, with an enormous and very rapid turnover of stock, this is not too surprising. Nevertheless, the prospect of Indian artwork depicting king cheetahs was sufficiently intriguing for the author to publish an article in the popular-format monthly magazine *Fortean Times* discussing this exciting possibility (Shuker, 2011), in the hope that it would elicit responses from readers who may be familiar with such art.

Fig. 1—"Akbar Hunts Near Lahore and Hamid Bakkari is Punished By Having His Head Shaved and Being Mounted on an Ass" (public domain, from Wikipedia)

On 21 February 2011, *Fortean Times* received an email from one reader, Laura Beaton, containing an image for the author to examine. It was followed a day later by an email from her sent directly to the author and containing details of the image—which is the painting with which this scientific paper is concerned.

Results

Consisting of opaque watercolor and gold on paper, the painting originally appeared in the Akbar Nama ('Book of Akbar'), which is the official chronicle of the reign of Akbar the Great, the third Mughal Emperor (reigned 1556-1605). Written by Akbar's court historian and biographer, Abu Fazl, between 1590 and 1596, the three-volume manuscript was illustrated with paintings produced by many different artists, some subsequent to the manuscript's preparation. The Victoria and Albert (V&A) Museum in London, U.K., purchased a partial copy of this manuscript in 1896 from Frances Clarke, the widow of Major General John Clarke, who bought it in India while serving as Commissioner of Oudh between 1858 and 1862 (V&A, 2013).

The painting is entitled 'Akbar Hunts Near Lahore and Hamid Bakkari is Punished By Having His Head Shaved and Being Mounted on an Ass', and is described as follows on the V&A's website (V&A, 2013):

> This illustration to the Akbarnama (Book of Akbar) is the right side of a double-page composition (see Museum no. IS.2:56-1896 for the left side). The entire composition depicts a ceremonial hunt that took place near

Lahore, in present-day north-east Pakistan, in 1567. The Mughal emperor Akbar (r.1556–1605) is shown in the centre of the painting mounted on horseback with his sword raised. At top right Hamid Bakkari is shown being punished for firing an arrow at one of the servants of the court by having his head shaved and being forced to ride backwards on an ass. The composition was designed by the Mughal court artist Miskina, who also painted the face of the emperor, and the rest was painted by Sarwan.

As can be seen from this paper's reproduction of the painting (Fig. 1), it portrays a number of familiar, readily-identifiable species of Indian mammal, including the cheetah, blackbuck *Antilope cervicapra*, and chital *Axis axis* (also known as the spotted or axis deer).

Also depicted is a single specimen of a large but hitherto-unidentified cat. Just in front of the mounted emperor is a pile of slain chital and other deer, to the immediate left of which is the cat in question, lying dead. As revealed in the close-up image of it (Fig. 2), this cat has a gracile body, slender with long lithe limbs, plus a long banded tail, and a pelage that is patterned not with spots but with short stripes. In addition, a dark, distinctive teardrop-like streak runs down from the cat's eye to its upper jaw.

Fig. 2—Close-up of the dead striped cat (top left), plus a normal spotted cheetah (bottom right) for direct comparison purposes (public domain, from Wikipedia)

Discussion

Despite its striped pelage, the dead cat bears no resemblance whatsoever to the only known species of habitually striped Asian cat, the tiger *Panthera tigris*. Nor does it correspond with any documented pelage variety of any other Asian cat species (Sunquist and Sunquist, 2002; Hunter and Barrett, 2011).

However, its entire morphology is distinctly cheetah-like, as substantiated by comparing its form with those of the normal spotted cheetahs portrayed in the painting taking part in the hunt. Even its facial 'teardrop' streak corresponds precisely with that of the spotted cheetahs present and is a unique feature of the cheetah species (Hunter and Barrett, 2011). Moreover, using the chital deer carcases directly alongside it as a scale, the cat is almost as large in size as they are, and is therefore much too big to belong to any of the smaller Asian cat species.

Although the dead cat's striped pelage is less ornate than that of most of the African examples on record, taking its striping together with its overall morphology the felid that it most closely resembles is the king cheetah. Bearing in mind that all other mammals in this painting are depicted in a painstakingly accurate, naturalistic manner, there is no reason to assume any different for the portrayal of the dead striped cat.

It is suggested, therefore, that the dead striped cat may represent an Asian equivalent of the African king cheetah, and, as such, the first example to have been recorded in the literature. The question is whether its striped pelage is directly homologous to that of the African king cheetah, i.e. induced by the same mutant gene allele responsible for the king cheetah morph; or whether it is due to an analogous mutant allele whose expression has produced a comparable (albeit somewhat more conservative) phenotype to that of the African king cheetah.

The Asian striped cat possesses the same degree of tail banding (Bottriell, 1987) as is typical for the African king cheetah. Whether it also possesses the characteristic series of longitudinal stripes running down the king cheetah's spine (Bottriell, 1987) is unknown, because the angle at which the Asian striped cat has been painted precludes any observation of its spine. Whereas the striped markings of the African king cheetah tend to be somewhat curled and irregular in shape (Bottriell, 1987), however, those of the Asian cat appear predominantly to be short vertical bars. This difference indicates that the Asian cat's striping may therefore be due to the expression of a different mutant allele from the version responsible for the African king cheetah's stripes. Consequently, it seems most likely that in terms of genetic identity, the Asian striped cat (if indeed a cheetah) and the African king cheetah are analogous rather than homologous striped morphs of the cheetah.

This in turn begs the question as to whether a new term should be created for the Asian striped cat—a quasi-king cheetah, perhaps, or a pseudo-king cheetah? Yet there is a notable precedent for not using separate terms for genetically analogous morphs in felids. Melanistic (all-black) specimens of the leopard *Panthera pardus* and the jaguar *Panthera onca*, for instance, both possess pelages with abnormally dark background coloration, making them look identical to one another in this respect. Yet the mutant alleles respectively responsible for inducing these two morphs are entirely different. In the leopard, it is a recessive mutant allele of the agouti (ASIP) gene; in the jaguar, it is a dominant mutant allele of the receptor gene MC1R (Shuker, 1989; Eizirik et al., 2003). Nevertheless, all-black leopards and all-black jaguars are both referred to as melanistic; separate terms for them have not been coined.

Only one previous putative example of a cheetah morph has been recorded for the Asian cheetah in the Indian subcontinent. This was a blue-spotted, white-furred cheetah specimen owned by the Mughal Emperor Jahangir (1569-1627; reigned 1605-1627), eldest surviving son of Akbar the Great (Divyabnanusinh, 1987). In contrast, a number of different cheetah morphs other than the king cheetah have been recorded in Africa, including melanistic specimens (Shuker, 1989), small-speckled specimens dubbed cheetalines (Pocock, 1921; Combes, 2011; Shuker, 2012), and the so-called woolly cheetah (Sclater, 1877; Shuker, 1989; Shuker, 2012). Consequently, the possibility that the cat depicted in the Mughal painting considered here is an Asian cheetah morph is of especial interest and potential zoological significance.

Moreover, this is not the first time that an examination of Mughal art has yielded important information relative to the appearance of morphs in felids. The earliest record of a white tiger can once again be found within the Akbar Nama, consisting of a painting entitled 'Akbar Slays a Tigress Which Attacked the Royal Cavalcade'. Of the five tigers depicted in this painting, two of them are white (Divyabhanusinh, 1986).

Clearly, therefore, continued examination of Mughal art may conceivably uncover further examples of ostensible cryptozoological interest, including additional representations of this paper's striped mystery cat (Raheel Mughal did allege seeing more than one Mughal painting depicting striped cheetahs). Such discoveries would be very beneficial in assessing the cat's zoological identity more precisely.

Acknowledgements

I wish to offer my grateful thanks to Raheel Mughal and Laura Beaton for their much-appreciated correspondence; and to the two reviewers for their valuable comments.

References

- Ammann, K. (1993). Close encounters of the furred kind. *BBC Wildlife*, 11 (July): 14-15.
- Bottriell, L.G. (1987). *King Cheetah: The Story of the Quest.* E.J. Brill (Leiden).
- Combes, G. (2011). The Phantom. http://guycombes.wordpress.com/2011/06/15/the-phantom/ 15 June.
- Cooper, A.L. (1926). A curious skin. *The Field,* 148 (14 October): 690.
- Divyabnanusinh (1986). The earliest record of a white tiger (*Panthera tigris*). *Journal of the Bombay Natural History Society*, 83 (December): 163-165.
- Divyabnanusinh (1987). Record of two unique observations of the Indian cheetah in Tuzuk-I-Jahangiri. *Journal of the Bombay Natural History Society*, 84 (August): 269-274.
- Eizirik, E., et al. (2003). Molecular genetics and evolution of melanism in the cat family. *Current Biology*, 13 (4 March): 448-453.
- Frame, G. (1992). First record of king cheetah outside southern Africa. *Cat News*, No. 16 (March).
- Hichens, W. (1937). African mystery beasts. *Discovery*, 18: 369-373.
- Hunter, L. and Barrett, P. (2011). *A Field Guide to the Carnivores of the World.* New Holland Publishers (London).
- Kaelin, C.B., et al. (2012). Specifying and sustaining pigmentation patterns in domestic and wild cats. *Science*, 337 (21 September): 1536-1541.
- Pocock, R.I. (1921). An interesting cheetah. *The Field*, 137 (19 March): 352.
- Pocock, R.I. (1927). Description of a new species of cheetah (*Acinonyx*). *Proceedings of the Zoological Society of London*, No. 17: 245-252.
- Pocock, R.I. (1939). *The Fauna of British India Including Ceylon and Burma*. Taylor and Francis (London).
- Sclater, P.L. (1877). Felis lanea. P*roceedings of the Zoological Society of London*, (19 June): 532-534.
- Shuker, K.P.N. (1989). *Mystery Cats of the World: From Blue Tigers to Exmoor Beasts.* Robert Hale (London).
- Shuker, K.P.N. (2011). Return of the king. *Fortean Times*, No. 271 (February): 58-59.

• Shuker, K.P.N. (2012). *Cats of Magic, Mythology, and Mystery: A Feline Phantasmagoria*. CFZ Press (Bideford).
• Sunquist, M. and Sunquist, F. (2002). *Wild Cats of the World*. University of Chicago Press (Chicago).
• V&A (2013). http://collections.vam.ac.uk/item/O9646/akbar-hunts-near-lahore-and-painting-miskina/ last updated 31 July 2013.
• Van Aarde, R.J. and Van Dyck, A. (1986). Inheritance of the king coat colour pattern in cheetahs *Acinonyx jubatus*. *Journal of Zoology* (A), 209: 573-587.

Appendix 2
Alleged Sightings of North American Black Pumas/Black Panthers Posted on *ShukerNature* by Readers

The internet is notorious for its ephemeral, uncertain nature, information appearing online then abruptly disappearing with frightening regularity. On 16 August 2012, I uploaded onto my *ShukerNature* blog an extensive article dealing with the highly contentious subject of melanistic pumas, entitled "The Truth About Black Pumas - Separating Fact From Fiction Regarding Melanistic Cougars." This article obviously struck a chord with readers because it attracted dozens of posted responses that included many alleged sightings of creatures variously claimed by their observers to have been black pumas or black panthers but which had never previously been published. In view of their cryptozoological value and the afore-mentioned unpredictable lifespan of online data, it would be very sensible to preserve these sightings in permanent, hard-copy format. Consequently, I am documenting them herewith, presented chronologically together with the respective names/usernames (when given) of those posting them.

As is so often true with cryptozoological reports, there is no confirmation that any of them are genuine, but they definitely deserve to be recorded, just in case.

Anonymous, 1 October 2012

I live in S. TX and heard several people talking about seeing a black panther in the area. I wondered if they were seeing a black jaguar, which would have been wonderful on several levels or perhaps the elusive/mythical black puma; cougars are known to live in this part of the state, although now rare.

So I kept my eyes peeled when I was in those places where I heard this mystery cat had been seen and was utterly surprised to see a large, very dark colored jaguarundi. The first I had ever seen period; not even on a TV nature show nor any wildlife books. I had to look it up on the internet.

I saw the cat several times over the course of about three years and it was completely understandable how one could mistake it for one of its larger cousins; being a good bit larger than the domestic cat, seen from a mid distance the mistake would be an easy one to make, especially if the cat were in tall grass, disguising the shorter limbs.

Of course to my shock I was told the animal could not possibly be a jaguarundi when I reported it [to] the TX Dept of Wildlife. They do not exist in the this [sic] area I was told. They said they did a two year study and found no evidence. Of course, they failed to say that since this is an endangered species if they had found one it may have pissed off more than a few hunters and those rancher[s] that set snares willy nilly, since it would probably have changed some laws involving those activities.

I doubt that the jaguarundi critters are still around now with all of the industrialization of our South TX plains and brush country. I am sure they have either moved on or perished under the onslaught. It's really too sad to contemplate.

Anonymous, 5 December 2012

I've lived in Sarasota county and Charlotte county Fl most of my life and people have seen black panthers more often than cougars including me. In 1966 when I was 12 I was on a school bus going to Englewood from North Port around 8 am when I seen a black panther run across the road right in front of the bus. It leapt across the road in about 3 leaps and was gone. It was totally black and had a long tail and like I said earlier there has been more sightings of black panthers than cougars. There was a story in the paper in the 80s about a mother with cubs went after a old person on a 3 wheeler and the person had to put the bike between the panther and himself until a car came by and chased it away.

Anonymous, 13 July 2013

I live in Humboldt County, Ca. in an area with an unusually high population of bears, mountain lions and bobcats. Our area of about five thousand acres of land closed to the public is bursting with predators of all local types and their prey.

My wife and I were standing next to my truck talking when I spotted a black mountain lion walking a brushy fence line with a freshly killed rabbit in its mouth. The cat was in full view at a range of 150 yards on a hillside that sloped downward towards us, allowing for an unobstructed view of his whole body.

We had seen so many mountain lions over the years that our first reaction was to note the color as it was the first black one we had seen. I used my 10x binoculars to survey its coloring. As far as I could see it was a nice shiny black except for the top of its head which was a very dark brown. Almost black. This cat was young and bold, weighing around 90# [90 lb] and very healthy.

We were fascinated as much by his hunting technique as his color. He had one rabbit in his mouth and was looking for more. As we were discussing how it might pull off another kill with one already in its mouth he winded two deer on the ridge about 50 yards above it. It crouched low and began a stalk in the open on the deer. That didn't work as the deer were watching him before he caught their scent. They left.

With the deer gone this young cat realized he was standing in the middle of a 4 acre pasture in ankle high grass. This made him nervous and he made his way into cover with his rabbit.

The next time he was seen was at close quarters by one of my employees about a week later.

My guy started a truck that was backed up very close to a black berry patch. As he stepped out of the truck to do his morning inspection of the vehicle the cat came storming from the rear area of the truck and ran by him "like his hair was on fire" was the way he described it. Straight line and over the hill.

The next man to see it was a professor from MIT that I had been working with for about two years.

I was to meet him less than a mile from where the two earlier sightings had been. He had no knowledge of this cat prior to his encounter.

Upon reaching the job site he exited his car and I noticed at once that he was pale and very agitated. I spoke to him and he ignored me as he began to pace back and forth. I figured someone had died. Suddenly he whirled around and said "OK, I'm going to ask you this. Have you ever seen a black Mountain Lion?"

I couldn't have been more relieved as I informed him this was the third sighting in as many weeks. He looked at me with a stunned expression as he told me that he had been sure he was losing his mind. Never seen a man so relieved.

He had come around a corner in his car and almost hit it. He was amazed at the leap it had made to escape a collision.

In talking to the old timers in the area I collected stories of two being killed in the valley over the last 100 years or so by sheep ranchers.

My family had no idea this cat was so unusual as we had seen many bears of different color phases over the years and figured it was the same sort of thing. Just didn't seem like a big deal until I stumbled onto Dr. Shuker's article.

Now I wonder years later if I had known about all the controversy; would I have shot it?

A total of four people saw this cat from all angles and no white was seen in its coat.

Unknown, 31 December 2013

I live in the Diablo range of San Bonito County CA. The farmers see and kill black pumas all the time. They hang circle hooks of meat from the trees to kill them so they don't attack the livestock. There are sooo many large cats around here. Just this week I saw a 50 lb bobcat that had no spots or any markings.

[I responded to this post by Unknown as follows (but received no reply):
If they kill black pumas all the time, why haven't they ever saved a carcase of one and submitted it for formal scientific identification and examination? To date, no scientifically confirmed black puma has EVER been documented from North America - fact. So if you can obtain a specimen of one of these black felids being regularly killed in your region, please do so and submit it for identification to a local museum, or even a vet. It would be a major scientific discovery if proven to be a genuine black puma.]

Unknown, 17 January 2014

My Mom told me that they have a black cougar near her place, eating dogs and deer. I looked up this article and she answered me back with the name of a gentlemen that lives near it and may have taken pictures of it.

They are near Poplar Bay, Pigeon Lake, Alberta, Canada.

If you would like contact info you can send me an email at: english.teacher.df@gmail.com

cheers,

Kiauhmitl

PS I once slept in a tent near a cougar in heat that did an amazing shriek howl for probably more than an hour. That was near the town of "Rosebud" in northern California, where they had a large population of cougars in 1993.

[I did send an email to Kiauhmitl, but I did not receive any reply]

Anonymous, 9 July 2014

In 2010 just outside Lyons Mississippi we came very close to running over a very large black cat. This is in the Delta area of Mississippi and within a few miles of the river. It was night but the cat was in full view in high beam. At first I thought it was two black labs playing in the hwy but as soon as my wife slammed on brakes we could tell it was one large black cat. From the tip of tail to the tip of its nose was over 6 feet. We raise and breed English Mastiffs so we are very used to seeing very big dogs, that being said we did not mistake a house cat or a dog for it. I can not say I saw a black puma but I can say we saw a 6 foot close to 200 pound black cat.

Patricia Rose, 8 October 2014
My mother and I saw a Black Panther near Bemidji Minnesota. It was in the winter it had just snowed. The giant black cat approached the road and we stopped the car. It walked around in front our car, pacing and whipping its tail. Its coat was shiny and black. It had yellow eyes. It seemed fearless. We turned on our headlights to illuminate it even more and it snarled and then bolted off. We often told people about the event and one day I was informed of another sighting about 4 miles away that same year. It was a huge, pitch black cat, at least 6 feet from its nose to its tail. Beautiful and scary.

I was really glad that there were two of us seeing this beautiful giant black cat. We can validate each other. We don't care what people want to call it officially but there was no confusion as to what we saw. It was a giant, all black cat that looked similar to a mountain lion, maybe a little sleeker, jet black and yellow eyes.

Unknown, 8 December 2014
Patricia Rose, Am I interested in your experience. In the mid 1970's, I had a similar experience near Pequot Lakes, Minnesota. I saw a 100 lb or so black cat at the edge of the woods sleeking; it was close enough that I could tell it wasn't a dog or other animal, but what looks to my [me] like a black panther. The way it was sleeking along was just like a cat... and the long tail is memorable along with the head. This is a lot of years ago, but the memory is so engrained in my mind, I can still see the scene. I was the passenger, so I had plenty of time to study the animal. Growing up on a farm with animals, I was pretty good at identifying form. For me, I was shocked to see this as I didn't know that black panthers could be around. We just thought that possibly a zoo animal got loose.

Siamese, 21 December 2014 *[not a sighting but an example from literary fiction that I hadn't encountered before, so it warrants recording here]*
In the first two books of the 'little house' autobiography series by Laura Ingalls Wilder panthers are mentioned. One in Wisconsin where it hunted Laura's grandfather, the tracks of another were seen by a stream - also in Wisconsin [-] and a third was seen by Laura's father and eventually was shot by a member of a nearby Native American tribe.

The panthers are described as having a high pitched scream that is very similar to a woman screaming and hunt by dropping down onto their prey from trees.

Here is the most detailed description in the books:

"One night Pa looked at Black Susan, stretching herself before the fire and running her claws out and in, and he said:

"Do you know that a panther is a cat, a great, big wild cat?"

"No," said Laura.

"Well, it is," said Pa. "Just imagine Black Susan bigger than Jack, and fiercer than Jack when he growls. Then she would be just like a panther."

He settled Laura and Mary more comfortably on his knees and he said, "I'll tell you about Grandpa and the panther."

"Your Grandpa?" Laura asked.

"No, Laura, your Grandpa. My father."

"Oh," Laura said, and she wriggled closer against Pa's arm. She knew her Grandpa. He lived far away in the Big Woods, in a big log house. Pa began:

The Story of Grandpa and the Panther.

"Your Grandpa went to town one day and was late starting home. It was dark when he came riding his horse through the Big Woods, so dark that he could hardly see the road, and when he heard a panther scream he was frightened, for he had no gun."

"How does a panther scream?" Laura asked.

"Like a woman," said Pa. "Like this." Then he screamed so that Laura and Mary shivered with terror.

Ma jumped in her chair, and said, "Mercy, Charles!"

But Laura and Mary loved to be scared like that.

"The horse, with Grandpa on him, ran fast, for it was frightened, too. But it could not get away from the panther. The panther followed through the dark woods. It was a hungry panther, and it came as fast as the horse could run. It screamed now on this side of the road, now on the other side, and it was always close behind.

"Grandpa leaned forward in the saddle and urged the horse to run faster. The horse was running as fast as it could possibly run, and still the panther screamed close behind.

"Then Grandpa caught a glimpse of it, as it leaped from treetop to treetop, almost overhead.

"It was a huge, black panther, leaping through the air like Black Susan leaping on a mouse. It was many, many times bigger than Black Susan. It was so big that if it leaped on Grandpa it could kill him with its enormous, slashing claws and its long sharp teeth.

"Grandpa, on his horse, was running away from it just as a mouse runs from a cat.

"The panther did not scream any more. Grandpa did not see it any more. But he knew that it was coming, leaping after him in the dark woods behind him. The horse ran with all its might.

"At last the horse ran up to Grandpa's house. Grandpa saw the panther springing. Grandpa jumped off the horse, against the door. He burst through the door and slammed it behind him. The panther landed on the horse's back, just where Grandpa had been.

"The horse screamed terribly, and ran. He was running away into the Big Woods, with the panther riding on his back and ripping his back with its claws. But Grandpa grabbed his gun from the wall and got to the window, just in time to shoot the panther dead.

"Grandpa said he would never again go into the Big Woods without his gun."

[My response:

Thank you so much for posting this, as its description of a mystery black panther shows that back in the days when these books were first published (early 1930s), such animals were known about and being seen. The creature's high-pitched scream is instantly reminiscent of a puma's, so perhaps it really was a melanistic puma. But without physical evidence, we can only ever speculate, sadly.]

Unknown, 2 July 2015
I live in the Okanagan valley of BC CANADA. I have photos of a large black cat the size of a mature black bear. The pictures are from a distance as I have enjoyed seeing such a rare large cat. Almost every morning he/she is in my sights.

Kolemann, 19 July 2015
My name is Kolemann and I work for a tree trimming company contracted with power companies in western Kansas. I was working in Russell Springs, a very rural nearly deserted settlement, this last week when I saw something I couldn't explain. One block in front of me I witnessed a very large black cat the approximate size of a mountain lion running across a vacant lot at high speed looking for cover. I saw it for a mere 30 seconds but it was close enough for me to see it had absolutely no pattern in its fur and was completely black from head to the tip of its tail and large enough for me to be somewhat concerned for my safety. I have been perplexed by this encounter since and seek some rational explanation to my sighting. It may be irrelevant but while driving to work that very day I saw two antelope and know mountain lions will follow these animals perhaps this animal was also. If anyone has suggestions or answers to my dilemma please email me at kolemannmp@gmail.com

Unknown, 13 November 2015
5 years ago we were travelling on highway 96 in northern California. It was evening my bright lights were on. On the side of the highway was high Brown dry grass. We saw in the grass huge bluish eyes. We slowed down and in the grass we all saw a huge head and shoulders crouched down of a huge black cat....in the rugged mountains of this area there have been stories going back to the 1800s of big black cat attacks.

Robyn, 10 December 2015
I grew up in Edmond Oklahoma, my mother and I, and a neighbor also on the road headed opposite direction, had a huge black panther looking animal cross in front of our cars in the headlights. Pure shiny black, with yellow greenish eyes. We were absolutely shocked. We were on our wooded neighborhood street in North Edmond Oklahoma back before it was fully developed the way it is now. I don't think my mother ever reported it. Don't know if the other person in the other car did.

[A few minutes after posting the above report, Robyn posted the following longer version of it:]
In Edmond Oklahoma, my mother and I had a black panther looking animal cross in front of our car in the headlights on our wooded neighbor hood street. It was a huge pure black shiny cat with yellow greenish eyes. We were extremely shocked. Another car approaching us from the opposite direction saw it as well. We were afraid to let our dogs out off leash for a very long time! I don't think my mother ever reported it. Not sure if the other person reported it either. I've always wondered exactly what it was and where it possibly could have come from. Looked like it should have been from the zoo. We had never heard of black cats in Oklahoma. It was huge, probably 6 ft long!

Unknown, 4 February 2016

I saw a black puma first hand in Leavenworth Washington in 2001-2, it was summer maybe June or July. It was maybe 100-150 yards from me across the river clear as day. I watched it walk down river, so big, so agile.

[A few minutes after posting this report, Unknown posted the following slightly reworded version it:]

I saw a black puma first hand in Leavenworth Washington on the summer of 2002. I was at the river and it was on the other side maybe 100-150 yards from me. It was mid day no shadows, I watched its full profile walking.

Unknown, 10 March 2016

I live in Chicago but was driving in through Montana three days ago, about 28 miles south east of Lewistown on west 200MT. As I drove, a black figure caught my eye in the field to the right of me. As I approached closer, I could see that it was a large cat only about 50 yards from the road. The cat was pitch black, had a thick tail over two feet long and about 55 to 60 pounds based off the size of our 40 pound dog at home. The cat was slowly stalking through the field as if it was sneaking up on prey. I knew that this was something significant and needed to be posted. I felt like I had just seen something on the level of Bigfoot, but the fact is this cat was right there! Anyone driving west bound would be able to see this super rare event!!!

Unknown, 1 June 2016

Enjoyed ur article very much. I have been an avid hunter and wild animal enthusiast over 35 years. Being from the south and my knowledge of its wild animals and birds (NC/SC stateline mountain area) I am constantly called about someone who has seen a black panther. I try to go check the area of the sightings if it's within an hour drive and give my opinion. I have set out trail cameras multiple times at these areas, checked tracks, scat and livestock/deer kills, etc. Unfortunately I have never seen a single piece of evidence to prove their existence. What I see or find is juvenile black bear, bobcat, coyotes, large dogs and large common house cats. When the day comes that I see one alive, dead or even a trail cam pic of my own.... I will possibly believe. However, chances are even then it will be a melanistic cougar, melanistic bobcat, someone's pet or zoo escapee. I've done a pile of research myself and have never even heard of scientists finding a true panther's remains or fossils in my area. The screams people hear at night are generally bobcats. "Do cougars exist here in my area?" is another question I'm asked constantly. There is no proof of this that I've seen either as far as them making this area their home. They did 100 years or so ago. It's possible that we get a roaming cougar from time to time. One was killed in central Georgia few years back that had came [come] all the way from the glades. The one that was killed on Pennsylvania highway few years ago had traveled all the way from the Dakotas. So yes they can roam but is not the norm. Most time it is juvenile male that has been drove [driven] from his area by a dominant male and his natural compass goes haywire. He basically just sets out in wrong direction to find a new life in search of food and a mate. 25 years ago there were no coyotes in my area or anywhere within couple hundred miles of here. Now they are everywhere. So yes anything can migrate, but I don't think it possible that a black panther has swam [swum] the

Atlantic Ocean from Africa and Asia to get here...

Unknown [Bill], 4 June 2016
I was chatting with my mother today about a cougar that was spotted near town recently. She then began telling me about a black cougar her and two friends saw in a remote area of northwestern California, Del Norte County, some 18 or so years ago. She remembered how they were all stunned that it was black. I did some research today to find out how common they are and found this post. I told her she had a better chance of spotting Bigfoot than a black cougar! Bill...

Unknown, 13 December 2016
My wife and I believe we saw a black puma cross near house in the Kansas Country side this past Sat. We know what we saw was not a canine and way too big to be a large domesticated cat. It moved like a big cat. Very unsettling to us as we have 3 young daughters.

Unknown, 12 January 2017
I had no idea Black Panthers were controversial. I've seen two in my life. One in San Augustine County, Texas around 1979. The other about two weeks ago in Kendall County, Texas. There were four of us that watched it walk across a field with a clear view in bright sunlight. All four of us had our phones and cameras and none of us thought to take a picture. I now understand they may not exist, but that hasn't stopped me from seeing two of them.

Oneder, 20 February 2017
My wife, son and I saw a large black cougar on Read Island in BC Canada several years ago. We did not know they were rare. We were very close as we were on a boat on a dock and the cat walked the rocks along the beach less than 50 yards away and did not see us inside our boat. We watched for about 2 minutes till it went out of sight. We saw its long tail and long body and were so surprised we forgot to even take a picture. Only later did we learn that they supposedly do not exist. There were three of us at close range in good light for a couple of minutes who saw this. Sorry no photo. Seems like almost a mystical beast.

Unknown, 7 June 2017
I live in Eatonville, WA and saw a black cougar run across the road close to where I live. I live in an isolated area and it scared the crap out of me. If I had to guess I would say the creature was 80-100 lbs. I didn't want to tell many people because I didn't want to sound crazy!

Unknown, 25 September 2017
I am sure I saw a huge black cat alongside the Nooksack River near the Washington/British Columbia border yesterday, also the size of a mature bear, and about three times the size of the male lab I was with.

Michael Blouch, 3 December 2017
My detailed account for the record... I've never publicly told my personal experience before.

Letter sent to the Florida Fish and Wildlife Conservation Commission:

To whom it may concern,

Florida Turnpike near Leesburg, as I was headed towards the Orlando International Airport. It was at 5:55 AM when I encountered a very large animal, which appeared to me to be a black panther (as strange as that may sound).

Here is what happened and what I observed:

The animal ran in full stride directly across the highway in front of my vehicle from the left side, while I was driving at a constant speed of 70 mph set via the cruise control. It first became visible via my left eye's peripheral vision. Instantaneously, I swerved my car towards the left, to avoid a collision. The fact is that it was so close to the front of my car that I actually lost sight of it momentarily, but somehow I missed it. Then in a split second, I could faintly see it lunged across the roadway to my right side.

The best description I can give:

The animal I observed was completely jet black, (not dark brown). It had a low profile physique, and a muscular looking torso and relatively muscular legs. It easily stood at a height of 1-1/2 - 2 ft. to the shoulder. There is absolutely no doubt in my mind that what I saw was a feline type of species, as opposed to some other type of cat. It was much too large and heavy, with big paws. I would estimate that the body alone was approximately 36 inches long, with a large head and relatively small ears. The tail was very long, a minimum of 18 inches, and like a heavy 3" thick mooring rope, like a typical cougar. I would conservatively estimate the weight at 110 - 120 lbs, based on my familiarity with other wild animals. For comparison of scale and weight, my wife had a 168 lb. St Bernard dog and it appeared smaller than that according to her. I didn't see it head on, but my wife did, and vividly recalls its muscular body and its bright yellow eyes.

When it ran in front of the car, the end of its thick, tail flipped above the hood on my car, from the change in the air, or perhaps the animal readjusted its balance to make the final leap.

Although I slowed down the car as we passed it, for a better look behind me, I lost sight of it. Not knowing the safety risk of an encounter, we decided not to stop the car or get out of the car. Frankly it scared us, and I wanted to get away from the area immediately.

I had my fog lights turned on so there was plenty of bright light, providing clear vision.

My wife observed exactly the same thing I did at the same time. She had a better view of the animal's face and eyes than I did as it approached our car. I would be willing to take a polygraph test to confirm I am speaking the truth.

The reason I am sending this message is that I feel compelled to notify the state authorities for the purpose of communicating my experience, so others may be on the lookout for this potentially dangerous and equally rare animal, and to substantiate what it actually is. — Mr. Michael Blouch

Mr Earl, 20 April 2018

Definite Black Cougar. Duchesne River in Utah at the confluence with Wolf Creek. Fishing midstream I look up and saw a very large pitch black cat with a long tail is padding along the river bank, ten yards away. The cat doesn't see me. I then whistle for my friend 15 yards downstream fishing to look up and see it. The cat stops and turns to me, slowly looks me up and down for 5 or 6 seconds. Definitely Cougar, I have seen many, round face, triangle ears. He decides he doesn't want to get wet and pads off upstream. Huge paw prints on bank show he extended claws when I whistled and he turned to me.

Follow up, a week later my friend is at Strawberry Reservoir watching Kokanee run, sees a DWR guy and tells him the story. DWR officer says, "You saw him? I saw him two years ago at the top of Wolf Creek drainage and nobody will believe me."

For what it's worth, he was ten yards away, looked right at me, walked slowly, I know what I saw. I have over 40 years living in the mountains of Utah and Montana.

Unknown, 11 June 2018

I live in Austin Texas and work as a farrier all around the city and have met several people who have or know those who have encountered large black cats in the area around Dale, Lytton springs and west of Lockart in the Mac Mahan area. One client whose neighbor saw a huge black panther cross her driveway at night from her car just feet away around the same time had all her cats killed and decapitated while her mastiff dog went missing for several days only to return severely mauled and with festering shoulder wound that I observed 2 weeks later. It had mostly healed by then but the hair had yet to grow back and the skin was still peeling. There were 4 long distinct claw marks spread 11 inches apart because I measured them. I have large hands and this was twice as wide as I could spread my four fingers. I'm no expert but I think this puts this in the Jaguar size category which jibes with the neighbor's description of the cat being as tall as the hood of her car and at least as long as the width of the car if not the driveway.

CET collective, 9 August 2018

In 1987 my youngest daughter and I were driving in the mountains east of Bountiful Utah and on our way down the mountain a very large jet-black male Puma came down the bank to cross the road about 60 feet in front of my car. He stopped and looked at us. I had stopped too. My daughter, about 8 at the time asked me, "What do we do?"

I said, "Roll up your window."

The big cat studied us for like 7 seconds and then went on his way down the trail on the other side of the road that he had just crossed. I didn't have a camera with me.

I didn't know at the time that black pumas were said to be a myth, but I found out soon enough. It could not have been a leopard or Jaguar, he stood taller and lighter in weight that one of those big cats. Could it have been a hybrid? Possibly. I just know he was startlingly big, had classic Puma lines and was coal black.

Unknown, 3 October 2018

When I was about 12, my father and I saw a huge black cat - the strangest part of this story is that it was in the city of Philadelphia. We were by the Delaware river; it crept out from some bushes next to a building and slowly walked out of sight. I absolutely know what I saw. It was maybe 50 yards away. And then, about 10 years later, I was watching a show about strange animal sightings and low and behold, there was a segment about big black cat sightings in the PA, NY and NJ area. I was dumbfounded. I indeed encountered a large and mysterious black cat in the eastern US.

Unknown, 12 December 2018
I seen a black cougar on the mountain just off Westsyde road towards Karindale road in Kamloops BC about 25 years ago. Told my folks I seen a panther. No one believed me but I tell yah what I seen it as a kid so there's no telling me otherwise.

Iane batot, 19 December 2018
I've been interested in cougar sightings in the East (I live in North Carolina) since childhood, and find that the Big Cat sighting phenomenon has a LOT in common with Bigfoot and giant water critter sightings! Although I think we WILL have a reproducing population of cougars here eventually (especially in the Appalachian Mountain chain, which is IDEAL cougar habitat), as they expand naturally Eastward, I don't think we do YET. As for the whole black cougar/panther notion, I have had the sad (so sad!) experience of seeing so-called professionals I work with (I work in a large zoo) who were called in to look at a very clear video taken of a supposed "black panther" nearby, by a woman who swore it was as big as a Labrador Retriever. I was asked to look at the video, which was a very nice one of an OBVIOUS black housecat (a BIG old wide-cheeked tom, but Felis domesticus, nonetheless!) It also AMAZED me that no one thought to look for TRACKS, as the film was taken during a recent snow. I was sent out to look for tracks, and of course all I found was housecat tracks! A HUGE cage trap was set (my influence and opinion not being valued particularly, and/or just not as interesting!) which eventually caught--you guessed it, a black housecat! But people still denied it and INSISTED they saw a "panther"! The local news crews and law enforcement turned it all into a 3-ring circus--and I learned that yes, many people DO see a black housecat and cry "panther"! Hard as that may be to believe! Even the so-called "experts"! This SAME phenomenon happened a few years later, with this time a tan-colored housecat claimed to be a LIONESS!! (tabby marking clearly visible on its tail!) It was filmed walking across a frozen pond in North Carolina (where virtually no pond ever freezes enough to support a 250+lb. lioness!) by a woman from her deck, and she was TERRIFIED! Another field day from the local zoo officials, press, and law enforcement, although local Wildlife Officials tried to tell them it was just a housecat. I was not asked my opinion at all this time, having been such a spoilsport during the last go-round!.....to be continued...

[Sadly, however, no continuation has been posted by Iane batot so far.]

Unknown, 4 August 2019
Still no update? [re analyses conducted upon the skull and pelt of a South American black mystery cat called the onça-canguçú, mentioned by me in my article] Reading about it here really peeked [piqued] my interest, as it reminds me of the one I saw, as impossible as it may be. Even though authorities claim the Eastern mountain lion is extinct, there are still mountain lions here, I can't count the number of people I know that's seen them. And even though science claims there is no "black Panthers" here, I know of 5 people that have seen them, most recently a neighbor seen a black one along with a litter of kittens. Years ago we had a game cam picture of what appeared to be a very large, black felid. Nothing concrete of course, but the muscly leg and long tail looked more feline than canine to me. Reading the above description of a black puma being white underneath makes me want to ask the neighbor more

specifics about the one he saw. It also made me more confident in what I had seen years ago, for when I had told my friends and family about they told me I must have been mistaken because the black Panthers are solid black and don't have any white on them.

As I said, Impossible as it may seem, this is what happened:
Approximately 20 years ago I was riding to town with my grandmother, there was a long straight stretch, maybe several hundred yards or so, and the house at the end of it had a bunch of pine trees planted along the road. As we approached the yard with the trees described, I saw it walking along the road toward us, between the road and the trees. We had to slow down because of sharp turns ahead so I got a pretty good look at it. It looked big, bulky, muscular, had a pretty large cat head, had a long slender tail, it was solid black, except for under his head, when he looked up at the car his neck/chest was white. As I said later on as I told people they said I must have been mistaken because of the white I saw. So reading this has peaked [piqued] my curiosity even as impossible as it may seem.

Austin, 27 December 2019
Captured a different looking mountain lion on my wildlife cam. The cat doesn't appear to be totally black, but much darker than all the other lions I've captured. Any thoughts on this? Here's the link:
https://www.youtube.com/watch?v=5xqw_mzodSc

[My two replies:
1) Hi Austin, Very interesting video, thanks for linking to it above in your comment! Although not widely realised, the puma (aka mountain lion aka cougar) occurs in two confirmed but very different color forms or morphs - the familiar brownish-red morph and the less familiar grey morph, which can appear darker than the brown one. I suspect that the first puma in your video is a dark grey puma and the second one the more familiar brown one. I saw on YouTube once a video of two puma cubs next to each other, one of which was brown, the other one dark grey, and they looked very different, with the grey one darker than the brown, and yet both forms can occur within a single litter (thereby proving that they are not separate species or subspecies, merely genetically-induced color forms). I'll see if I can find that video, and if I do I'll post a link to it here. All the best, Karl
2) Found it, or one very similar to it, featuring a brown puma cub with a grey puma cub, both still possessing spots (which are lost as the cubs grow older): https://www.youtube.com/watch?v=PntYVa-wst4]

John Todd, 24 April 2020
Biologists are quite mistaken to say that black pumas (*Felis* [now *Puma*] *concolor*) (we've always called them cougars or mountain lions) do not exist. They may be EXTREMELY rare, but at least ONE existed. In 1975 my kid brother Charlie, and his friend, Anthony took my dog for a walk and encountered one. This was in the woods less than 1/2 mile from a subdivision (housing tract) Fairwood West near Renton, King County, Washington State, USA. Not far from Seattle. We frequently heard Cougars from the forested hills not far away. My brother and his friend encountered a black cougar. My dog stood between my brother and his friend and growled and barked. The cougar decided that discretion was the better part of valour

and retreated. I'd never heard of a black cougar and neither had my brother. He and his friend were both 7 at the time and I was 21. He did not say "Gray" or "Silver-Gray", or "Dark Brown" he said "Black". I know he was telling the truth because he was TERRIFIED. There are many cougar sightings in our area because humans are building further and further out into the woods. I've never heard of a cougar attacking a human locally. If you e-mail me I will give you a more detailed account. If you send me a mailing address I will send you maps of the area, photos (at least of my brother at 7 and the dog) and an account from my brother in his words. At the time, I didn't know that there was anything remarkable about a black cougar. Never heard of a gray one. Only in the last couple of years have I had any interest in Cryptozoology and learned they were controversial. Your article said they may have existed once. No doubt a few modern Cougars have inherited some genes from this cat, which though usually unexpressed, rarely produce a throwback. I recently read an article about how many unexpressed genes we have, some even from viruses. Someday, unless someone captures a live example, when the total genome of *Felis concolor* has been sequenced, the matter will be settled. Jon T. Todd (jonttodd@yahoo.com)

[I have emailed John Todd for further details as promised by him above, and am currently awaiting a response.]

Dave, 22 August 2020

About 1985, I saw for several days 2 adolescent cougars, they were jet black. 200 yards, through 10 power binoculars. They were playing together, in the open, in the sun. Location. Los Altos Hills CA.

In Tuolumne County CA, I believe I have seen the same lion twice in 2 years. It was dark brown, like Schreber's drawing above [the illustration of *Felis discolor*, see Chapter 6]. Both sightings about a mile apart. Last sighting 8/19/20 Location Lyons lake road/trail from Confidence, the old rail bed. I biked up behind the lion both times, about 200 feet away.

Say what you want, this is what I saw.

Reiterating what I suggested in Chapter 6, the above reports, claiming time and again that what were seen were pumas but wholly coal-black or jet-black in color, make me wonder whether the three so-called black puma specimens documented elsewhere in this book (the Brazilian example shot by William Thomson, Chapter 5; the Costa Rican example shot by Miguel Ruiz Herrero, Chapter 6; and London Zoo's 19th-century live example, Appendix 3), all of which had pale underparts, were nothing more than exceptionally dark examples of the puma's grey morph, rather than actual melanistic specimens.

After all, normal leopards and jaguars have pale underparts too but their melanistic versions are black all over. Surely, therefore, a genuine melanistic puma should be the same. So perhaps North America's mystery "black-all-over" panthers really are genuine melanistic pumas—but if that is true, why has not a single specimen ever been brought to scientific attention? For me, this remains the single greatest mystery of all concerning such felids. Also, we shouldn't forget that identical creatures are also being reported in regions of the world (Britain, continental Europe, Australia) where pumas are not even native, and where the occurrence of any escapee/released melanistic pumas is wholly unrealistic, as no such cats are known to exist in captivity to begin with.

Clearly, there is still much to make sense of when seeking an explanation for reports of alleged black pumas.

Appendix 3
A Black Puma at the London Zoo?

About 23 years ago, I purchased an extremely interesting hand-colored copper engraving from 1862 that I had never seen before and whose subject appears to have been a bona fide black puma that was exhibited alive for a while at the London Zoo. I later discovered a slightly different version of this fascinating engraving online, and I subsequently documented both of these remarkable, highly-unexpected illustrations and their potential cryptozoological significance in *Cats of Magic, Mythology, and Mystery* (2012). Here is what I wrote, together with the two engravings:

As far as I was aware...no such animal [black puma] had ever been kept in captivity, at least not in Europe. But all that changed a while ago during one of my numerous visits to one of my all-time favourite places—Hay On Wye, Herefordshire's world-famous 'Town of Books', nestling on the Welsh border.

In addition to around 30 bookshops at present, this small town also has shops devoted to antiquarian prints. As an avid collector of such items, I was browsing in one of these shops one sunny Saturday afternoon during the late 1990s when I came upon a truly remarkable example—remarkable because it is not often that an antiquarian print depicts a cryptozoological cat!

The print in question, which was an original hand-colored copper engraving dating from 1862 (as written in pencil on its reverse), and which I naturally lost no time in purchasing, is duly reproduced here (its previous appearance, in an article of mine published by the now defunct British monthly magazine *Beyond*, where it was reproduced in its original full-color format, may well have been the first time that it had ever been published anywhere), and appears to portray a bona fide black puma.

My black puma engraving (Karl Shuker)

Certainly, it comes complete with jet-black upperparts, slaty-grey underparts, and white chest—very different from normal pumas, which are either tawny brown-rufous or silver-grey (the puma exhibits two distinct color morphs), but matching precisely those few confirmed black puma specimens. Most interesting of all, however, is the engraving's caption: "The Puma. In the Gardens of the Zoological Society". This means that if the puma in the engraving has been colored accurately, and there is no reason why it should not have been, a black puma, that most mysterious of mystery cats, was once actually on display at London Zoo!

When I first discovered this engraving, I wondered whether its astonishing black puma was exhibited at London Zoo at the same time as the zoo's unique captive woolly cheetah [see Chapter 4], bearing in mind that the engraving was dated 1862. Who knows, if so, it may even have been in the enclosure next door!

In February 2011, however, I discovered a second version of the same puma engraving, but this one was dated 1825. Moreover, the hand-coloring on this latter version, reproduced here, is much more skilful.

So which (if either) is the correct date for it? The mystery deepens, and darkens—which is very apt for anything featuring a black puma at its core!

A second version of my black puma engraving (public domain)

GLOSSARY

ABUNDISM: Abnormal multiplication of body markings.

AGOUTI GENE: Gene responsible, with grey or golden fur, for each hair of such fur being banded with yellow and black stripes.

ALBINISM: Abnormally pale colouration, resulting from absence or limited distribution of eumelanin and/or other melanin pigments; animals so affected are termed albinos. Adj. = albinistic.

ALLELE: One of a pair or series of alternative forms exhibited by a given gene.

ALLOPATHIC SPECIATION: When populations of one original species become geographically isolated from one another, with each population eventually evolving into a new, separate species (cf. **SYMPATRIC SPECIATION**).

BIFURCATE: Having two prongs, forked, or branching off into two separate parts.

BIG CAT: In scientific sense, term denoting any cat species with a hyoid apparatus not composed wholly of bone, thereby enabling the cat to roar, but not to purr continuously, e.g. lion, tiger, leopard, jaguar.

BIPEDAL: Walking on the hind limbs only, as with humans, birds, bears on occasion, etc.

B.P.: Before the present day.

CANID: Any member of the taxonomic family Canidae, e.g. dogs, wolves, foxes, jackals, bush-dog.

CANINE: (1) Adjective, applied to anything dog-like in appearance, behavior, etc; (2) Noun, referring to the single upper and lower pair of fang-like teeth possessed by many mammals, and located just behind the incisors; used for seizing and tearing prey, and highly developed in most (although not all) carnivorous mammals.

CARNASSIALS: Premolars and molars bearing sharp blade-like edges for slicing meat.

CHINCHILLA: A phenotype-describing term alluding to the characteristic pale colouration of certain mutant forms, such as the Rewa white tigers and the Timbavati white lions, resulting from absence of phaeomelanin, but not necessarily indicating genetic homology between such mutants; sometimes loosely referred to as partial albinism.

CROSS-BREEDING: Mating between animals of different species, subspecies, or breeds; also known as hybridization.

CRYPTIC MARKINGS: Dark pelage markings highly visible in normal individuals but largely obscured in melanistic individuals by their pelage's abnormally dark background colouration.

CRYPTID: Noun applied to any animal whose existence or identity has yet to be formally ascertained, e.g. mystery cats, lake monsters, bigfoot, etc.

CRYPTOZOOLDGY: The scientific investigation of animals whose existence or identity has yet to be formally ascertained, e.g. this book's mystery cats, Loch Ness monster and other lake monsters, sea serpents, yeti, bigfoot, Congolese mokele-mbembe, etc.

CURSORIAL: Mode of existence based around fleet-footed activity, and characteristic of long-limbed animals, e.g. cheetah, wolf, antelopes.

DASYURE: Name assigned to six species of Australian carnivorous, weasel-like marsupial belonging to the genus *Dasyurus*.

DENSE PIGMENTATION GENE: Gene responsible for inducing a mammal's normal degree of coat color density.

DILUTE: In genetic sense, the Dense Pigmentation gene's recessive mutant allele partly responsible for blue-colored fur in mammals.

DIMORPHISM: The occurrence of two distinct morphs within a species.

DISTAL: The part of any structure furthest away from its point of attachment (cf. **PROXIMAL**).

DOMINANT: In genetic sense, term applied to any allele of a given gene that can prevent the action of another allele of that same gene.

DORSAL: Appertaining to the upper side or upper surface of an animal (cf. **VENTRAL**).

ERYTHRISM: Abnormally red colouration, resulting from excess of red-brown pigment phaeomelanin and/or chestnut-red pigment erythromelanin. Adj. = erythristic.

EUMELANIN: Blackish-brown pigment responsible for black colouration in mammals; often referred to non-specifically as melanin.

EUPLERID: Any member of the taxonomic family Eupleridae, e.g. Madagascan mongooses, fossa, fanaloka, falanoucs.

EUTHERIAN: Any modern-day mammal other than monotremes and marsupials; characterized by development of placenta from embryo's waste-products sac (allantois), and a relatively long gestation period (during which the young can attain an advanced degree of development prior to birth.

EXOTIC: Noun or adjective, loosely applied to any animal currently existing in captivity and/or in the wild state within an area not contained in its normal, native range of distribution.

EXTENSION OF BLACK GENE: Gene responsible for normal distribution of eumelanin present in mammals' fur.

FELID: Noun applied to any member of the taxonomic family Felidae, consisting of the cats, e.g. lion, tiger, lynx, puma, domestic cat, cheetah.

FELINE: Adjective applied to anything cat-like in appearance, mode of movement, etc.

FERAL: Appertaining to animals that have lapsed from a domesticated into a wild condition (cf. **NATURALIZED**).

FULL COLOR GENE: Gene responsible for inducing a mammal's normal complement of pelage pigment.

GRACILITY: The possession of a notably slender, graceful body and long, lithe limbs.

HETEROZYGOUS: Possessing two different alleles of a given gene.

HOMOZYGOUS: Possessing a pair of the same allele of a given gene.

HYBRID: The offspring of a mating between organisms of two different species, subspecies, or breeds. Often designated by the names of the two parent forms separated by an "x," e.g. wildcat x domestic cat = hybrid of a mating between a wildcat and a domestic cat.

HYOID APPARATUS: Bony structure in throat region, whose composition influences the sounds made by cat species.

INTERGENERIC: Appertaining to comparisons, interactions, etc, between individuals belonging to different genera.

INTERSPECIFIC: Appertaining to comparisons, interactions, etc, between individuals belonging to different species.

MACHAIRODONTIDS: Saber-toothed cats—a taxonomic (sub) family of felids characterized by extra-large upper canines and undersized lower canines; extinct since the Pleistocene, according to the current fossil record.

MANDIBLE: The lower jaw in cats and other mammals.

MARSUPIAL: Any member of the mammalian taxon Marsupialia; most marsupial species are characterized by the possession of a pouch (and sometimes by a placenta derived from the embryo's yolksac), and by a notably short gestation period that results in the young being born in a very immature state, after which they make their way to the mother's pouch in which they then continue their development and suckle.

MELANISM: Abnormally dark background colouration, as exemplified by that of the fur of the black panther, black jaguar, and other cat forms displaying this condition (cf. **PSEUDO-MELANISM**); caused by excess of eumelanin. Adj. = melanistic or melanic.

METATHERIAN: Alternative name for marsupial.

MONOTREME: Any modern-day egg-laying mammal, i.e. platypus and echidnas.

MORPH: One of two or more physically different (but not taxonomically different) forms of individual exhibited by a given species, e.g. the black panther is a melanistic morph of the leopard *Panthera pardus.*

MORPHOLOGY: The physical appearance and structure of an animal.

MORPHOMETRIC: Appertaining to analyses involving detailed measurements derived from an organism's morphology.

MOZAICISM: Abnormal condition in which aberrant patches of pigment are distributed over an animal's body surface.

MUSTELID: Any member of the taxonomic family Mustelidae, e.g. weasels, martens, otters, badgers, wolverine.

MUTANT: An individual exhibiting heritable characteristics that have arisen spontaneously (as opposed to gradually) and differ from corresponding characteristics possessed by normal individuals of the species concerned.

MUTATION: A change in the physical characteristics of an organism that results from a change in the structure or amount of that organism's genetic material (DNA).

NASAL: Appertaining to the nose or nostrils.

NATURALIZED: Appertaining to any wild (but not domesticated) animal species that by escaping from captivity and/or by deliberate introduction has become established in an area to which it is not normally native (cf. **FERAL**).

NIGRISM: Abnormal fusion of body markings.

NOMEN NUDUM: A published scientific name for a given species or subspecies that for various reasons is no longer accepted by science as legitimate, and therefore should not be used, e.g. *Smilodon neogaeus.*

NON-AGOUTI: In genetic sense, the Agouti gene's recessive mutant allele responsible for melanism in many mammals.

PELAGE: The furry coat of mammals.

PERS. COMM.: Abbreviation of personal communication—term denoting that the information preceding it in this book's text is material obtained via personal communication with a correspondent, rather than from some published source or from personal research.

PHAEOMELANIN: Reddish-brown melanin pigment present in the pelage of many mammals.

PLANTAR PAD: The pad present on the palm of a mammal's paw.

PLEIOTROPIC: Appertaining to any gene responsible for a range of quite different effects upon an organism.

POLYMORPHISM: The occurrence of more than two distinct morphs within a given species.

PRIMARY GUARD HAIRS: Very long, coarse, stiff hairs (longer than normal aka awn guard hairs) scattered sparingly throughout a mammal's pelage, and sometimes differing markedly in color from the predominant pelage colouration present.

PROXIMAL: The part of any structure closest to its point of attachment (cf. **DISTAL**).

PSEUDO-MELANISM: Condition mimicking melanism to some degree, but in which abnormally dark colouration results from abnormal fusion and multiplication of body markings (cf. **MELANISM**).

PUG MARK: The impression left behind by any pad on a mammal's paw that makes contact with the ground.

RACE: Loose term, popularly used to designate local populations within a subspecies.

RECESSIVE: Appertaining to any allele whose action can be prevented by that of another allele of the same gene.

RETRACTILE: Capable of being drawn back or withdrawn (actively and/or passively), as with the claws of most cat species.

SMALL CAT: In scientific sense, term denoting any cat species with a hyoid apparatus composed wholly of bone, thereby enabling it to purr continuously, but not to roar, e.g. puma, lynxes, domestic cat, wildcat, ocelot, jaguarundi.

SPOOR: Track or trail constituting an animal's paw prints.

SUBSPECIES: A collection of morphologically similar populations of a species, inhabiting a geographical subdivision of that species' total distribution range, and differing taxonomically from other collections of populations of that species.

SYMPATRIC SPECIATION: When populations of one original species evolve into new, separate species without first having been isolated geographically from one another (cf. **ALLOPATRIC SPECIATION**).

TAXONOMIC: Anything appertaining to the scientific classification of organisms; any characteristics of, or differences between, organisms that are considered to have bearing upon those organisms' evolutionary relationships to one another.

TAXONOMY: The science of classification, as applied to organisms.

THYLACOLEONID: Any member of the taxonomic family Thylacoleonidae, consisting of the supposedly extinct marsupial lions of Australia.

VENTRAL: Appertaining to the underside or lower surface of an animal (cf. **DORSAL**).

VIBRISSAE: A mammal's whiskers (often touch-sensitive).

VIVERRID: Any member of the taxonomic family Viverridae, e.g. civets, genets, binturong.

SELECTED BIBLIOGRAPHY

To preserve this book's historical nature and worth, the great majority of this bibliography consists of the references that appeared in this book's original 1989 edition:

SHUKER, K.P.N. *Mystery Cats of the World: From Blue Tigers to Exmoor Beasts* (Robert Hale: London, 1989).

However, I have also included a representative selection of the numerous more recent, essential references utilized in preparing this 2020 edition's updated sections of main text.

For ease of direct cross-referencing between main text and bibliography, all references for each chapter (and introduction) are listed in this bibliography in the order that they are first referenced within that chapter's (and introduction's) main text.

Abbreviations: *J.B.N.H.S.* = *Journal of the Bombay Natural History Society*; *P.Z.S.L.* = *Proceedings of the Zoological Society of London*

Introduction

DENIS, A. *Cats of the World* (Constable: London, 1964).
BOORER, M. *Wild Cats* (Hamlyn: London, 1969).
GUGGISBERG, C.A.W. *Wild Cats of the World* (David & Charles: Newton Abbot, 1975).
GRZIMEK, B. (Ed.) *Grzimek's Animal Life Encyclopedia, Vol 12* (Van Nostrand Reinhold: London, 1975).
RICCIUTI, E. *The Wild Cats* (Windward: Leicester, 1979).
MACDONALD, D. (Ed.) *The Encyclopaedia of Mammals, Vol 1* (Allen & Unwin: London, 1984).
KITCHENER, A. *The Natural History of the Wild Cats* (Christopher Helm: London, 1991).
CORBET, G.B. & HILL, J.E. *A World List of Mammalian Species* (3rd Edit.) (Natural History Museum Publications/Oxford University Press: London, 1991).
ALDERTON, D. *Wild Cats of the World* (Blandford: London, 1993).
SUNQUIST, M. & SUNQUIST, F. *Wild Cats of the World* (University of Chicago Press: Chicago, 2002).
HUNTER, L. & BARRETT, P.A. *Field Guide to the Carnivores of the World* (New Holland: London, 2011).
WALKER, W.F. *A Study of the Cat* (W.B. Saunders: London, 1967).
SAVAGE, R.J.G. Evolution in carnivorous mammals. *Palaeontology* Vol 20, pp 237-271 (1977).
LEYHAUSEN, P. *Cat Behavior* (Garland: New York, 1979).
LAWRENCE, M.J. & BROWN, R.W. *Mammals of Britain: Their Tracks, Trails and Signs* (Revised Edit.) (Blandford: London, 1973).
BANG, P. & DAHLSTROM, P. *Collins Guide to Animal Tracks and Signs* (Collins: London, 1974).
NELSON, E.W. *Wild Animals of North America* (Revised Edit.) (National Geographic Society: Washington, 1930).
SETON, E.T. *Animal Tracks and Hunter Signs* (Edmund Ward: London, 1959).

SMITHERS, R.H.N. *Land Mammals of Southern Africa* (Macmillan: Johannesburg, 1986).
TABOR, R. *The Wildlife of the Domestic Cat* (Arrow Books: London, 1983).
GONYEA, W. & ASHWORTH, R. The form and function of retractile claws in the Felidae and other representative carnivorans. *Journal of Morphology* Vol 145, pp 229-238 (1975).
ROMER, A.S. *Vertebrate Paleontology* (3rd Edit.) (University of Chicago Press: Chicago, 1966).
STEEL, R. & HARVEY, A.P. (Eds) *The Encyclopaedia of Prehistoric Life* (Mitchell Beazley: London, 1979).
SAVAGE, R.J.G. & LONG, M.R. *Mammal Evolution: An Illustrated Guide* (British Museum (Natural History): London, 1986).
BRYANT, H.N. & CHURCHER, C.S. All sabretoothed carnivores aren't sharks. *Nature* Vol 325, p 488 (Feb 5 1987).
MACDONALD, D. *The Velvet Claw: A Natural History of the Carnivores* (BBC Books: London, 1992).
TURNER, A. *The Big Cats and Their Fossil Relatives* (Columbia University Press: New York, 1997).
PROTHERO, D.R. *The Princeton Field Guide to Prehistoric Mammals* (Princeton University Press: Princeton, 2017).
KURTEN, B. *Pleistocene Mammals of Europe* (Weidenfeld & Nicolson: London, 1968).
SUTCLIFFE, A. *On the Track of Ice Age Mammals* (British Museum (Natural History): London, 1985).
AGUSTÍ, J. & ANTÓN, M. *Mammoths, Sabertooths, and Hominids: 65 Million Years of Mammalian Evolution in Europe* (Columbia University Press: New York, 2002).
KURTEN, B. & ANDERSON, E. *Pleistocene Mammals of North America* (Columbia University Press: New York, 1980).
GRAYSON, D.K. *Giant Sloths and Sabertooth Cats: Extinct Mammals and the Archaeology of the Ice Age Great Basin* (University of Utah Press: Salt Lake City, 2016).
MAYR, E. *Populations, Species, and Evolution* (Belknap Press: Massachusetts, 1970).
MAYR, E. *Principles of Systematic Zoology* (Tata McGraw-Hill: New Delhi, 1971).
ROBINSON, R. *Genetics For Cat Breeders* (Pergamon: Oxford, 1971).
ROBINSON, R. Homologous genetic variation in the Felidae. *Genetica* Vol 46, pp 1-31 (1976).
ROBINSON, R. Homologous coat color variation in *Felis*. *Carnivore* Vol 1, pp 68-71 (Jan 1978).
ROBINSON, R. The breeding of spotted and black leopards. *Journal of the Bombay Natural History Society* Vol 66, pp 423-429 (1969).
O'GRADY, R.J.P. Melanism in breeding wild cats. *Proceedings of the 4th Symposium of the Association of British Wild Animal Keepers* Vol 4, pp 32-41 (1979).
VELLA, C.M. *et al. Robinson's Genetics for Cat Breeders and Veterinarians* (4th Edit.) (Butterworth-Heinemann: Oxford, 1999).

Chapter 1

KURTEN, B. *Pleistocene Mammals of Europe* (Weidenfeld & Nicolson: London,

1968).
SUTCLIFFE, A. *On the Track of Ice Age Mammals* (British Museum (Natural History): London, 1985).
STUART, A.J. *Pleistocene Vertebrates in the British Isles* (Longman: Essex, 1982).
NAISH, D. Europe, where the sabre-tooths, lions and leopards are. *Tetrapod Zoology* https://scienceblogs.com/tetrapodzoology/2008/03/12/european-cats-part-I Mar 12 2008.
NAISH, D. Pumas of South Africa, cheetahs of France, jaguars of England. *Tetrapod Zoology* https://scienceblogs.com/tetrapodzoology/2008/03/13/european-cats-part-ii Mar 13 2008.
NAISH, D. Britain's lost lynxes and wildcats. *Tetrapod Zoology* https://scienceblogs.com/tetrapodzoology/2008/04/21/european-cats-part-iii Apr 21 2008.
BISHOP, M.J. The mammal fauna of the Early Middle Pleistocene Cavern infill site of Westbury-Sub-Mendip, Somerset. *Special Papers in Palaeontology* No 28, pp 1-108 (1984).
GREEN, H.S. Pontnewydd Cave. A lower Paleolithic hominid site in Wales. The first report. *National Museum of Wales Quaternary Studies Monographs* No 1, pp 171-180 (1984).
CLAIR, C. *Unnatural History* (Abelard-Schuman: London, 1967).
DENT, A. *Lost Beasts of Britain* (Harrap: London, 1974).
GOSS, M. Phantom felines: Parts 1-5, *The Unknown* No 13, pp 5-10 (Jul 1986); No 14, pp 16-21 (Aug 1986); No 15, pp 70-75 (Sept 1986); No 16, pp 62-67 (Oct 1986); No 17, pp 11-17 (Nov 1986).
ANON. The wild dog of Ennerdale. *Chambers's Journal* (Ser 6) Vol 7, pp 470-472 (1904).
FORT, C. *The Books of Charles Fort* (Henry Holt: New York, 1941).
ROBERTS, A. *Cat Flaps!* (Brigantia Books: Brighouse, 1986).
COBBETT, W. *Rural Rides, Vol 1* (William Cobbett [1st Edit. self-published]: London, 1830).
ANON. Mystery "tiger of the north". *Daily Express* (London) Jan 14 1927.
BLAKE, M. *et al.* Multidisciplinary investigation of a 'British big cat': a lynx killed in southern England c. 1903. *Historical Biology* Vol 26(4), pp 441-448 (2014) [online Apr 23 2013].
NAISH, D. A lynx, shot dead in England in c. 1903. *Tetrapod Zoology* https://blogs.scientificamerican.com/tetrapod-zoology/edwardian-lynx-from-england/ Apr 24 2013.
ROBERTS, I. "Ghoulies and ghosties". *The Field* Vol 171, p 677 (Mar 19 1938); plus sequence of responses by other authors. *The Field* Vol 171 (Apr 16, 23, May 7, 28 1938).
BORD, J. & BORD, C. *Alien Animals* (Revised Edit.) (Panther: London, 1985).
McEWAN, G. *Mystery Animals of Britain and Ireland* (Robert Hale: London, 1986).
ANON. Safari in Shooters Hill. *Kentish Mercury* (Greenwich) Jul 19 1963.
ANON. The cheetah hunt goes on – with snarls at dawn. *Kentish Mercury* (Greenwich) Jul 26 1963.
RICKARD, R.J.M. If you go down to the woods today... *INFO Journal* No 13, pp 3-18 (May 1974).
CLARK, E. Surrey hunts its mystery puma. *The Observer* (London) Oct 11 1964.

HEAD, V. Trailing the Surrey puma. *The Field* Vol 225, p 627 (Apr 8 1965).
BURTON, M. Is this the Surrey puma? *Animals* Vol 9, pp 458-461 (Dec 1966).
MIDDLETON, C. Claws. On the trail of the big cats. *Radio Times* Vol 249, pp 100-101 (May 17 1986).
ANON. Puma is seen stalking a rabbit. *The Times* (London) Jul 5 1966.
ROGERS, E.J. One or two pumas. *The Field* Vol 225, p 770 (Apr 29 1965).
ANON. Pumas proof claimed. *Western Morning Times* (Plymouth) Jan 14 1985.
RICKARD, R.J.M. The Surrey puma & friends: More mystery animals. *The News* No 14, pp 3-8, 17, 24 (Jan 1976).
ANON. [Reports regarding Nottingham lioness.] *Nottingham Evening Post* (Nottingham) Jul 29, 30, 31, Aug 2, 3, 4, 5, 6 1976.
ANON. The Nottingham lion saga. *Fortean Times* No 18, pp 25-26 (1976).
ANON. Scottish mystery cat updates. *Fortean Times* No 20, pp 18-19 (Feb 1977).
RICKARD, R.J.M. The Scottish 'lioness'. *Fortean Times* No 26, pp 43-44 (summer 1978).
ANON. [Reports regarding Wolverhampton puma.] *Express & Star* (Wolverhampton) Jul 23, 24, 25, 26, 28, 31 1980.
RICKARD, R.J.M. Wolverhampton wonders. *Fortean Times* No 34, pp 21-22 (winter 1980).
RICKARD, R.J.M. Scottish puma – Saga or farce? *Fortean Times* No 34, pp 24-25, 36 (winter 1980).
DOW, A. & AIRS, G. Trapped! *Daily Record* (Glasgow) Oct 30 1980.
DOW, A., AIRS, G. & MOWA, B. Experts fall out over puma. *Daily Record* (Glasgow) Oct 31 1980.
ANON. Expert's hoax claim shattered. *Daily Record* (Glasgow) Nov 6 1980.
DOW, A. Vet's verdict on the big cat. *Daily Record* (Glasgow) Nov 11 1980.
MOWAT, B. [Report re second Cannich puma.] *Daily Record* (Glasgow) Nov 10 1980.
ANON. [Death of Felicity.] *The Scotsman* (Edinburgh) Feb 7 1985.
McGOVERN, B. The prowling menace at Janet's door. *Sunday Express* (London) Jan 27 1985.
BORD, J. & BORD, C. Strange creatures in Powys. *Fortean Times* No 34, pp 18-20 (winter 1980).
ANON. Island Monster shot. *Daily Mirror* (London) Feb 17 1940.
ANON. The Island "Monster" a fox. *Isle of Wight County Press* (Newport) Feb 24 1940).
ANON. (1983). Big cat riddle. *The News* (Portsmouth) Aug 20 1983.
ANON. (1983). 'Big cats' sightings puzzle. *The News* (Portsmouth) Sept 12 1983.
COPPELL, D. Gruesome Island find sparks hunt. *The News* (Portsmouth) Sept 15 1983.
ANON. 'Puma' mystery deepens. *The News* (Portsmouth) Oct 20 1983.
ANON. (1984). Cat scare near dump. *The News* (Portsmouth) Mar 15 1984.
ELSDON-DEW, M. Island riddle of a giant cat. *Sunday Express* (London) Apr 29 1984.
ANON. Out of place. *Fortean Times* No 42, pp 40-42 (autumn 1984).
ANON. 'Paws' lives...alive and kicking in the island. *The News* (Portsmouth) Dec 20 1984.
ANON. Kill sparks new 'big cat' alert. *The News* (Portsmouth) Jun 17 1985.

ANON. Island 'big cat' on prowl again. *The News* (Portsmouth) Jul 22 1985.
RICKARD, R.J.M. Once more with felines. *Fortean Times* No 44, pp 28-31 (summer 1985).
ANON. Spotty cat story turns out to have an old twist in tail. *Isle of Wight County Press* (Newport), Jan 14 1994.
RICKARD, R.J.M. The Exmoor Beast and others. *Fortean Times* No 40, pp 52-61 (summer 1983).
BEER, T. *The Beast of Exmoor: Fact or Legend?* (Countryside Productions: Barnstaple, 1984).
BRIELY, N. *They Stalk By Night: Big Cats of Exmoor and the South-West* (Yeo Valley Productions: Bishop Nympton, 1989).
RICKARD, R.J.M. And now...the Dartmoor panther. *Fortean Times* No 35, p 45 (summer 1981).
HOWE, R. The black beast strikes once more. *Western Morning News* (Plymouth) Apr 22 1983.
ANON. Marine sights cat-like beast. *Western Morning News* (Plymouth) May 5 1983.
HART-DAVIS, D. On the trail of the Beast of Exmoor. *Sunday Telegraph* (London) Jul 10 1983.
ANON. £1,000 reward. *Daily Express* (London) May 7 1983.
ANON. Back on the trail of the Beast. *Western Morning News* (Plymouth) May 19 1983.
ANON. 'Cunning canine' hunt. *Daily Express* (London) Jun 7 1983.
ANON. A beastly encounter. *Western Morning News* (Plymouth) May 31 1983.
CHARLESTON, M. Marines end hunt for Beast of Exmoor. *Daily Express* (London) Jul 9 1983.
COATES, J. & TAFT, A. The Exmoor Beast is dead, but... *Sunday Times* (London) Jul 31 1983.
SETON, C. Sheep-killing Beast of Exmoor still at large. *The Times* (London) Aug 1 1983.
ANON. New victim. *Western Morning News* (Plymouth) Aug 16 1983.
FLETCHER, K. Making of a monster myth. *Sunday Times* (London) Jul 10 1983.
ANON. Exmoor Beast killed claim. *Western Morning News* (Plymouth) Sept 17 1983.
HENDERSON, D. Now, the beasts of Exmoor. *North Devon Journal-Herald* (Barnstaple) Nov 1 1984.
ANON. Return of the Beast: new signs. *North Devon Journal-Herald* (Barnstaple) Oct 4 1984.
BEER, T. Personal communications (1987-1988).
SIMPSON, D. 'More than one' Exmoor Beast. *Western Morning News* (Plymouth) Nov 13 1985.
ANON. Moor beasts. *The Mail on Sunday* (London) Dec 1 1985.
METCALFE, T. Wolverine theory is dismissed. *Western Morning News* (Plymouth) Dec 10 1985.
SHUKER, K.P.N., Who's afraid of the big bad wolverine? *Fortean Times* No 85, pp 36-37 (Feb-Mar 1996).
CLOUGH, M. Schoolboy sees The Beast. *Western Morning News* (Plymouth) Jan 17 1986.

ANON. Beast is a lynx. *The Sun* (London) Nov 6 1986.
TURK, F.A. Nature scene in Cornwall. *West-Briton* (Truro) Dec 31 1986.
JEWELL, L. (Producer) *The Natural History Programme* (B.B.C. Radio Bristol radio programme: Bristol, first broadcast Jan 22 1987).
STADDON, G. Dabs of the Beast. *North Devon Gazette* (Barnstaple) Jan 30 1987.
ANON. 'Beast' alert. *Daily Telegraph* (London) May 5 1987.
ANON. Exmoor searchers see big black cat. *BBC Wildlife* Vol 5, p 610 (Nov 1987).
RICHEY, C. This is the Exmoor Beast says mystery 'Jungle Man'. *Mid Devon Star* (Bideford) Jan 1 1988.
SHUKER, K.P.N. Feline clues on the moors. *Fortean Times* No 52, pp 26-27 (summer 1989).
ANON. Farmer loses 30 lambs to a beast. *Western Morning News* (Plymouth) Jan 23 1988.
STADDON, G. Beast is wrongly accused. *North Devon Gazette* (Barnstaple) Feb 5 1988.
ANON. 'Beast' kills foal. *Express & Star* (Wolverhampton) Jan 30 1988.
ANON. Beast of the moor. *Sunday Independent* (Plymouth) Jan 31 1988.
SAGE, A. Youths find mystery cat's skull. *Western Morning News* (Plymouth) Jan 31 1988.
REYNOLDS, D. Killer beast of moor could be leopard – Expert's opinion. *Sunday Independent* (Plymouth) Feb 21 1988.
ANON. Exmoor beast mystery lives on. *The Times* (London) Apr 22 1988.
ANON. Mystery as rare cat shot on moors. *Express & Star* (Wolverhampton) Apr 22 1988.
SHUKER, K.P.N. Exotic moorland cats are former captives. *Western Morning News* (Plymouth) May 3 1988.
PORTER, M. Scots Leopard [Cat] Shot'. *Sunday Telegraph* (London), Mar 6 1988.
ANON. *The "Beast" Strikes Again* [booklet in the 'Legends of Exmoor' series] (Exmoor National Park: Dulverton, 1988).
STADDON, G. Expert arrives to hunt Beast. *North Devon Gazette* (Barnstaple) Sept 2 1993.
STADDON, G. [Re Dr Karl Shuker's Beast-seeking Exmoor visit.] *Sidmouth Herald* (Sidmouth) Sept 4 1993.
BAKER, S.J. & WILSON, C.J. *The Evidence For the Presence of Large Exotic Cats in the Bodmin Area and Their Possible Impact on Livestock* (MAFF Publications: London, 1995).
SHUKER, K.P.N. The Kellas cat: reviewing an enigma. *Cryptozoology* Vol 9, pp 26-40 (1990).
BOWERS, A. Kellas cats, scotching the myth. *The Scottish Big Cat Trust* http://scotcats.online.fr/abc/identification/kellascataron.html (n.d. – first accessed Jan 29 2004).
KERRIDGE, R. The big cat cover-up. *Sunday Express* (London), Dec 15 1996.
CAMPBELL, John G. *Superstitions of the Highlands and Islands of Scotland* (J. MacLehose: Glasgow, 1900).
SHUKER, K.P.N. *Cats of Magic, Mythology, and Mystery: A Feline Phantasmagoria* (CFZ Press: Bideford, 2012).
TABOR, R. *The Wildlife of the Domestic Cat* (Arrow Books: London, 1983).
WOOD, G.L. *The Guinness Book of Pet Records* (Guinness Superlatives: Middlesex,

1984).
TOMKIES, M. *My Wilderness Wildcats* (Macdonald & Jane's: London, 1977).
CORBET, G.B. & SOUTHERN, H.N. (Eds) *The Handbook of British Mammals* (2nd Edit.) (Blackwell: London, 1977).
FRENCH, D.D. *et al.* Morphological discriminants of Scottish wildcats (*Felis silvestris*), domestic cats (*F. catus*) and their hybrids. *Journal of Zoology* Vol 214, pp 235-259 (1988).
ROBINSON, R. Hybrids of wild and domestic cats. *Carnivore Genetics Newsletter* Vol 2, pp 93-94 (1972).
DENIS, A. *Cats of the World* (Constable: London, 1964).
BOURNE, H.L. *Living on Exmoor* (Galley Press: London, 1963).
ROBINSON, H.W. Wild cats on Westmorland-Lancashire border. *The Naturalist* No 849, p 292 (1927).
FREETHY, R. *Man and Beast: The Natural and Unnatural History of British Mammals* (Blandford: Dorset, 1983).
GUGGISBERG, C.A.W. *Wild Cats of the World* (David & Charles: Newton Abbot, 1975).
WEIGEL, I. Other small cats. In: GRZIMEK, B. (Ed.) *Grzimek's Animal Life Encyclopedia, Vol 12* (Van Nostrand Reinhold: London, 1975), pp 312-332.
RICCIUTI, E.R. *The Wild Cats* (Windward: Leicester, 1979).
BAKER, S.J. Escaped exotic mammals in Britain. *Mammal Revue* Vol 20, pp 75-96 (1990).
YOUNG, S.P. & GOLDMAN, E.A. *The Puma: Mysterious American Cat* (American Wildlife Institute: Washington, 1946).
BJÄRVALL, A. & ULLSTRÖM, S. *The Mammals of Britain and Europe* (Croom Helm: London, 1986).
ELGMORK, K. Et uvanlig gaupespor og lit om gaupe i Vassfartraktene 1960-1969. *Fauna* Vol 25, pp 258-264 (1972).
SMITHERS, R.H.N. *Land Mammals of Southern Africa* (Macmillan: Johannesburg, 1986).
LEVER, C. *The Naturalized Animals of Britain and Ireland* (New Holland: London, 2009).
LEVER, C. *Naturalized Mammals of the World* (Longmans: London, 1985).
LEVER, C. *Naturalized Birds of the World* (Longmans: London, 1987).
RICKARD, R.J.M. The end of a caper. *Fortean Times* No 20, p 20 (Feb 1977).
ANON. Escaped rare cat shot dead. *Kentish Gazette* (Canterbury) Apr 2 1976.
ANON. Trail of death and destruction at 120 mph. *Daily Mail* (London) Oct 17, 30 (1987).
MOORE, R. Affluent cats. *Cryptozoology* Vol 5, p 146 (1986).
STEPHENSON, C. A puma hunt in Surrey. *Wide World Magazine* Vol 11, pp 511-515 (1903).
ANON. Leopard hunt. *Blackburn Standard* (Blackburn), Jun 8 1836.
OSWALD, A. Chillington Hall, Staffordshire – II. *Country Life* Vol 103, pp 378-381 (Feb 20 1948).
ANON. That's no lion – it's just an old softie called Finn. *Daily Mail* (London) Dec 21 1988.
ANON. Farmer shot dog. *Mid Devon Star* (Bideford) Jan 1 1988.
POWELL, J. Moor sheep face killer dogs peril. *Western Morning News* (Plymouth)

Jan 5 1988.
ANON. Sheep farmer plans vigil. *Western Morning News* (Plymouth) Jun 1 1988.
INNES, S. Beware of the dog. *The Unexplained* No 86, pp 1710-1713 (1983).
INNES, S. Tracking down black dogs. *The Unexplained* No 87, pp 1730-1733 (1983).
ANON. Cat seven feet in length. *Reynolds's Illustrated Newspaper* (London) Jul 7 1929.
RICKARD, R.J.M. Hyena in Sussex. *Fortean Times* No 19, pp 9-10 (Dec 1976).
DOYLE, P. Paws for thought on cat muddle. *Express & Echo* (Exeter) Feb 7 1983.
ANON. Big cat? No – it's a badger. *Express & Echo* (Exeter) Feb 8 1983.
HEUVELMANS, B. Annotated checklist of apparently unknown animals with which cryptozoology is concerned. *Cryptozoology* Vol 5, pp 1-26 (1986).
JENKINSON, R.D.S. A rapid but short lived colonisation of the British Isles by the Northern lynx. In: GILBERTSON, D.D. & JENKINSON, R.D.S. (Eds) *In the Shadow of Extinction: A Quaternary Archaeology and Palaeoecology of the Lake, Fissures and Smaller Caves at Creswell Crags SSSI* (Department of Prehistory St Archaeology, University of Sheffield: Sheffield, 1984), pp 111-115.
HETHERINGTON, D.A. *et al.* New evidence for the occurrence of Eurasian lynx (*Lynx lynx*) in medieval Britain. *Journal of Quaternary Science* Vol 21, pp 3-8 (2005).
PULLAR, P. The missing lynx – should it be reintroduced? *Scottish Field* https://www.scottishfield.co.uk/outdoors/wildlifeandconservation/the-missing-lynx-should-it-be-reintroduced/ Sept 13 2019.
SCREETON, P. Diary of a cat flap. *Folklore Frontiers* No 4, pp 19-25 (Jan 1987).
BELL, E. The Durham cat – the story so far. *Folklore Frontiers* No 7, pp 11-14 (1988).
ANON. Mystery beast sighted! *Forres Gazette* (Forres) Mar 14 1984.
ANON. Playmates disturb mystery big cat. *Forres Gazette* (Forres) Apr 10 1985.
ANON. Moray's 'big cat' moves upcountry! *Forres Gazette* (Forres) Jan 20 1988.
ANON. Hit-and-run driver kills mystery wildcat. *The News* (Portsmouth) Jul 28 1988.
ANON. 'Killer' found dead. *South Shropshire Journal* (Ludlow) 10 Feb 1989.

Chapter 2

SATUNIN, C. The black cat of Transcaucasia. *P.Z.S.L.* Vol II, pp 162-163 (1904).
SMIRNOV, N. [Paper re *Felis daemon.*] *Bulletin du Musée de Caucase* Vol 11, pp 84-86 (1917).
OGNEV, S.I. [*Mammals of the U.S.S.R. and Adjacent Countries. Vol III: Carnivora*] (Israel Program for Scientific Translations: Jerusalem, 1962).
POCOCK, R.I. *Catalogue of the Genus Felis* (British Museum (Natural History): London, 1951).
HENDEY, Q.B. The Late Cenozoic Carnivora of the South-Western Cape Province. *Annals of the South African Museum* Vol 63, pp 1-369 (Jan 1974).
DESMAREST, A.G. *Mammalogie ou Description des Espèces des Mammifères* (Imprimeur-Libraire: Paris, 1820).
CUVIER, G. *Dictionnaire des Sciences Naturelles, Vol 8* (Le Normant: Paris, 1817).
ROBINSON, R. Homologous genetic variation in the Felidae. *Genetica* Vol 46, pp

1-31 (1976).
KITCHENER, A. *et al.* A revised taxonomy of the Felidae: The final report of the Cat Classification Task Force of the IUCN Cat Specialist Group. *Cat News* Special Issue 11, pp 1-80 (winter 2017).
SCHWARZ, E. Die Wildkatze der Balearen. *Zoologischer Anzeiger* Vol 91, pp 223-224 (1930).
PARRACK, J.D. *The Naturalist in Majorca* (David & Charles: Newton Abbot, 1973).
HEUVELMANS, B. Annotated checklist of apparently unknown animals with which cryptozoology is concerned. *Cryptozoology* Vol 5, pp 1-26 (1986).
VIGNE, J.-D. Zooarchaeology and the biogeographical history of the mammals of Corsica and Sardinia since the last ice age. *Mammal Review* Vol 22(2), pp 87–96 (1992).
BATE, D.M.A. On the mammals of Crete. *P.Z.S.L.* pp 315-323 (1905).
LAVAUDEN, L. Sur le chat sauvage de la Corse. *Comptes Rendus de l'Académie des Sciences* Vol 189, pp 1023-1024 (Dec 2 1929).
ARRIGHI, J. & SALOTTI, M. Le chat sauvage (*Felis silvestris* Schreber, 1977) en Corse. Confirmation de sa presence et approche taxonomique. *Mammalia* Vol 52, 123-125 (1988).
SAMUEL, H. French claim discovery of new 'cat-fox' species in Corsican forest. *Daily Telegraph* (London) https://www.telegraph.co.uk/news/2019/06/20/french-claim-discovery-new-cat-fox-species-corsican-forest/ Jun 20 2019.
SCHARFF, R.F. On the former occurrence of the African wild cat (Felis ocreata, Gmel.) in Ireland. *Proceedings of the Royal Irish Academy B* Vol 26, pp 1-12 (1906).
CHALMERS, P. Cat of eighteen lives. *The Field* Vol 178, p 672 (Dec 6 1941).
MacDERMOTT, A. Wild cats in Ireland. *The Field* Vol 179, p 16 (Jan 3 1942).
KINAHAN, J.R. [Note re Andrews's nomenclatural findings for Irish wildcat.] *Proceedings of the Natural History Society of Dublin* Vol 1, p 69 (Dec 9 1853).
WOOD, J.G. *Illustrated Natural History: Mammalia* (Routledge, Warne, & Routledge: London, 1859-1863).
HAMILTON, E. Remarks upon the supposed existence of the wild cat (*Felis catus*) in Ireland. *P.Z.S.L.* pp 211-214 (Mar 3 1885).
LLOYD, A.T. The population genetics of cats in northern Ireland. *Carnivore Genetics Newsletter* Vol 3, pp 373-377 (1979).
TEGETMEIER, W.B. Exhibition of a specimen of the wild cat (*Felis catus*) obtained in Donegal. *P.Z.S.L.* p 3 (1885).
SCHARFF, R.F. The wild cat in Ireland. *Irish Naturalist* Vol 14, p 79 (Apr 1905).
WELCH, R. The wild cat in Ireland. *Irish Naturalist* Vol 14 (May 1905).
KANE, W.F. de V. Wild cats formerly indigenous in Ireland. *Irish Naturalist* Vol 14, pp 165-166 (Jul 1905).
WARREN, R. Supposed wild cat in Ireland. *Irish Naturalist* Vol 14, pp 135-136 (Jun 1905).
WARREN, R. The wild cat in Ireland. *Irish Naturalist* Vol 14, pp 183-184 (Aug 1905).
SCHARFF, R.F. The wild cat in Ireland. *Irish Naturalist* Vol 14, p 184 (Aug 1905).
BARRETT-HAMILTON, G.E.H. The wild cat in Ireland. *Irish Naturalist* Vol 20, p 55 (Mar 1911).

MOFFAT, C.B. The mammals of Ireland. *Proceedings of the Royal Irish Academy B* Vol 44, pp 61-128 (1937-1938).
STELFOX, A.W. Notes on the Irish "wild cat". *Irish Naturalists' Journal* Vol 15, pp 57-60 (Jul 1965).
MONTGOMERY, W.I. *et al.* Origin of British and Irish mammals: disparate post-glacial colonisation and species introductions. *Quaternary Science Reviews* Vol 98, pp 144-165 (Aug 15 2015).
HOLIDAY, F.W. *The Goblin Universe* (Llewellyn Publishers: Minnesota, 1986).
CUNNINGHAM, G. & COGHLAN, R. *The Mystery Animals of Ireland* (CFZ Press: Bideford, 2010).
SUTCLIFFE, A.J. *On the Track of Ice Age Mammals* (British Museum (Natural History): London, 1985).
NAISH, D. Europe, where the sabre-tooths, lions and leopards are. *Tetrapod Zoology* https://scienceblogs.com/tetrapodzoology/2008/03/12/european-cats-part-I Mar 12 2008.
NAISH, D. Pumas of South Africa, cheetahs of France, jaguars of England. *Tetrapod Zoology* https://scienceblogs.com/tetrapodzoology/2008/03/13/european-cats-part-ii Mar 13 2008.
FERRANT, V. & FRIANT, M. Quelques caractères de tigre chez le *Felis spelea* Goldf. *Bulletin. Museum National d'Histoire Naturelle (Paris)* (Ser 2) Vol 11, pp 508-512 (1939).
FISHER, J., SIMON, N., & VINCENT, J. *The Red Book* (Collins: London, 1969).
JAEKEL, 0. Prähistorische Löwen aus dem Formenkreis der Felis spelaea. *Zoologischer Anzeiger* Vol 70, pp 225-236 (1927).
MAZAK, V. On a supposed prehistoric representation of the Pleistocene scimitar cat, *Homotherium* Fabrini, 1890 (Mammalia; Machairodontidae). *Zeitschrift für Säugetierkunde* Vol 35, pp 359-362 (1970).
KURTEN, B. *Pleistocene Mammals of Europe* (Weidenfeld & Nicolson: London, 1968).
AGUSTÍ, J. & ANTÓN, M. *Mammoths, Sabertooths, and Hominids: 65 Million Years of Mammalian Evolution in Europe* (Columbia University Press: New York, 2002).
ROUSSEAU, M. Un félin à canine-poignard dans l'art paléolithique. *Archéologia* No 40, pp 81-82 (May-Jun 1971).
ROUSSEAU, M. Une Machaïrodonte dans l'art aurignacien? *Mammalia* Vol 35, pp 648-657 (Dec 1971).
REUMER, J. *et al.* Late Pleistocene survival of the saber-toothed cat *Homotherium* in northwestern Europe. *Journal of Vertebrate Paleontology* Vol 23, 260-262 (2003).
BERLITZ, C. *Atlantis* (Macmillan: London, 1984).
CAMPION-VINCENT, V. *Des Fauves Dans Nos Campagnes: Légendes, Rumeurs et Apparitions* (Imago: Paris, 1992).
BRODU, J.-L. & MEURGER, M. *Les Félins-Mystère: Sur Les Traces d'Un Mythe Moderne* (Pogonip: Paris, 1984).
TAAKE, K.-H. *Gévaudan Tragedy: The Disastrous Campaign of a Deported 'Beast'* (Kindle e-book: Seattle, 2015).
SHUKER, K.P.N. The Beast of Gévaudan – Wolf, Man...or Wolf-Man? *ShukerNature* http://karlshuker.blogspot.com/2015/08/

the-beast-of-gevaudan-wolf-manor-wolf.html Aug 22 2015.
BEGOUEN, H. A propos du crane de panthère, dit de Malarnaud. *Bulletin de la Société d'Histoire Naturelle (Toulouse)* Vol 56, pp 469-471 (1927).
ASTRE, G. Le crane de panthère, dit de Malarnaud, et sa non-fossilisation. *Bulletin de la Société d'Histoire Naturelle (Toulouse)* Vol 56, pp 471-474 (1927).
ANON. Land beasties. *Doubt* No 17, p 260 (1947).
ANON. British report. *Doubt* No 18, p 269 (1947).
CHARTRAIN, D. La bête du Cezallier. *Amazone* No 1, p 17 (Oct 1982).
MAGIN, U. Continental European big cats. *Pursuit* Vol 17, pp 114-115 (autumn 1984).
ANON. Saint-Raphaël: un félin (puma ou panthère) égaré. *Matin* (Nice) Feb 16 1983.
ANON. La "bête de Valescure" toujours introuvable. *Matin* (Nice) Feb 18 1983.
ANON. French news. *Fortean Times* No 43, p 47 (spring 1985).
ANON. Other big news. *Fortean Times* No 42, pp 41-42 (autumn 1984).
ANON. Black panther found prowling roofs in French town. *BBC News* (London) https://www.bbc.co.uk/news/world-europe-49753154 Sept 19 2019.
ANON. Recaptured wandering black panther stolen from French zoo. *BBC News* (London) https://www.bbc.co.uk/news/world-europe-49819815 Sept 24 2019.
LÜPS, P. Zwei neue Nachweise der Wildkatze im Kanton Bern. *Mitteilungen der Naturforschenden Gesellschaft in Bern* Vol 28, pp 71-73 (1971).
LÜPS, P. Nachweis der Wildkatze *Felis s. silvestris* Schreber 1777 im Berner Jura. *Jahrbuch des Naturhistorisohen Museums Bern* Vol 7, pp 195-199 (1981).
LÜPS, P. Daten zu einer am 1.1.1983 am Strassenrand zwischen Sonceboz und La Heutte tot aufgefundenen: Wildkatze *Felis s. silvestris* Schreber 1777. Unpublished report (1983).
RAGNI, B. Riuscira la lince a ricolonizzare il territorio italiano? *Natura Montagna* Vol 30, pp 41-43 (1983).
ANON. A panther at large. *The Field* Vol 82, p 263 (Aug 12 1893).
ANON. Tiger in the Orloff Government. *The Field* Vol 82, p 302 (Aug 19 1893).
BURTON, R.G. A wild beast in Russia. *The Field* Vol 82, p 882 (Dec 9 1893).
PIETILA, A. Raids by starving tigers strike terror in Soviet towns. *Sun* (Baltimore) Feb 15 1987.
KESWAL. Mysterious wild beast in Russia. *The Field* Vol 83, p 127 (Jan 27 1894).

Chapter 3

CORBET, G.B. & HILL, J.E. *A World List of Mammalian Species* (3rd Edit.) (Natural History Museum Publications/Oxford University Press: London, 1991).
ANON. Cheetahs spotted. *Independent* (London) Jun 10 1988.
ANON. Cheetah sighting in India. *Oryx* Vol 25(2), p 69 (Apr 1991).
SHUKER, K.P.N. Blue tigers, black tigers, and other Asian mystery cats. *Cat World*, No 214, pp 24-25 (Dec 1995).
GEE, E.P. The white tigers. *Animals* Vol 3, pp 282-286 (1964).
THORNTON, I.W.B., YEUNG, K.K., & SANKHALA, K.S. The genetics of the white tigers of Rewa. *Journal of Zoology* Vol 152, pp 127-135 (1967).
ROBINSON, R. The white tigers of Rewa and gene homology in the Felidae. *Genetica* Vol 40, pp 198-200 (1969).

THORNTON, I.W.B. White tiger genetics – further evidence. *Journal of Zoology* Vol 185, pp 389-394 (1978).
GEE, E.P. Albinism and partial albinism in tigers. *J.B.N.H.S.* Vol 56, pp 581-587 (1959).
XIAO, X. *et al.* The genetic basis of white tigers. *Current Biology* Vol 23(11), pp 1031-1035 (Jun 3 2013 [May 23 2013 online]).
WIEGEL, I. Big felids. In: GRZIMEK, B. (Ed.), *Grzimek's Animal Life Encyclopedia, Vol 12* (Van Nostrand Reinhold: London, 1975), p 348.
PREMKUMAR, R. 'White tiger' spotted in the Nilgiris. *The Hindu* https://www.thehindu.com/news/national/tamil-nadu/white-tiger-in-the-nilgiris-is-a-first/article19217223.ece Jul 5 2017.
CUVIER, G. *Le Règne Animal... [The Animal Kingdom...], 16 Vols* (Roulin & Valenciennes: Paris, 1836-49).
ROBINSON, R. Homologous genetic variation in the Felidae. *Genetica* Vol 46, pp 1-31 (1976).
WOOD, J.G. *The Illustrated Natural History: Mammalia* (Routledge, Warne, & Routledge: London, 1859-1863).
LYDEKKER, R. *The Game Animals of India, Burma and Tibet* (Rowland Ward: London, 1907).
NARAYAN, V.N. Notes on man eating tigers. *J.B.N.H.S.* Vol 28, p 1124 (1922).
POCOCK, R.I. Tigers. *J.B.N.H.S.* Vol 33, pp 505-541 (1929).
SHUKER, K.P.N. *Cats of Magic, Mythology, and Mystery: A Feline Phantasmagoria* (CFZ Press: Bideford, 2012).
VELLA, C.M. *et al. Robinson's Genetics for Cat Breeders and Veterinarians* (4th Edit.) (Butterworth-Heinemann: Oxford, 1999).
CARTER, W.H. Black tigers. *The Times* (London) Oct 16 1936.
SCHROETER, W.K.H. Farbanomalien und Streifenreduktionen beim Tiger (*Panthera tigris* Linnaeus, 1758). *Sitzungsberichte der Gesellschaft Naturforschender Freunde zu Berlin* Vol 12, pp 154-158 (1973).
SCHROETER, W. On color, color deviation, stripe diminuation and color-brightening in the tiger, *Panthera tigris. Säugetierkundliche Mitteilungen* Vol 29(4), pp 1-8 (1981).
CALDWELL, H.R. *Blue Tiger* (Duckworth: London, 1925).
HEUVELMANS, B. Annotated checklist of apparently unknown animals with which cryptozoology is concerned. *Cryptozoology* Vol 5, pp 1-26 (1986).
CALDWELL, J.C. *Our Friends the Tigers* (Hutchinson: London, 1954).
XU, D.C. *Mystery Creatures of China: The Complete Cryptozoological Guide* (Coachwhip Publications: Greenville, 2018).
CASE, D. Letter. *Mainly About Animals* No 36, p 22 (Jan 1988).
STONOR, C. Rare sightings of black tigers. *Country Life* Vol 136, p 691 (Sept 17 1964).
HUGHES-HALLETT, N.M. Black tigers. *Country Life* Vol 136, p 926 (Oct 8 1964).
ANON. [Black tiger report.] *The Observer* (London) Jan 27 1844.
BUCKLAND, C.T. A black tiger. *The Field* Vol 73, p 422 (Mar 23 1889).
HAUXWELL, T.A. Possible occurrence of a black tiger (*Felis tigris*). *J.B.N.H.S.* Vol 22, pp 788-789 (1914).
EDITORS OF THE FIELD. Black tigers. *The Field* Vol 152, p 657 (Oct 25 1928).

HIGGINS, J.C. [Black tiger specimen was bear.] *J.B.N.H.S.* Vol 35, pp 673-674 (1932).
DOLLMAN, G. Black tigers. *The Times* (London) Oct 19 1936.
DOLLMAN, G. Black tigers – evidence for existence. *The Times* (London) Feb 4 1937.
CAPPER, S. Black tigers. *The Times* (London) Oct 20 1936.
GRIFFITHS, R.G. "Black tigers". *The Times* (London) Jan 19 1937.
PRATER, S.H. Black tigers. *J.B.N.H.S.* Vol 39, pp 381-382 (1937).
BRANDER, A.A.D. Black tigers. *The Field* Vol 169, p 80 (Jan 9 1937).
GEE, E.P. *The Wild Life of India* (Collins: London, 1964).
BURTON, R.G. Black tigers. *The Field* Vol 152, pp 656-657 (Oct 25 1928).
ANON. White tiger cub in Indian zoo turns black. *Daily Telegraph* (London) https://www.telegraph.co.uk/news/newsvideo/weirdnewsvideo/7973055/White-tiger-cub-in-Indian-zoo-turns-black.html Aug 31 2010.
SINGH, L.A.K. Colour variation in tiger: preliminary insight into evolution and genetics. *Zoos' Print* Vol 12, pp 9-11 (May 1997).
SHUKER, K.P.N. Black is black...isn't it? *Fortean Times* No 109, p 44 (Apr 1998).
SINGH, L.A.K. *Born Black: The Melanistic Tiger in India* (WWF-India: New Delhi, 1999).
BASU, M. Bright and black. *DownToEarth* https://www.downtoearth.org.in/news/wildlife-biodiversity/bright-and-black-57520 Apr 14 2017.
SHUKER, K.P.N. *ShukerNature Book 1: Antlered Elephants, Locust Dragons, and Other Cryptic Blog Beasts* (Coachwhip Publications: Greenville, 2019).
TURNBULL-KEMP, P. *The Leopard* (Bailey Bros. & Swinfen: London, 1967).
GUGGISBERG, C.A.W. *Wild Cats of the World* (David & Charles: Newton Abbot, 1975).
LOCKE, A. *The Tigers of Trengganu* (Museum Press: London, 1954).
DAY, D. *The Doomsday Book of Animals* (Ebury Press: London, 1981).
WOOD, G.L. *The Guinness Book of Animal Facts and Feats* (3rd Edit.) (Guinness Superlatives: Middlesex, 1982).
SITWELL, N. Comment. *Animals* Vol 14, p 389 (Sept 1972).
ANON. No trace of Bali tiger. *Wildlife* Vol 17, p 187 (Apr 1975).
ANON. Bali may still have large cats. *Wildlife* Vol 22, p 17 (Feb 1980).
SODY, H.J.V. The Balinese tiger, *Panthera tigris balica* (Schwartz). *J.B.N.H.S.* Vol 36, pp 233-235 (1932).
HORNER, T. (Ed.). Tiger resurrection? [Javan Tiger.] *WWF Update* p 3 (spring 1994).
BAMBANG, M. In search of 'extinct' Javan tiger'. *Jakarta Post* (Jakarta), Oct 30 2002.
ANON. New mammal discovered. *Animals* Vol 10, pp 501-503 (Mar 1968).
IMAIZUMI, Y. A new genus and species of cat from Iriomte, Ryukyu Islands. *Journal of the Mammalogical Society of Japan* Vol 3, pp 75-106 (1967).
WURSTER-HILL, D.H. *et al.* Banded chromosome study of the Iriomote cat. *Journal of Heredity* Vol 78, pp 105-107 (Mar 1987).
SWANCER, B. The yamapikarya – Japan's mystery cat, Part 1. *Cryptomundo* http://www.cryptomundo.com/cryptozoo-news/yamapikarya1/ Feb 6 2009.
SWANCER, B. The yamapikarya – Japan's mystery cat, Part 2. *Cryptomundo* http://www.cryptomundo.com/cryptozoo-news/yamapikarya2/ Feb 7 2009.

SWANCER, B. Japanese mystery cats – revisiting the yamapikarya. *Cryptomundo* http://www.cryptomundo.com/cryptotourism/japanese-mystery-cats/ Feb 24 2012.
POCOCK, R.I. *Catalogue of the Genus Felis* (British Museum (Natural History): London, 1951).
HARRISSON, T. The large mammals of Borneo. *Malayan Nature Journal* Vol 4, pp 70-76 (1949).
DAVIES, A.G. & PAYNE, J.B. *A Faunal Survey of Sabah* (WWF Malaysia: Kuala Lumpur, 1982).
RABINOWITZ, A. *et al.* The clouded leopard in Malaysian Borneo. *Oryx* Vol 21, pp 107-111 (Apr 1987).
CHARLES, J.K. *et al.* Announcement!!! The melanistic Bornean clouded leopard – first ever photograph. *Heart of Borneo – Brunei Sungai Ingei Faunal Expedition* http://ingeiexpedition.blogspot.com/2011/03/announcement-melanistic-bornean-clouded.html Mar 3 2011.
HICKS, F.C. *Forty Years Among the Wild Animals of India* (Pioneer: Allahabad, 1910).
FRASER, C. A hermaphrodite tiger. *J.B.N.H.S.* Vol 43, p 99 (1942).
GRAY, A.P. *Mammalian Hybrids* (2nd Edit.) (Commonwealth Agricultural Bureaux: Slough, 1972).
ANON. Back from extinction. *Wildlife* Vol 17, p 232 (May 1975).
ROBINSON, R. The breeding of black and spotted leopards. *J.B.N.H.S.* Vol 66, pp 423-429 (1970).
0'GRADY, R.J.P. Melanism in breeding wild cats. *Proceedings of the Symposium of the Association of British Wild Animal Keepers* Vol 4, pp 32-41 (1979).
FOOKS, H.A. A pale colour form of the panther. *J.B.N.H.S.* Vol 42, pp 435-436 (1941).
POCOCK, R.I. The panthers and ounces of Asia. *J.B.N.H.S.* Vol 34, pp 64-82, 307-336 (1930).
LYDEKKER, R. On an abnormal leopard skin. *P.Z.S.L.* pp 1-3 (Jan 14 1908).
COLLINS, H.O. Is it unique? *Bulletin of the South Californian Academy of Science* Vol 14, pp 49-51 (1915).
PRATER, S.H. *The Book of Indian Animals* (3rd Edit.) (Bombay Natural History Society: Bombay, 1971).
ANON. MA – Hong Kong. *Fortean Times* No 25, p 35 (spring 1978).
DIVYABNANUSINH. Record of two unique observations of the Indian cheetah in *Tuzuk-I-Jahangiri*. *Journal of the Bombay Natural History Society* Vol 84, pp 269-274 (Aug 1987).
SHUKER, K.P.N. Return of the king [putative king cheetahs in Asia]. *Fortean Times* No 271, pp 58-59 (Feb 2011).
SHUKER, K.P.N. A historical depiction of a king cheetah in Asia? *Journal of Cryptozoology* Vol 2, pp 31-39 (2013).
SHUKER, K.P.N. A white sabre-tooth alive and well and living in China? *ShukerNature* http://karlshuker.blogspot.com/2014/09/a-white-sabre-tooth-alive-and-well-and.html Sept 30 2014.
HERKLOTS, G.A.C. *The Hong Kong Countryside* (South China Morning Post: Hong Kong, 1951).
ANON. The story of the Shing Mun tiger. *Animals* Vol 9, pp 16-17 (May 1966).

Chapter 4

CORBET, G.B. & HILL, J.E. *A World List of Mammalian Species* (3rd Edit.) (Natural History Museum Publications/Oxford University Press: London, 1991).

SHUKER, K.P.N. African mystery cats. *Cat World* No 213, pp 14-15 (Nov 1995).

SHUKER, K.P.N. *Cats of Magic, Mythology, and Mystery: A Feline Phantasmagoria* (CFZ Press: Bideford, 2012).

HICHENS, W. African mystery beasts. *Discovery* Vol 18, pp 369-371 (Dec 1937).

COOPER, A.L. A curious skin. *The Field* Vol 148, p 690 (Oct 14 1926).

HILLS, D. & SMITHERS, R.H.N. The "king cheetah": a historical review. *Arnoldia Zimbabwe* Vol 9, pp 1-23 (1980).

POCOCK, R.I. Variation of the pattern in leopards. *The Field* Vol 148, p 707 (Oct 21 1926).

POCOCK, R.I. Description of a new species of cheetah (*Acinonyx*). *P.Z.S.L.* pp 245-252 (1927).

POCOCK, R.I. The Rhodesian cheetah. *The Field* Vol 151, p 593 (Apr 15 1928).

CHAPMAN, A. The Rhodesian cheetah. *The Field* Vol 151, p 654 (Apr 19 1928).

POCOCK, R.I. The *Fauna of British East India Including Ceylon and Burma: Mammalia, Vol 1 - Primates & Carnivora* (2nd Edit.) (Taylor & Francis: London, 1939).

JAMES, A. The puzzle of king cheetahs. *The Field* Vol 219, pp 1018-1019 (May 24 1962).

BOTTRIELL, L.G. *King Cheetah: The Story of the Quest* (E.J. Brill: Leiden, 1987).

BOTTRIELL, L.G. The king of the cats? *Excellence* Vol 3, pp 52-56 (Apr 1987).

ANON. Schutz durch Streifen. *GEO* No 3, pp 179-182 (Mar 1987).

de GRAAF, G. A familiar pattern deviation of the cheetah (*Acinonyx jubatus*). *Custos* Vol 3, p 2 (1974).

ANON. The king cheetah puzzle. *Wildlife* Vol 24, p 73 (Feb 1982).

OAKES, P. Spot the difference. *The Mail on Sunday - YOU Magazine* (London) pp 36-37, May 10 1987.

HALLET, J-P. & PELLE, A. *Animal Kitabu* (Elek Books: London, 1967).

SAYRE, R. Can the leopard change its spots? *Audubon* Vol 86, pp 26-27 (1984).

van AARDE, R.J. & van DYCK, A. inheritance of the king cheetah coat colour pattern in cheetahs *Acinonyx jubatus*. *Journal of Zoology A* Vol 209, pp 573-578 (1986).

KAELIN, C.B. *et al.* Specifying and sustaining pigmentation patterns in domestic and wild cats. *Science* Vol 337, pp 1536-1541 (2012).

STANFORD UNIVERSITY MEDICAL CENTER. How the sub-Saharan cheetah got its stripes: Californian feral cats help unlock biological secret. *ScienceDaily* https://www.sciencedaily.com/releases/2012/09/120920141147.htm Sept 20 2012.

BOTTRIELL, L.G. & BOTTRIELL, P. Paper presented at the ISC Annual Members Meeting, Royal Museum of Scotland, Jul 26 1987.

SHUKER, K.P.N. The cat who would be king. *All About Cats* Vol 3, pp 54-55 (Sept 1996).

SHUKER, K.P.N. A historical depiction of a king cheetah in Asia? *Journal of Cryptozoology* Vol 2, pp 31-39 (2013).

FRAME, G. First record of king cheetah outside southern Africa. *Cat News* No 16 (Mar 1992).
SHUKER, K.P.N. The mpisimbi – a now-extinct king cheetah strain in East Africa? *ShukerNature* http://karlshuker.blogspot.com/2014/07/the-mpisimbi-undiscovered-but-now.html Jul 28 2014.
HOLDSWORTH, E.W. Report: The Secretary on additions to the menagerie. *P.Z.S.L.* pp 532-534 (Jun 19 1877).
SCLATER, P.L. Mr P.L. Sclater on *Felis lanea. P.Z.S.L.* pp 655-656 (Jun 18 1878).
MIVART, St. G. *The Cat* (John Murray: London, 1881).
SCLATER, P.L. The woolly cheetah. *P.Z.S.L.* p 476 (Nov 4 1884).
GANDAR DOWER, K.C. *The Spotted Lion* (William Heinemann: London, 1937).
GOSS, M. In search of...Africa's spotted lion. *Fate* Vol 39, pp 78-91 (Jun 1986).
SHUKER, K.P.N. Lion spotting: on the trail of the mystifying Marozi. *All About Cats* Vol 5, pp 42-43 (Nov-Dec 1998).
GUGGISBERG, C.A.W. *Wild Cats of the World* (David & Charles: Newton Abbot, 1975).
CAMPBELL, L. Moorland cheetah. *East African Natural History Society Bulletin* pp 34-35 (May-Jun 1983).
POCOCK, R.I. An interesting cheetah. *The Field* Vol 137, p 352 (Mar 19 1921).
COMBES, G. The Phantom. *Guy Combes Artist* https://guycombes.wordpress.com/2011/06/15/the-phantom/ Jun 15 2011.
AMMANN, K. Close encounters of the furred kind [Qattara cheetah]. *BBC Wildlife* Vol 11, pp 14-15 (Jul 1993).
HOATH, R. A deader desert [Qattara cheetah]. *BBC Wildlife* Vol 14 (Sept 1996).
BRYNER, J. Elusive Saharan cheetah captured in photos. *Live Science* https://www.livescience.com/10926-elusive-saharan-cheetah-captured-photos.html Dec 24 2010.
O'BRIEN, S.J. *et al.* The cheetah is depauperate in genetic variation. *Science* Vol 221, pp 459-462 (Jul 29 1983).
ANON. ...But the cheetah's future is rosier. *New Scientist* Vol 114, p 27 (Apr 9 1987).
McGUINNESS, C.J. *Nomad* (Methuen: London, 1934).
GANDAR DOWER, K.C. A spotted lion? *The Field* Vol 165, p 388 (Feb 23 1935).
HEUVELMANS, B. *On the Track of Unknown Animals* (Rupert Hart-Davis: London, 1958).
GANDAR DOWER, K.C. In quest of the spotted lion. *The Field* Vol 166, p 21 (Jul 6 1935).
POLLARD, J.R.T. *African Zoo Man* (Robert Hale: London, 1963).
POCOCK, R.I. Note on the spotted lion of the Aberdares. In: GANDAR DOWER, K.C. *The Spotted Lion* (William Heinemann: London, 1937), pp 317-321.
FOWLE, A. Spotted lions. *The Field* Vol 165, p 1361 (Jun 1 1935).
RICHARDSON, B.V. Spotted lions. *The Field* Vol 170, p 1515 (Dec 11 1937).
HAMILTON-SNOWBALL, G. Spotted lions. *The Field* Vol 192, p 412 (Oct 9 1948).
POLLARD, J.R.T. Spotted lions. *The Field* Vol 192, p 553 (Nov 13 1948).
FLETT, G. Spotted lion. *The Field* Vol 193, p 76 (Jan 15 1949).
FORAN, W.R. Legendary spotted lion. *The Field* Vol 196, p 535 (Sept 30 1950).

SIMON, N. *Between the Sunlight and the Thunder* (Collins: London, 1962).
HUXLEY, E. *On the Edge of the Rift: Memories of Kenya* (William Morrow: New York, 1962).
ANON. The lion who just can't change his spots. *Daily Mirror* (London) Jan 28 1970.
FLORIO, P.L. Birth of a lion x leopard hybrid in Italy. *International Zoo News* Vol 30, pp 4-6 (1983).
TURNBULL-KEMP, P. *The Leopard* (Bailey Bros. & Swinfen: London, 1967).
SCHERREN, H. Some feline hybrids. *The Field* Vol 111, p 711 (Apr 25 1908).
ANON. Supposed hybrid lion and leopard. *The Field* Vol 111, p 711 (Apr 25 1908).
POCOCK, R.I. Striped lions and stripeless tigers. *The Field* Vol 159, p 149 (Jan 30 1932).
SHUKER, K.P.N. Behold, Uneeka! Uncovering a hitherto-forgotten photograph of the only living lijagupard ever to be exhibited in Britain – a ShukerNature world-exclusive! *ShukerNature* http://karlshuker.blogspot.com/2018/02/behold-uneeka-uncovering-hitherto.html Feb 19 2018.
SHUKER, K.P.N. *ShukerNature Book 2: Living Gorgons, Bottled Homunculi, and Other Monstrous Blog Beasts* (Coachwhip Publications: Greenville, 2020).
KINGDON, J. *East African Mammals: An Atlas of Evolution in Africa, Vol IIIA* (Academic Press: New York, 1977).
HALTENORTH, T. & DILLER, H. *A Field Guide to the Mammals of Africa Including Madagascar* (Collins: London, 1980).
HEUVELMANS, B. Annotated checklist of apparently unknown animals with which cryptozoology is concerned. *Cryptozoology* Vol 5, pp 1-26 (1986).
PRINCE WILLIAM OF SWEDEN. *Among Pygmies and Gorillas* (Gyldendal: London, 1923).
LINDBLOM, G. *The Akamba in British East Africa* (2nd Edit.) (Appelbergs Botryckeri Aktiebolag: Uppsala, 1920).
GUGGISBERG, C.A.W. *Simba: The Life of the Lion* (Bailey Bros. & Swinfen: London, 1962).
PITMAN, C.R.S. *A Game Warden Among His Charges* (Nisbet: London, 1931).
TEMPLE-PERKINS, E.A. *Kingdom of the Elephant* (Andrew Melrose: London, 1955).
HEUVELMANS, B. *Sur la Piste des Bêtes Ignorées, Vol 2* (Plon: Paris, 1955).
McBRIDE, C. *The White Lions of Timbavati* (Paddington Press: London, 1977).
McBRIDE, C. *Operation White Lion* (Collins & Harvill Press: London, 1981).
SMUTS, G.L. *Lion* (Macmillan: London, 1982).
STEVENSON-HAMILTON, J. Albinism among lions. *The Field* Vol 197, p 358 (1951).
BRYNER, J. Rare white lions released to wild. *NBC News* http://www.nbcnews.com/id/27245016/ns/technology_and_science-science/t/rare-white-lions-released-wild/#.XlcF2fTgq00 Oct 17 2008.
ROBINSON, R. & de VOS, V. Chinchilla mutant in the lion. *Genetica* Vol 60, pp 61-63 (1982).
ROBINSON, R. New mutant in the lion. *Carnivore Genetics Newsletter* Vol 2, p 236 (1974).
SPEIGHT, W.L. Mystery monsters in Africa. *Empire Review* Vol 71, pp 223-228

(1940).
LAYARD, H. Early *Adventures in Persia, Susiana and Babylon* (John Murray: London, 1887).
KAY, J. *Okavango* (Hutchinson: London, 1962).
ADAMSON, G. *My Pride and Joy: An Autobiography* (Harvill Press: London, 1986)
SHUKER, K.P.N. Black lions – manipulation, melanism, and mozaicism. *ShukerNature* http://karlshuker.blogspot.com/2012/06/black-lions-manipulation-melanism-and.html Jun 12 2012.
SHUKER, K.P.N. Exposing another black lion photograph as a fake. *ShukerNature* http://karlshuker.blogspot.com/2012/10/exposing-another-black-lion-photograph.html Oct 1 2012.
SHUKER, K.P.N. Exposing yet another fake black lion photograph. *ShukerNature* http://karlshuker.blogspot.com/2017/02/exposing-yet-another-fake-black-lion.html Feb 15 2017.
ANON. Zoo set to breed the first black lion. *The Sun* (London) Nov 3 1975.
SHUKER, K.P.N. A hitherto-unseen photograph of Ranger – Scotland's (nearly) black lion. *ShukerNature* http://karlshuker.blogspot.com/2013/05/a-hitherto-unseen-photograph-of-ranger.html May 21 2013.
HERAN, I. *Animal Coloration: The Nature and Purpose of Colours in Vertebrates* (Hamlyn: London, 1976).
HEUVELMANS, B. *Les Félins Encore Inconnus d'Afrique* [*The Still-Unknown Cats of Africa*] (L'Oeil du Sphinx: Paris, 2007).
SHUKER, K.P.N. Green lion...or green leopard? On the track of Heuvelmans's unknown mystery cat. *ShukerNature* http://karlshuker.blogspot.com/2012/11/green-lionor-green-leopard-on-track-of.html Nov 18 2012.
ROBINSON, R. Homologous genetic variation in the *Felidae*. *Genetica* Vol 46, pp 1-31 (1976).
ANON. A leopard which "changed its spots"! *Illustrated London News* Vol 188, p 1012 (Jun 6 1936).
ANSELL, W.F.H. An aberrant leopard from Rhodesia. *Arnoldia Rhodesia* Vol 3, pp 1-6 (1967).
PILFOLD, N.W. *et al.* Confirmation of black leopard (*Panthera pardus pardus*) living in Laikipia County, Kenya. *African Journal of Ecology* https://doi.org/10.1111/aje.12586 Jan 29 2019.
BURTON, W. Exhibition of some melanistic and black leopard skins. *P.Z.S.L.* p 346 (Apr 28 1908).
M.A.B. Black leopards. *Journal of the East African Natural History Society* Vol 16, pp 223-224 (1942).
GÜNTHER, A. Note on a supposed melanotic variety of the leopard, from South Africa. *P.Z.S.L.* pp 243-245 (Mar 3 1885).
GÜNTHER, A. Second note on the melanotic variety of the South African leopard. *P.Z.S.L.* pp 203-205 (Apr 6 1886).
SHUKER, K.P.N. The dark side of leopards. *All About Cats* Vol 4, pp 46-47 (May-Jun 1997).
SKEAD, C.J. *Historical Mammal Incidence in the Eastern Cape (2 Vols)* (Department of Nature and Environmental Conservation of the Provincial Administration of the Cape of Good Hope: Cape Town, 1987).
MOISER, C.M. The melanotic leopards of Eastern Cape, South Africa. In:

DOWNES, J. (Ed.), *The CFZ Yearbook 1997* (CFZ Publications: Exwick, 1997), pp 43-50.
SCLATER, W.L. *The Mammals of South Africa, Vol 1: Primates, Carnivora and Ungulata* (R.H. Porter: London, 1900).
POCOCK, R.I. The story of the 'Nandi bear'. *Natural History* Vol 2, pp 162-169 (1930).
LANE, F.W. *Nature Parade* (4th Edit.) (Jarrolds: London, 1955).
STEERE, E. *Swahili Tales, As Told by Natives of Zanzibar* (Bell & Daldy: London, 1870).
"FULAHN" [HICHENS, W.]. On the trail of the brontosaurus & co. *Chambers's Journal* (Ser 7) Vol 17, pp 692-695 (1927).
HARDY, I.W. Golden cat in the Aberdares National Park. *East African Natural History Society Bulletin* pp 111-112 (Sept-Oct 1979).
PITMAN, C.R.S. *Annual Report of the Game Department for the Year Ended 31st Dec, 1950* (Government Printer: Entebbe, 1951).
DURRELL, G. *The Bafut Beagles* (Rupert Hart-Davis: London, 1954).
JOHNSTON, H. *Liberia* (Hutchinson: London, 1906).
HICHENS, W. Africa's mystery beasts. *Wide World Magazine* Vol 62, pp 171-176 (Dec 1928).
VON BUOL, P. 'Buffalo lions': a feline missing link? *Swara* Vol 23(2), pp 20-25 (Jul-Sept 2001).
GNOSKE, T.P. & KERBIS PDETERHANS, J. Cave lions: the truth behind biblical myths. *In The Field* Vol 71, pp 2-6 (2000).
MORANT FORÉS, A. The mngwa: an unknown subspecies of African lion? *cz@yahoogroups.com* Jul 30 2001.
NAISH, D. Buffalo lions. *cz@yahoogroups.com* Jul 31 2001.
SUNQUIST, M. & SUNQUIST, F. *Wild Cats of the World* (University of Chicago Press: Chicago, 2002).
von HEUGLIN, T. *Reise noch Abessinien, den Gala-Landern, Ost-Sudan, und Chartum in den Jahren 1861 u. 1862* (H. Costenoble: Jena, 1868).
von HEUGLIN, T. *Reise in Nordostafrika, Vol 2* (G. Westermann: Braunschweig, 1877).
PARKYNS, M. *Life In Abyssinia* (2nd Edit.) (John Murray: London, 1868).
DERANIYALAGA, P.E.P. Does the tiger inhabit the Sudan? *Spolia Zeylanica* Vol 26, p 159 (1951).
MANDELA, N. *Long Walk To Freedom* (Little, Brown: London, 1994).
MEEK, J. On the trail of the Borneo cat-fox [also includes details re Africa's Rungwe tiger]. *Guardian* (London), Dec 7 2005.
RYAN, J. [Report re Beast of Mayanja.] *San Francisco Examiner & Chronicle* (San Francisco) Feb 17 1974.
ANON. [Report re Beast of Bungoma.] *Daily Mirror* (London) Apr 24 1974.
SHUKER, K.P.N. When did the sabre-tooths really die out? *All About Cats* Vol 4, pp 50-51 (Jul-Aug 1997).
SHUKER, K.P.N. *Still In Search Of Prehistoric Survivors: The Creatures That Time Forgot?* (Coachwhip Publications: Greenville, 2016).
RAYNAL, M. Personal communications (1987).
VINCENT, J.-F. *Le Pouvoir et le Sacré Chez les Hadjeray du Tchad* (Anthropos: Paris, 1975).

DE BURTHE D'ANNELET, A.J.V. *À Travers l'Afrique Française...* (Roger: Paris, 1932).
MERFIELD, F.G. & MILLER, H. *Gorillas Were My Neighbours* (Longmans, Green: London, 1956).
von NOLDE, I. Der Coje ya Menia: ein Sagenhaftes Tier Westafrikas. *Deutsche Kolonialzeitung* Vol 51, pp 123-124 (1939).
KRUMBIEGEL, I. Was ist der 'Löwe des Wassers'? *Kosmos* Vols 42-43, pp 143-146 (1947).
EVANS-PRITCHARD, E.E. Notes on some animals in Zandeland. *Man* Vol 63, pp 139-142 (Sept 1963).
HEUVELMANS, B. *Les Derniers Dragons d'Afrique* (Plon: Paris, 1978).
RAYNAL, M. "Lions d'eau" Africains et machairodontes (Dernière Partie). *Cryptozoologia* pp 3-9 (Sept 1996).
JOYE, E. Le Mourou-ngou se porte bien!! *Cryptozoologia*, No 6, pp 1-5 (Sept 1994).
WHITFIELD, P. *The Hunters* (Hamlyn: London, 1978).
STOW, G.W. & BLEEK, D. *Rock Paintings in South Africa* (Methuen: London, 1930).
HUNTER, J.A. *Hunter* (Hamish-Hamilton: London, 1952).
FROBENIUS, L. & FOX, D.C. *African Genesis* (Turtle Island Foundation: Berkeley, 1983).
BRONSON, E.B. *In Closed Territory* (A.C. McClurg: Chicago, 1910).
JORDAN, J.A. *Elephants and Ivory: True Tales of Hunting and Adventure* (Rinehart: London, 1956).
MAHAUDEN, C. *Kisongokimo: Chasse et Magie Chez les Balubas* (Flammarion: Paris, 1965).
JORDAN, J.A. *The Elephant Stone* (Nicholas Kaye: London, 1959).
ANON [quoting JORDAN, J.A.]. The brontosaurus. Hunter's story of tusked and scaly beast. *Daily Mail* (London) Dec 16 1919.
HAGENBECK, C. *Beasts and Men* (Longmans, Green: London, 1909).
HUGHES, J.E. *Eighteen Years on Lake Bangweulu* (The Field: London, 1933).
ANON. A "burrowing" cat. *Leeds Mercury* (Leeds) Jan 26 1925.
LAVAUDEN, L. Animaux disparus et légendaries de Madagascar. *Revue Scientifique* pp 1-12 (May 1931).
CAZARD, P. Le lion des rocs. *Le Chasseur Français* p 664 (Oct 1939).
MERY, F. *The Life, History, and Magic of the Cat* (Hamlyn: London, 1967).
HOSEY, G. Madagascan cats. *Fortean Times* No 176, p 72 (Nov 2003).
DECARY, R. *La Faune Malgache* (Payot: Paris, 1950).
NAISH, D. New mammal for Madagascar. *cz@yahoogroups* May 19 2003.
SAUTHER, M.L. *et al.* Taxonomic identification of Madagascar's free-ranging "forest cats". *Conservation Genetics* https://doi.org/10.1007/s10592-020-01261-x (Feb 28 2020).
SOKOL, J. Madagascar's mysterious, murderous cats identified. *Science* Vol 367, p 1178 (Mar 10 2020).
BORGERSON, C. The *fitoaty*: an unidentified carnivoran species from the Masoala peninsula of Madagascar. *Madagascar Conservation & Development* Vol 8(2), pp 81-85 (Nov 2013).
FARRIS, Z.J. *et al.* Feral cats and the *fitoaty*: first population assessment of the black

forest cat in Madagascar's rainforests. *Journal of Mammalogy* Vol 20(10), pp 1-8 (Dec 2015).
MURPHY, A. Makira lessons: the fitoaty (aka the creature with seven livers). *Medium* https://medium.com/@Asia_Murphy/makira-lessons-the-fitoaty-aka-the-creature-with-seven-livers-36da668baf2b Feb 14 2017.

Chapter 5

CORBET, G.B. & HILL, J.E. *A World List of Mammalian Species* (3rd Edit.) (Natural History Museum Publications/Oxford University Press: London, 1991).
ANON. Florida panther crisis. *Oryx* Vol 19, p 172 (Jul 1985).
FLORIDA FISH & WILDLIFE CONSERVATION COMMISSION. Florida panther program. *Myfwc* https://myfwc.com/wildlifehabitats/wildlife/panther/ (2017).
FISHER, J., SIMON, N., & VINCENT, J. *The Red Book* (Collins: London, 1969).
BANFIELD, A.W.F. *The Mammals of Canada* (University of Toronto Press: Toronto, 1974).
DOWNING, R.L. The search for cougars in the Eastern United States. *Cryptozoology* Vol 3, pp 31-49 (1984).
WRIGHT, B.S. *The Eastern Panther* (Clark, Irwin: Toronto, 1971).
GOLLEY, F.B. *Mammals of Georgia* (University of Georgia Press: Athens, 1962).
RANDOLPH, V. *We Only Lie to Strangers* (Greenwood Press: Westport, 1951).
WHITE, K. Mountain lions in Missouri? Now there's proof. *News-Leader* (Springfield) Feb 1 2017.
WRIGHT, B.S. The latest specimen of the Eastern puma. *Journal of Mammalogy* Vol 42, pp 278-279 (May 1961).
HARDISON, S. *Wildlife in North Carolina* Vol 40, pp 15-17 (1976).
SASS, H.R. The panther prowls the East again. *Saturday Evening Post* Vol 226, pp 31, 133-134, 136 (Mar 13 1954).
NOWAK, R.M. *The Cougar in the United States and Canada* (1976). Cited in: DOWNING, R.L. The search for cougars in the Eastern United States. *Cryptozoology* Vol 3, pp 31-49 (1984).
WRIGHT, B.S. Rediscovering the Eastern panther. *Animals* Vol 6, p 85 (Mar 23 1965).
YOUNG, S.P. & GOLDMAN, E.A. *The Puma: Mysterious American Cat* (American Wildlife Institute: Washington, 1946).
LEWIS, J.C. Evidence of mountain lions in the Ozarks and adjacent areas, 1948-1968. *Journal of Mammalogy* Vol 50, pp 371-372 (1969).
FEGELY, T. Rare animal sightings occur. *Latrobe Bulletin* (Latrobe) Aug 12 1987.
RICCIUTI, E.R. *The Wild Cats* (Windward: Lebester, 1979).
DENIS, A. *Cats of the World* (Constable: London, 1964).
ANON. Here, kitty kitty. *The News* No 9, p 16 (Apr 1975).
ANON. Animal sightings. *Res Bureaux Bulletin* No 40, pp 7-8 (Nov 9 1978).
RICKARD, R.J.M. Large cats. *Fortean Times* No 28, p 49 (winter 1979).
COLEMAN, L. On the trail: phantom panthers. *Fortean Times* No 30, pp 47-49 (autumn 1980).
GREENWELL, J.R. (Ed.) "Mystery cats" stalk again. *ISC Newsletter* Vol 1, pp 6-7 (spring 1982).

COHEN, D. *The Encyclopedia of Monsters* (Dodd, Mead: New York, 1982).
GORDON, S. Cougars in PA. *PA Woods & Waters* (Feb 1987).
HAMILTON, R. Big-cat comeback. [A Connecticut quarterly periodical, title unknown to me] p 5 (summer 1987).
ANON. Cougar controversy continues throughout West. *West Virginia Advocate* (Cacapon Bridge) Sept 14 1987.
INGRAM, B. A big cat in Virginia? Do mountain lions/cougars roam the Commonwealth? *Virginia Country* pp 24-25 (fall 1987).
ANON. Cougars in Hardy County. *West Virginia Advocate* (Cacapon Bridge) Jan 14 1988.
ANON. Cougar report. *West Virginia Advocate* (Cacapon Bridge) Feb 15 1988.
DOWNING, R.L. Techniques used in the search for Eastern cougars. Paper presented at the ISC Annual Members Meeting, University of Maryland, May 14 1988.
TISCHENDORF, J.W. & ROPSKI, S.J. (Eds) *Proceedings of the Eastern Cougar Conference, 1994* (American Ecological Research Institute: Aerie, 1996).
U.S. FISH & WILDLIFE SERVICE. U.S. Fish and Wildlife Service concludes Eastern cougar extinct. *U.S. Fish & Wildlife Service* https://www.fws.gov/northeast/ecougar/ecougar_newsrelease.html March 2 2011.
U.S. FISH & WILDLIFE SERVICE. Long extinct Eastern cougar to be removed from Endangered Species List correcting lingering anomaly. *U.S. Fish & Wildlife Service* https://www.fws.gov/northeast/ecougar/pdf/Cougar_News_Bulletin_Final_1_18.pdf Jan 22 2018.
COLEMAN, L. *Mysterious America* (Faber & Faber: London, 1983).
MAYES, M. *Shadow Cats: The Black Panthers of North America* (Anomalist Books: San Antonio, 2018)
ANON. Land beasties. *Doubt* No 17, p 260 (1947).
SHUKER, K.P.N. An American mystery black panther depicted in a famous modern-day painting. *ShukerNature* http://karlshuker.blogspot.com/2016/04/an-american-mystery-black-cat-depicted.html Apr 15 2016.
COLEMAN, L. Mystery animals in Illinois. *Fate* Vol 24, pp 48-54 (Mar 1971).
CLARK, J. & COLEMAN, L. On the trail of pumas, panthers and ULAs (Unidentified Leaping Animals). *Fate* Vol 25, pp 72-82 (Jun 1972) & pp 92-102 (Jul 1972).
COLEMAN, L. California odyssey: observations on the western para-panther. *Anomaly Research Bulletin* No 23, pp 5-8 (1978).
COLEMAN, L. Black "mountain lions" in California? *Pursuit* Vol 12, pp 61-62 (Apr 1979).
CLARK, J. & COLEMAN, L. *Creatures of the Outer Edge* (Warner Books: New York, 1978).
WRIGHT, B.S. *The Ghost of North America* (Vantage Press: New York, 1959).
ANON. Big black cat roams S.J. hills. *San Jose News* (San Jose) Dec 15 1973.
ANON. Black African panther is roaming Bay area. *Arkansas Gazette* (Little Rock) Dec 16 1973.
COLEMAN, L. Phantom panther on the prowl. *Fate* Vol 30, pp 62-67 (Nov 1977).
HALE, D. Clues to Van Etten 'swamp monster' aren't easy prey. *Daily Telegram* (Elmira) Jul 21 1977.

ANON. Van Etten swamp monster print believed discovered. *Daily Telegram* (Elmira) Jul 25 1977.
ANON. Conservation officer urges calm in Van Etten search. *Daily Telegram* (Elmira) Jul 28 1977.
ANON. Biologists doubt authenticity of swamp monster print. *Daily Telegram* (Elmira) Aug 18 1977.
HARTLEY, T. 'Swamp monster' mystery unsolved. *Sunday Telegram* (Elmira) Jan 8 1978.
CROOKS, S. Swamp monster is just a memory now. *Sunday Telegram* (Elmira) May 13 1979.
FULLER, C. Hunting black panthers. *Fate* Vol 38, pp 26, 28 (Jan 1985).
SCHAFFNER, R. Phantom panthers in suburbia. *Fate* Vol 38, pp 71-73 (May 1985).
ANON. 'Big cat' has Nauvoo excited. *Tuscaloosa News* (Tuscaloosa) May 29 1987.
JENKINS, R. Mysterious cat-like creature stalking pets, terrorizing area of South Jersey. *Star-Ledger* (Newark) Jun 23 1987.
ANON. Mystery animal roaming South Jersey. *Asbury Park Press* (Asbury Park) Jun 24 1987.
SIPRESS, A. 'Mystery animal' stalking near S. Jersey communities. *Philadelphia Inquirer* (Philadelphia) Jun 29 1987.
SCHAFFNER, R. Panther pandemonium. *Creature Chronicles* No 12, pp 4-10 (Feb 1988).
ALBERT, V. Personal communication, Nov 26 1987.
HALL, E.R. *The Mammals of North America, 2 Vols* (2nd Edit.) (Wiley: New York, 1981).
NELSON, E.W. *Wild Animals of North America* (Revised Edit.) (National Geographic Society: Washington, 1930).
LOTT, V. A fisher in Florida. *INFO Journal* No 44, p 9 (May 1984).
LANGFORD, C. *Winter of the Fisher* (Manor Books: New York, 1975).
IVES, K.E. Black panther. *INFO Journal* No 41, p 17 (Oct 1982).
KELSEY, P. Response. *INFO Journal* No 41, p 17 (Oct 1982).
THOMAS, A.C. Personal communication. Oct 26 1984.
ARMENT, C. *Varmints: Mystery Carnivores of North America* (Coachwhip Publications: Landisville, 2010).
ANON. Jaguarundi survives in Texas. *Oryx* Vol 21, p 123 (Apr 1987).
LEVER, C. *Naturalized Mammals of the World* (Longmans: London, 1985).
ULMER, F.A. Melanism in the Felidae, with special reference to the genus Lynx. *Journal of Mammalogy* Vol 22, pp 285-288 (1941).
PARADISO, J. Melanism in Florida bobcats. *Florida Scientist* Vol 36, pp 215-216 (1973).
ARMENT, C. *Cryptozoology: Science & Speculation.* Coachwhip Publications: Landisville, 2004).
SHUKER, K.P.N. The truth about black pumas – separating fact from fiction regarding melanistic cougars. *ShukerNature* http://karlshuker.blogspot.com/2012/08/the-truth-about-black-pumas-separating.html Aug 16 2012.
ROBINSON, R. Homologous genetic variation in the Felidae. *Genetica* Vol 46, pp 1-31 (1976).
COLEMAN, L. Melanistic phases of Felidae in captivity: preliminary survey results.

Carnivore Genetics Newsletter Vol 2, pp 209-211 (1974).
CAHALANE, V.H. *Mammals of North America* (Macmillan: London, 1961).
GUGGISBERG, C.A.W. *Wild Cats of the World* (David & Charles: Newton Abbot, 1975).
McCABE, R.A. The scream of the mountain lion. *Journal of Mammalogy* Vol 30, pp 305-306 (1949).
ANON. Black pumas. *Pursuit* Vol 5, p 12 (Jan 1972).
GOLDMAN, E.A. The jaguars of North America. *Proceedings of the Biological Society of Washington* Vol 45, pp 143-146 (1932).
MERRIAM, H. Is the jaguar entitled to a place in the California fauna? *Journal of Mammalogy* Vol 1, pp 38-40 (1919).
DAY, D. *The Doomsday Book of Animals* (Ebury Press: London, 1981).
NOWAK, R. Retreat of the jaguar. *National Parks Conservation Magazine* Vol 49, pp 10-13 (Dec 1975).
TAYLOR, W.P. Recent record of the jaguar in Texas. *Journal of Mammalogy* Vol 28, p 66 (1947).
HOCK, R.J. Southwestern exotic felids. *American Midland Naturalist* Vol 53, pp 324-328 (Apr 1955).
BROWN, D.E. On the status of the jaguar in the southwest. *Southwestern Naturalist* Vol 28, pp 459-460 (1983).
FISHER, J. The Zoo's leopards and jaguars. *The Field* Vol 176, p 224 (Aug 17 1940).
DITTRICH, L. Die Vererbung des Melanismus beim Jaguar (*Panthera onca*). *Zoologishe Garten* Vol 49, pp 417-428 (1979).
RABINOWITZ, A. *Jaguar* (Collins London, 1987).
EIZIRIK, E. *et al.* Molecular genetics and evolution of melanism in the cat family. *Current Biology* Vol 13(5), pp 448-453 (Mar 4 2003).
COLEMAN, L. On the trail: maned mystery cats. *Fortean Times* No 31, pp 24-27 (spring 1980).
BRANDON, J. *Weird America* (E.P. Dutton: New York, 1978).
COLEMAN, L. On the trail: an answer from the Pleistocene. *Fortean Times* No 32, pp 21-22 (summer 1980).
ANON. Lion at large? *Newsweek* (New York) Mar 27 1961.
ANON. Animal appearances. *Res Bureaux Bulletin* No 27, p 6 (1977).
ANON. [Report regarding alleged lion sighted in Florida.] *Palm Beach Post* (Palm Beach) Jan 24 1978.
ANON. More animal stories. *Res Bureaux Bulletin* No 50, p 4 (Aug 1979).
COLEMAN, L. *Curious Encounters* (Faber & Faber: London, 1985).
ANON. The big white hunters of Texas. *Express & Star* (Wolverhampton) Apr 22 1988.
ANON. Hunters lured by Texas ranch safaris. *Sunday Times* (London) Jun 12 1988.
SWANCER, B. Mysterious phantom lions in America. *Mysterious Universe* https://mysteriousuniverse.org/2018/09/mysterious-phantom-lions-in-america/ Sept 5 2018.
COLEMAN, L. Maned mystery lion on Norwalk video, *Cryptozoonews* http://www.cryptozoonews.com/maned-la/ Aug 4 2014.
KURTEN, B. & ANDERSON, E. *Pleistocene Mammals of North America* (Columbia University Press: New York, 1980).

SIMPSON, G.G. Large Pleistocene felids of North America. *American Museum Novitates* No 1136, pp 1-27 (1941).
BARNETT, R. *et al.* Phylogeography of lions (*Panthera leo* ssp) reveals three distinct taxa and a late Pleistocene reduction in genetic diversity. *Molecular Ecology* Vol 18(8), pp 1668-1677 (Apr 2009).
GRAYSON, M. Pleistocene panthers. *Fortean Times* No 36, pp 58-59 (winter 1982).
HEMMER, H. Fossil history of living Felidae. *Worlds Cats* Vol 3, pp 1-14 (1976).
WEST, P.M. The lion's mane. *American Scientist* Vol 93, pp 226-235 (May-Jun 2005).
MARTIN, L.D. & GILBERT, B.M. An American lion, *Panthera atrox*, from Natural Trap cave, North Central Wyoming. *Contributions to Geology, University of Wyoming* Vol 16, pp 95-101 (1978).
CHIMENTO, N.R. & AGNOLIN, F.L. The fossil American lion (*Panthera atrox*) in South America: Palaeobiogeographical implications. *Comptes Rendus Palevol* Vol 16, pp 850-864 (Nov-Dec 2017).
MERRIAM, J.C. & STOCK, C. *The Felidae of Rancho La Brea* (Carnegie Institution, Publication No 422: Washington, 1932).
ANON. Sabre-tooth and lion-like cats: skeletons found in asphalt. *Illustrated London News* Vol 182, pp 222-223 (Feb 18 1933).
CAPPARELLA III, A. The santer: North Carolina's own mystery cat? Parts I & II. *Shadows* No 4, pp 1-3 (Jan 1977) & No 5, pp 1-3 (Feb 1977).
LONG, B. Personal communication, Jun 30 1998.
SHUKER, K.P.N. *Still In Search Of Prehistoric Survivors: The Creatures That Time Forgot?* (Coachwhip Publications: Greenville, 2016).
SCHWARZ, E. Blue or dilute mutation in Alaskan lynx. *Journal of Mammalogy* Vol 19, p 376 (1938).
POLAND, H. *Fur-Bearing Animals in Nature and Commerce* (Gurney & Jackson: London, 1892).
SCHAFFNER, R. [Report regarding Russelville "tiger".] *Creature Chronicles* No 6, p 6 (spring 1983).
PIPES, G.H. *Strange Customs of the Ozark Hillbilly* (Hobson Books: New York, 1947).
SHUKER, K.P.N. Feline folklore of the wild, Wild West. *All About Cats* Vol 6, pp 30-31 (May-Jun 1999).
COLEMAN, L. Outing the Ozark howler. *Cryptozoonews* http://www.cryptozoonews.com/howler-redux/ Mar 19 2008.
GOSS, M. Personal communication, Mar 4 1986.
ELDIN, P. *Amazing Hoaxes and Frauds* (Octopus Books: London, 1987).

Chapter 6

GREENWELL, J.R. Is this the beast the Spaniard saw in Montezuma's zoo? *BBC Wildlife* Vol 6, pp 354-359 (Jul 1987).
GREENWELL. J.R. (Ed.). Onza specimen obtained – identity being studied. *ISC Newsletter* Vol 5, pp 1-6 (spring 1986).
MARSHALL, R. *The Onza* (Exposition Press: New York, 1961).
GREENWELL, J.R. (Ed.). Two new onza skulls found. *ISC Newsletter* Vol 4, pp 6-7 (winter 1985).

CARMONY, N.B. *Onza! The Hunt For a Legendary Cat* (High-Lonesome: Silver City, 1995).
DOBIE, J.F. *Tongues of the Monte* (Doubleday, Doran: New York, 1935).
CLARK, J. & COLEMAN, L. *Creatures of the Outer Edge* (Warner Books: New York, 1978).
ORR, P.C. *Felis trumani*, a new radiocarbon dated cat skull from Crypt Cave, Nevada. *Bulletin of the Santa Barbara Museum of Natural History, Department of Geology* Vol 2, pp 1-8 (1969).
SAVAGE, D.E. A survey of various late Cenozoic vertebrate faunas of the Panhandle of Texas, III: Felidae. *University of California Publications in Geological Science* Vol 36, pp 317-344 (1960).
MARTIN, L.D., GILBERT, B.M., & ADAMS, D.B. A cheetah-like cat in the North American Pleistocene. *Science* Vol 195, pp 981-982 (Mar 11 1977).
ADAMS, D.B. The cheetah: Native American. *Science* Vol 205, pp 1155-1158 (Sept 14 1979).
HEMMER, H. The onza as a paleocheetah – an example of possible Pleistocene persistence. Paper presented at the Third International Congress of Systematic and Evolutionary Biology, University of Sussex, England, Jul 7 1985.
BAILEY, M. Experts wild about onza. *The Observer* (London) Nov 16 1987.
ANON. Legendary cat shot. *Oryx* Vol 21, p 56 (Jan 1987).
GREENWELL, J.R. & BEST, T. The onza: its history and biology. Paper presented at the ISC Annual Members Meeting, Royal Museum of Scotland, Jul 26 1987.
COLARUSSO, J. Cryptoletter. *ISC Newsletter* Vol 6, p 10 (autumn 1987).
MOLNAR, R. Personal communications (1986-1987).
SHUKER, K.P.N. Cryptoletter. *ISC Newsletter* Vol 5, p 11 (winter 1986).
GREENWELL, J.R. (Ed.). Onza identity still unresolved. *ISC Newsletter* Vol 7, pp 5-6 (winter 1988).
SHUKER, K.P.N. *The Encyclopaedia of New and Rediscovered Animals: From the Lost Ark to the New Zoo – and Beyond* (Coachwhip Publications: Landisville, 2012).
SHUKER, K.P.N. Unmasking the onza – from Aztecs to Arizona. *All About Cats* Vol 5, pp 44-45 (Jan-Feb 1998).
DRATCH, P.A. *et al.* Molecular genetic identification of a Mexican onza specimen as a puma (*Puma concolor*). *Cryptozoology* Vol 12, pp 42-49 (1993-1996 [published 1998]).
PALMEROS, R.A.L. Was an onza shot in early 1995? In: DOWNES, J. (Ed.). *CFZ Yearbook 1996* (CFZ: Exwick, 1995), pp 167-168.
KURTEN, B. & ANDERSON, E. *Pleistocene Mammals of North America* (Columbia University Press: New York, 1980).
BARNETT, R. *et al.* Evolution of the extinct sabretooths and the American cheetah-like cat. *Current Biology* Vol 15, pp R589-R590 (2005).
GRAYSON, D.K. *Giant sloths and Sabertooth Cats: Extinct Mammals and the Archaeology of the Ice Age Great Basin* (University of Utah Press: Salt Lake City, 2016).
WENDT, H. *Out of Noah's Ark* (Weidenfeld & Nicolson: London, 1959).
SANDERSON, I.T. More new cats? *Pursuit* Vol 6, pp 35-36 (Apr 1973).
WILKINS, H. *Secret Cities of Old South America* (Library Publishers: New York, 1952).

DOWNES, J. Onza = normal puma. *cz@onelist.com* Jun 1 1998.
COLBERT, E.H. *Wandering Lands and Animals* (Hutchinson: London, 1973).
SIMPSON,G.G. *Splendid Isolation: The Curious History of South American Mammals* (Yale University Press: London, 1980).
CORBET, G.B. & HILL, J.E. *A World List of Mammalian Species* (3rd Edit.) (Natural History Museum Publications/Oxford University Press: London, 1991).
SHUKER, K.P.N. South American mystery cats: speckled tigers, striped tigers and sabre-toothed tigers? *Cat World* No 215, pp 36-37 (Jan 1996).
GUGGISBERG, C.A.W. *Wild Cats of the World* (David & Charles: Newton Abbot, 1975).
TURNBULL-KEMP, P. *The Leopard* (Bailey Bros. & Swinfen: London, 1967).
RABINOWITZ, A. *Jaguar* (Collins: London, 1987).
de AZARA, F. *Apuntamientos Para la Historia Natural de los Quadrupedos de Paraguay y Rio de la Plata, 2 Vols* (Ibarra: Madrid, 1802).
RENGGER, J.R. *Naturgeschichte der Säugethiere von Paraguay* (Schweighauser: Basle, 1830).
DOHERTY, R. Meet the world's first white jaguar cubs born in captivity. *AOL News* https://tinyurl.com/th6hj8e [original URL was exceedingly lengthy!] May 9 2012.
BUFFON, G.-L.L. *Histoire Naturelle, 36 Vols* (Imprimerie Nationale: Paris, 1749-1804).
PENNANT, T. *Pennant's History of Quadrupeds, Vol 1* (B. White: London, 1781).
BEWICK, T. & HODGSON, S. *A General History of Quadrupeds* (E. Walker: Newcastle, 1807).
VON SCHREBER, J.C.D. *Die Säugetiere in Abbildungen nach der Natur mit Beschreibungen* (Wolfgang Walther: Erlangen, 1774-1804)
ROBINSON, R. Homologous genetic variation in the Felidae. *Genetica* Vol 46, pp 1-31 (1976).
TINSLEY, J.B. *The Puma: Legendary Lion of the Americas* (Texas Western Press: El Paso, 1987).
GREENWELL, J.R. The puma plot thickens. *BBC Wildlife* Vol 13, p 26 (May 1995).
SHUKER, K.P.N. *The Beasts That Hide From Man: Seeking The World's Last Undiscovered Animals* (Paraview Press: New York, 2003).
HOCKING, P.J. Large Peruvian mammals unknown to zoology. *Cryptozoology* Vol 11, pp 38-50 (1992).
CONAN DOYLE, A. The story of the Brazilian cat. *Strand Magazine* Vol 16, pp 603-615 (Dec 1898).
HOCKING, P.J. Further investigation into unknown Peruvian mammals. *Cryptozoology* Vol 12, pp 50-57 (1993-1996 [published 1998]).
JARDINE, W. *The Natural History of the Felinae* (W.H. Lizars: Edinburgh, 1834).
BROCK, S.E. *Hunting in the Wilderness* (Robert Hale: London, 1963).
PERRY, R. *The World of the Jaguar* (Taplinger: New York, 1970).
POLAND, H. *Fur-Bearing Animals in Nature and Commerce* (Gurney & Jackson: London, 1892).
SHUKER, K.P.N. *Cats of Magic, Mythology, and Mystery: A Feline Phantasmagoria* (CFZ Press: Bideford, 2012).

HEMMER, H. Untersuchungen zur Stammesgeschichte der Pantherkatzen (Pantherinae). Teil 1. *Veröffentlichungen der Zoologischen StaatsSammlung* Vol 11, pp 1-121 (1966).
HALTENORTH, T. Ein Leopard-Puma Bastard. *Zeitschrift für Säugetierkunde* Vol 11, pp 349-352 (1936).
SCHERREN, H. Some feline hybrids. *The Field* Vol 111, p 711 (Apr 25 1908).
BIOCCA, E. *Yanoáma: The Narrative of a White Girl Kidnapped by Amazonian Indians* (E.P. Dutton: New York, 1970).
BRIDGES, W. It's the "fearsome warracaba tiger". *Animal Kingdom* Vol 57, pp 25-28 (1957).
KIRKE, H. *Twenty-Five Years in British Guiana* (Sampson Low: London, 1898).
BROWN, C.B. *Canoe and Camp Life in British Guiana* (Edward Stanford: London, 1876).
im THURN, E. *Among the Indians of Guiana* (Kegan Paul: London, 1883).
BUELER, L.E. *Wild Dogs of the World* (Constable: London, 1973).
WALKER, E.P. *Mammals of the World* (3rd Edit.) (Johns Hopkins Press: Baltimore, 1975).
KLEINMAN, D.G. Social behavior of the maned wolf (*Chrysocyon brachyurus*) and bush dog (*Speothos venaticus*): a study in contrast. *Journal of Mammalogy* Vol 53, pp 791-806 (1972).
FAWCETT, P.H. *Exploration Fawcett* (Hutchinson: London, 1953).
MALLINSON, J. *Travels In Search of Endangered Species* (David & Charles: Newton Abbot, 1989).
MACKAL, R.P. *Searching For Hidden Animals* (Doubleday: Garden City, 1980).
LEITE-PITMAN, M.R.P. & WILLIAMS, R.S.R. *Atelocynus microtis. The IUCN Red List of Threatened Species 2011* https://www.iucnredlist.org/species/6924/12814890#bibliography (2011).
MERCHANT, M. Personal communication, Jun 19 2019.
MORANT FORÉS, A. An investigation into some unidentified Ecuadorian mammals. *Virtual Institute of Cryptozoology* http://cryptozoo.pagesperso-orange.fr/expeditions/ecuador_eng.htm [no longer online] Oct 12 1999.
MORANT FORÉS, A. Maned cougars. *cz@yahoogroups.com* Aug 31 2001.
MATTHIESSEN, P. *The Cloud Forest* (Pyramid Books: New York, 1966).
ANON. Le félin aux dents de sabre. *Science Illustrée* No 9, p 62 (Sept 1998).
MILLER, G.J. Man and Smilodon: a preliminary report on their possible coexistence at Rancho La Brea. *Los Angeles County Museum Contributions to Science* Vol 163, pp 1-8 (1969).
COUTO, C. de P. "Tigre-dentes-de-sabre" do Brasil. *Boletin Conselho Nacional de Pesquisas* No 1, pp 1-30 (1955).
CROFT, D.A. *Horned Armadillos and Rafting Monkeys: The Fascinating Fossil Mammals of South America* (Indiana University Press: Bloomington, 2016).
MARTIN, P.S. & KLEIN, R.G. (Eds) *Quaternary Extinctions: A Prehistoric Revolution* (University of Arizona Press: Tucson, 1984).
MARTIN, P.S. The discovery of America. *Science* Vol 179, pp 969-974 (Mar 9 1973).
GOWLETT, J. *Ascent to Civilization* (Collins: London, 1984).
MANZUETTI, A. *et al.* An extremely large saber-tooth cat skull from Uruguay (late Pleistocene-early Holocene, Dolores Formation): body size and

paleobiological implications. *Alcheringa* https://doi.org/10.1080/03115518.2019.1701080 (Mar 2 2020 [online]).
BLAKE, C.C. Stone cells from Chiriqui. *Transactions of the Ethnological Society* Vol 2, pp 166-170 (1863).
CHALMERS, P. The terrible sabre-tooth tiger. *The Field* Vol 180, p 228 (Aug 29 1942).
CHATWIN, B. *In Patagonia* (Jonathon Cape: London, 1977).
SANCHEZ ROMERO, G. Personal communication, Sept 12 2001.
'BRADYPUS TAMIAS'. Personal communications, Jun 9, 11 2019.
SCHULTZ, C.B., SCHULTZ, M.P., & MARTIN, C.D. A new tribe of saber-toothed cats (Barbourofelini) from the Pliocene of North America. *Bulletin of the University of Nebraska State Museum* Vol 9, pp 1-31 (1970).
HEUVELMANS, B. Annotated checklist of apparently unknown animals with which cryptozoology is concerned. *Cryptozoology* Vol 5, pp 1-26 (1986).
RIGGS, E.S. A new marsupial sabre-tooth from the Pliocene of Argentina and its relationships to other South American predacious marsupials. *Transactions of the American Philosophical Society, New Series* Vol 24, pp 1-32 (1934).
KURTEN, B. *The Age of Mammals* (Weidenfeld & Nicolson: London, 1971).
MORANT FORÉS, A. Personal communication, Sept 26 1994.
NAISH, D. *et al.* 'Mystery big cats' in the Peruvian Amazon: morphometrics solve a cryptozoological mystery. *PeerJ* Vol 2(1), e291 https://peerj.com/articles/291/ (Mar 6 2014).
HEUVELMANS, B. *On the Track of Unknown Animals* (Rupert Hart-Davis: London, 1958).
LEHMANN-NITSCHE, R. Zur Vorgeschichte der Entdeckung vom Grypotherium bei Ultima Esperanza. Cited in: HEUVELMANS, B. *On the Track of Unknown Animals* (Rupert Hart-Davis: London, 1958).
DOBRIZHOFFER, M. *An Account of the Abipones, an Equestrian People of Paraguay, 3 Vols* (John Murray: London, 1822).
FALKNER, T. *A Description of Patagonia, and the Adjoining Parts of South America* (C. Pugh: London, 1774).
MUSTERS, G.C. *At Home With the Patagonians* (John Murray: London, 1871).
RICATTE, R. *De Vile du Diable aux Tumuc-Humac* (Seusée Universelle: Paris, 1979).
CHAPELLE, R. *J'ai Vécu l'Enfer de Raymond Maufrais* (Flammarion: Paris, 1970).
DOWNES, J. & DOWNES, C. (Eds). *CFZ Expedition Report 2007 Guyana* (CFZ Press: Bideford, 2007).
NAISH, D, Personal communication, Jun 17 1998.
ANON. Black pumas. *Pursuit* Vol 5, p 12 (Jan 1972).

Chapter 7

TYNDALE-BISCOE, H. *Life of Marsupials* (Edward Arnold: London, 1973).
SUTCLIFFE, A. *On the Track of Ice Age Mammals* (British Museum (Natural History): London, 1985).
RICH, P.V. *et al. Kadimakara: Extinct Vertebrates of Australia* (Pioneer Design Studio: Lilydale, 1985).
SCHOUTEN, P. *The Antipodean Ark* (Angus & Robertson: London, 1987).
VICKERS-RICH, P. & RICH, T.H. *Wildlife of Gondwana* (Reed: Chatswood,

1993).
LONG, J. *et al. Prehistoric Mammals of Australia and New Guinea: One Hundred Million Years of Evolution* (University of New South Wales Press: Sydney, 2002).
JOHNSON, C. *Australia's Mammal Extinctions: A 50000 Year History* (Cambridge University Press: Melbourne, 2006).
ANON. Tigers, devils, monsters and things. *Wildlife in Australia* Vol 6, p 54 (Jun 1969).
LUMHOLTZ, C. *Among Cannibals: an Account of Four Years Travel in Australia and of Camp Life With the Aborigines of Queensland* (Charles Scribner's Sons: New York, 1889).
BRANDL, E.J. *Australian Aboriginal Paintings in Western and Central Arnhem Land* (Australian Aboriginal Studies No 52, Australian Institute of Aboriginal Studies: Canberra, 1973).
CLEGG, J. Pictures of striped animals: which ones are thylacines? *Archaeological & Physical Anthropology of Oceania* Vol 13, pp 19-29 (1978).
HEUVELMANS, B. *On the Track of Unknown Animals* (Rupert Hart-Davis: London, 1958).
MARCHANT, R.A. *Beasts of Fact and Fable* (Phoenix House: London, 1962).
SHERIDAN, B.G. Notice on the existence in Queensland of an undescribed species of mammal. *P.Z.S.L.* pp 629-630 (Nov 7 1871).
SCOTT, W.T. Letter addressed to the Secretary, respecting the supposed "native tiger" of Queensland. *P.Z.S.L.* p 355 (Mar 5 1872).
SCOTT, W.T. Second letter, on the existence of a "native tiger" in Queensland. *P.Z.S.L.* p 796 (Nov 5 1872).
WHITLEY, G. Mystery animals of Australia. *Australian Museum Magazine* Vol 7, pp 132-139 (Mar 1 1940).
MAKEIG, P. Is there a Queensland marsupial tiger? *North Queensland Naturalist* Vol 37, pp 6-8 (1970).
SOULE, G. *The Mystery Monsters* (G.P. Putnam's Sons: New York, 1965).
McGEEHAN, J. Description of wild animal seen on Atherton Tableland. *North Queensland Naturalist* Vol 6, pp 3-4 (1938).
TROUGHTON, E. *Furred Animals of Australia* (8th Edit.) (Angus & Robertson: London, 1965).
GOUGER [IDRIESS, I.]. [Article documenting Idriess's two Queensland tiger incidents.] *Morning Bulletin* (Rockhampton) Jun 8 1922.
SMITH, M. The great North Queensland tiger hunt of 1923. *Malcolm's Musings*, http://malcolmscryptids.blogspot.co.uk/2013/11/the-great-north-queensland-tiger-hunt.html Nov 8 2013.
LE SOUEF, A.S. & BURRELL, H. *The Wild Animals of Australasia, Embracing the Mammals of New Guinea and the Nearer Pacific Islands* (George Harrap: London, 1926).
TATE, G.H.H. Mammals of Cape York Peninsula, with notes on the occurrence of rain forest in Queensland. *Bulletin of the American Museum of Natural History* Vol 98, pp 563-616 (1925).
BURTON, M. The supposed "tiger-cat" of Queensland. *Oryx* Vol 1, pp 321-326 (1952).
ANON. Naturalist team to hunt a tiger in N.Q. *Courier-Mail* (Brisbane) Aug 2

1970.
ANON. Hunt for tiger link. *Courier-Mail* (Brisbane) Aug 31 1970.
ANON. 100-year-old print a clue to tiger. *Courier-Mail* (Brisbane) Oct 25 1970.
ANON. Another strange beast sighted in Tableland. *Cairns Post* (Cairns) Jan 17 1973.
ANON. Aloomba woman reports seeing Tableland beast. *Cairns Post* (Cairns) Jan 18 1973.
ANON. [Queensland tiger sighting.] *West Australian* (Perth) Aug 24 1982.
ANON. [Supposed Queensland tiger killed was dog.] *Maryborough Chronicle* (Maryborough) Feb 2 1983.
SHUKER, K.P.N. *Cats of Magic, Mythology, and Mystery: A Feline Phantasmagoria* (CFZ Press: Bideford, 2012).
SHUKER, K.P.N. *Still In Search Of Prehistoric Survivors: The Creatures That Time Forgot?* (Coachwhip Publications: Greenville, 2016).
HEALY, T. & CROPPER, P. *Out of the Shadows: Mystery Animals of Australia* (Pan Macmillan Australia: Chippendale, 1994).
SMITH, M. *Bunyips & Bigfoots: In Search of Australia's Mystery Animals* (Millennium: Alexandria, 1996).
GUILER, E. *Thylacine: The Tragedy of the Tasmanian Tiger* (Oxford University Press: London, 1985).
GUILER, E & GODARD, P. *Tasmanian Tiger: A Lesson To Be Learnt* (Abrolhos Publishing: Perth, 1998).
GREENWELL, J.R. (Ed.) Thylacine reports persist after 50 years. *ISC Newsletter* Vol 4, pp 1-5 (winter 1985).
PARK, A. Is this toothy relic still on the prowl in Tasmania's wilds? *Smithsonian* Vol 16, pp 117-131 (Aug 1985).
GOSS, M. Tracking Tasmania's mystery beast. *Fate* Vol 36, pp 34-43 (Jul 1983).
BERESFORD, Q. & BAILEY, G. *Search For the Tasmanian Tiger* (Blubber Head Press: Sandy Bay, 1981).
BROWN, R. Has the thylacine really vanished? *Animals* Vol 15, pp 416-419 (Sept 1973).
FRAUCA, H. The riddle of the kangaroo wolf. *Animal Life* No 31, pp 32-33 (Mar 1965).
PYCRAFT, W.P. The fate of the thylacine. *Illustrated London News* Vol 196, p 564 (Apr 27 1940).
MOREY, G.W. A prehistoric beast still living. *The Field* Vol 151, p 822 (Nov 28 1931).
LOWRY, J.W.J. & MERRILEES, D. Age of the desiccated carcase of a thylacine (Marsupialia; Dasyuroidea) from Thylacine Hole, Nullarbor Region, Western Australia. *Helictite* Vol 7, pp 15-16 (1969).
MERRILEES, D. A check on the radiocarbon dating of desiccated thylacine (marsupial 'wolf') and dingo tissue from Thylacine Hole, Nullarbor Region, Western Australia. *Helictite* Vol 8, pp 39-42 (1970).
ACKMAN, P. *et al.* Wild dog pack could be 'extinct' native tigers. *Sunday Telegraph* (Sydney) Aug 8 1976, Mar 27 1977.
DOUGLAS, A.M. Tigers in Western Australia? *New Scientist* Vol 110, pp 44-47 (Apr 24 1986).
RICKARD, R.J.M. The return of the tiger? *Fortean Times* No 49, pp 5-7 (winter

1987).
HARRIS, S. Hold that tiger! *Walkabout* Vol 34, pp 28-31 (Jun 1968).
SHUKER, K.P.N. Is the Ozenkadnook Tiger a cardboard cryptid? *ShukerNature*, http://karlshuker.blogspot.co.uk/2017/07/is-ozenkadnook-tiger-cardboard-cryptid.html Jul 8 2017.
'JACK THE INSIDER' [HOYSTED, P.]. The prank that took 53 years to debunk. *The Australian* (Surry Hills) http://www.theaustralian.com.au/opinion/blogs/jack-the-insider-the-prank-that-took-53-years-to-debunk/news-story/e95ff01c0bdc0b023f5539aa4447769e [this article no longer online without subscription] Mar 24 2017.
FULLER, C. Bears on two legs? *Fate* Vol 30, p 34 (May 1977).
KOLIG, E. Aboriginal man's best foe? *Mankind* Vol 9, pp 122-123 (Dec 1973).
ALBERT, V.A. The Queensland tiger: Evidence for the possible survival of the marsupial lion, *Thylacoleo*, into recent times. Paper presented at the ISC Annual Members Meeting, Royal Museum of Scotland, Jul 26 1987.
KREFFT, G. Fabulous Australian animals. *Annals and Magazine of Natural History* (Ser 4) Vol 11, pp 315-316 (1873).
SMITH, M. The Queensland tiger: further evidence on the 1871 footprint. *Journal of Cryptozoology* Vol 1, pp 19-24 (2012).
SMITH, M. Review of the thylacine (Marsupialia, Thylacinidae). In: ARCHER, M. (Ed.) *Carnivorous Marsupials. Vol 1* (Royal Zoological Society of New South Wales: Sydney, 1982), pp 237-253.
HEUVELMANS, B. Annotated checklist of apparently unknown animals with which cryptozoology is concerned. *Cryptozoology* Vol 5, pp 1-26 (1986).
van DEUSEN, H.M. First New Guinea record of *Thylacinus*. *Journal of Mammalogy* Vol 44, pp 279-280 (1963).
PLANE, M. The occurrence of thylacines in tertiary rocks from Papua New Guinea. *Journal of Australian Geology and Geophysics* Vol 1, pp 78-79 (1976).
ALBERT, V.A. A bungle in the jungle. *Cryptozoology* Vol 6, pp 119-120 (1987).
OWEN, R. On the fossil mammals of Australia. Part 1. Description of a mutilated skull of the large marsupial carnivore (*Thylacoleo carnifex* Owen), from a calcareous conglomerate. *Philosophical Transactions of the Royal Society* Vol 149, pp 309-322 (1859).
FINCH, M.E. The discovery and interpretation of *Thylacoleo carnifex* (Thylacoleonidae, Marsupialia). In: ARCHER, M. (Ed.) *Carnivorous Marsupials. Vol 2* (Royal Zoological Society of New South Wales: Sydney, 1982), pp 537-551.
COPE, E.D. The ancestry and habits of *Thylacoleo*. *American Naturalist* Vol 16, pp 520-522 (1882).
FLOWER, W. On the affinities and probable habits of the extinct Australian marsupial *Thylacoleo carnifex* Owen. *Quarterly Journal of the Geological Society of London* Vol 24, pp 307-319 (1868).
KREFFT, G. On the dentition of *Thylacoleo carnifex* (Ow.). *Annals and Magazine of Natural History* (Ser 3), Vol 18, pp 148-149 (1866).
WELLS, R.T. *et al.* *Thylacoleo carnifex* Owen (Thylacoleonidae): marsupial carnivore? In: ARCHER, M. (Ed.) *Carnivorous Marsupials. Vol 2* (Royal Zoological Society of New South Wales: Sydney, 1982), pp 573-586.
ARCHER, M. & DAWSON, L. Revision of marsupial lions of the genus *Thylacoleo*

Gervais (Thylacoleonidae, Marsupialia) and thylacoleonid evolution in the late Cainozoic. In: ARCHER, M. (Ed.) *Carnivorous Marsupials. Vol 2* (Royal Zoological Society of New South Wales: Sydney, 1982), pp 477-494.

DAILY, B. *Thylacoleo*, the extinct marsupial lion. *Australian Museum Magazine* Vol 13, pp 163-166 (1960).

CASE, J.A. Differences in prey utilization by Pleistocene marsupial carnivores, *Thylacoleo carnifex* (Thylacoleonidae) and *Thylacinus cynocephalus* (Thylacinidae). *Australian Mammalogy* Vol 8, pp 45-52 (1985).

ROUSE, R. Are we being stalked by a prehistoric tiger? *Gold Coast Journal* (Gold Coast) Aug 5 1987.

GROVES, C.P. Review of: *Mystery Cats of the World: From Blue Tigers to Exmoor Beasts* by Karl P.N. Shuker. *Cryptozoology* Vol 11, pp 119-121 (1992).

OWENS, B.L. The strange saga of...the Emmaville panther. *Australian Outdoors and Fishing* pp 17-19, 83 (Apr 1977).

CROPPER, P. The panthers of southern Australia. *Fortean Times* No 32, pp 13-21 (summer 1980).

ANON. [Report re Emmaville panther.] *Sunday Express* (London) Mar 2 1969.

PASH, B. [Report re Kulja panthers.] *Sunday Times* (Perth) Nov 12, 18 1972.

O'REILLY, D. *Savage Shadow: The Search For the Australian Cougar* (Creative Research: Perth, 1981).

O'REILLY, D. *Savage Shadow: The Search For the Australian Cougar* (New Edit.) (Strange Nation Publishing: Hazelbrook, 2011).

WILLIAMS, M. & LANG, R. *Australian Big Cats: An Unnatural History of Panthers* (Strange Nation Publishing: Hazelbrook, 2010).

WALDRON, D. & TOWNSEND, S. *Snarls from the Tea-tree: Big Cat Folklore* (Arcadia: North Melbourne, 2012).

ANON. 'Puma' sighted. *Sun* (Sydney) Apr 28 1977.

FORREST, B. Panther-like animal stalks Vic outback. *Sun* (Sydney) Nov 25 1977.

O'REILLY, D. Is the killer cat a feral puma? *The Australian* (Sydney) Jun 5, 6 1978.

ANON. Something loose in Emerald, Australia. *Pursuit* Vol 13, p 94 (spring 1980).

HIGGINS, J. A mystery drawing to an end in two States. *Maryborough Advertiser* (Maryborough) Jul 15 1987.

ANON. Cat hunt in the St Arnaud area/Snares to catch a panther? *Maryborough Advertiser* (Maryborough) Jul 1987.

ANON. CF&L staff see huge cat near Maryborough. *Maryborough Advertiser* (Maryborough) Jul 1987.

ANON. CF & L memo gives cat control details. *Maryborough Advertiser* (Maryborough) Jul 1987.

ANON. US puma expert to see local evidence. *Maryborough Advertiser* (Maryborough) Jul 1987.

DAVIES, W. & PRENTICE, R. The feral cat in Australia. *Wildlife in Australia* Vol 17, pp 20-26, 32 (1980).

FIELD, A. Youth shoots stalking 'killer puma' on a mountain. *Daily Mirror* (Sydney) Nov 15 1977.

NAISH, D. Australia's new feral mega-cats. *Tetrapod Zoology* http://scienceblogs.com/tetrapodzoology/2007/03/04/australias-new-feral-mega-cats/ Mar 4 2007.

NAISH, D. Williams and Lang's Australian Big Cats: do pumas, giant feral cats and mystery marsupials stalk the Australian outback? *Tetrapod Zoology* https://

blogs.scientificamerican.com/tetrapod-zoology/williams-and-langs-australian-big-cats/ Feb 13 2012.
HEALEY, K. DNA tests conclude it is a (very) big cat. *Herald Sun* (Melbourne) Nov 27 2005.
KNOX, G. Big cat cover-up? *Hawkesbury Gazette* (Hawkesbury) Feb 17 2005.
GILROY, R. *Mysterious Australia* (Nexus: Mapleton, 1995).
ANON. [Lion-like beast sighting.] *Melbourne Argus* (Melbourne) Nov 3 1933.
GILROY, R. Mystery lions in the Blue Mountains. *Nexus* pp 25-28 (Jun-Jul 1992).
GILROY, R. & GILROY, H. *Out of the Dreamtime: The Search For Australasia's Unknown Animals* (URU Publications: Katoomba, 2006).
SMITH, M. Alien Big Cats in New Guinea? *Malcolm's Musings* http://malcolmscryptids.blogspot.com.au/2011/11/alien-big-cats-in-new-guinea.html Nov 11 2011.
LAWSON, J.A. *Wanderings in the Interior of New Guinea* (Chapman & Hall: London, 1875).
ANON. Taller-than-Everest adventures. *Sunday Herald* (Sydney) Aug 23 1953.
BALLARD, C. Parodic precision: the wanderings of John Lawson. In: JOLLY, M. *et al.* (Eds) *Oceanic Encounters: Exchange, Desire, Violence* (ANU E Press: Canberra, 2009), pp 221-258 http://press-files.anu.edu.au/downloads/press/p60461/mobile/ch08s06.html [excerpt].
ANON. [Report re Mangere lion.] *New Zealand Herald* (Auckland) Jul 8 1977.
ANON. [Early morning news item re Kaiapoi tiger.] *1YA Radio* (Auckland) Jul 18 1977.
ANON. NZ police hunt tiger's owner. *Sydney Morning Herald* (Sydney) Jul 23 1977.
ANON. NZ 'tiger' search goes on. *Sun-Herald* (Sydney) Jul 24 1977.
ANON. MA – New Zealand. *Fortean Times* No 25, p 35 (spring 1978).
KUBOTA, G. State sets traps for leopard-size cat near Olinda. *Star-Bulletin* (Honolulu) http://starbulletin.com/2003/06/17/news/index4.html Jun 17 2003.
FITZGERALD, B.M. Feeding ecology of feral house cats in New Zealand forest. *Carnivore Genetics Newsletter* Vol 4, pp 67-72 (1980).

Chapter 8

LEVER, C. *Naturalized Mammals of the World* (Longmans: London, 1985).
HANSEN, J. The wild dogs of Italy. *New Scientist* Vol 97, pp 590-591 (Mar 3 1983).
BOITANI, L. Wolf and dog competition in Italy. *Acta Zoologica Fennica* No 174, pp 259-264 (1983).
BECK, A.M. *The Ecology of Stray Dogs* (York Press: Baltimore, 1973).
SHUKER, K.P.N. Letter [re African mystery cats]. *Swara* p 36 (Jan-Feb 1988).
HEUVELMANS, B. *On the Track of Unknown Animals* (Rupert Hart-Davis: London, 1958).
SHUKER, K.P.N. *The Encyclopaedia of New and Rediscovered Animals: From the Lost Ark to the New Zoo – and Beyond* (Coachwhip Publications: Landisville, 2012).

ACKNOWLEDGEMENTS

Over 30 years have passed since this book's first edition was published, and I am very sad that so many of those kind persons who so generously assisted and steadfastly encouraged me back then and since then, including, above all, my own family, are no longer here, but their names and memory live on in the following list, in all of the other pages of this book's two editions, and far, far beyond them too. I am truly grateful and greatly honored to have known them and to have been a part of their lives as they were a part of mine.

I owe a great debt of gratitude to a considerable number of persons, societies, organizations, and publications for their very kind interest and most generous assistance in relation to my mystery cat researches and preparation of this book in both of its editions; in particular, I would like to express my most sincere thanks to the following:

Peter Adamson; Dr Victor Albert; *All About Cats*; *Animals and Men*; ASSAP Library; *Athene*; Australian News Summary; Bill Bailey; Simon J. Baker; Dr Michael Bassett; *BBC Wildlife*; Laura Beaton; Endymion Beer; Trevor Beer; Dr David Beeston; Martin Belderson; Phil Bennett; Matthew Bille; *Birmingham Evening Mail*; Birmingham Libraries; Janet and Colin Bord; Lena and Paul Bottriell; Chris Brack; Bradypus Tamias; Sandy Brander; Bill and Ken Brewer; Bristol, Clifton & West England Zoological Society; British Library; Markus Bühler; Owen Burnham; *Cairns Independent*; *Cat World*; *Cats*; Centre for Fortean Zoology (CFZ); Mark Chorvinsky; Paul Clacher; D.N. Clark-Lowes; Luke Cockle; Coffee Books (Stratford-upon-Avon); Loren Coleman; Guy Combes; Mary Cotterill; *Courier Mail* (Brisbane); Paul Cropper; *Daily Express*; *Daily Mail*; *Daily Mirror*; *Daily Record*; Dr Mike Dash; Sophy Day/The Tiger Trust; Jonathan and Corinna Downes; Dudley & West Midlands Zoological Society; Dudley Libraries; S. Ealham; Norman Evans; Dr Anthony Eve; Excalibur Books; *Express & Echo* (Exeter); *Express & Star* (Sandwell/Wolverhampton); Kevin Farley; *Fate*; Miroslav Fišmeister; *Folklore Fro*ntiers; Angel Morant Forés; *Forres Gazette*; Fortean Picture Library; *Fortean Times*; Keith Foster; Richard Freeman; *Glasgow Herald*; Michael Goss; Geoffrey Greed; J. Richard Greenwell; Edith and Florence Griffin; Adam Hart-Davis; Bob Hay; Heads 'N' Tails Taxidermists; Markus Hemmler; David Heppell; Dr Bernard Heuvelmans; Highland Wildlife Park; Daphne Hills; Peter Houston; *INFO Journal*; International Society of Cryptozoology

(ISC); *Journal of Cryptozoology*; Shereen Karmali; Clinton Keeling; Rt Rev. Mgr Michael Kirkham; Dr James Kirkwood; J.M. Knowles; Rebecca Lang; Graham Law; Shane Lea; Michael Andrew Leigh; Ted Leonard; Gerard van Leusden; Jorge A. de Lima Jr; Lorna Lloyd; Robert Lohman; Bryan Long; Dr Peter Lüps; Prof. Roy P. Mackal; Carl Marshall; Debbie Martyr; Marwell Zoological Park; Marcus Matthews; Ruth McArthur; Bill McKee; David McKinley; Michael Merchant; Dr Ralph Molnar; Steve Moore; David Morgan; Dr Desmond Morris; Tim Morris; Raheel Mughal; Richard Muirhead; Dr Darren Naish; Nate; National Museum of Wales; Natural History Museum London; *The News* (Portsmouth); James Nicholls; Edward Noble; Scott T. Norman; North Queensland Naturalists Club; Hodari Nundu; Prof. Stephen J. O'Brien; Richard J.P. O'Grady; *The Observer*; Edward R.J. Orbell; David Pepper; Michael Playfair; *Pursuit*; Queensland Museum; Radio Devon; Radio WM; Michel Raynal; William M. Rebsamen; Bob Rickard; John Rimmer; Roy Robinson; Gustavo Sanchez Romero; Dr Marc van Roosmalen; Royal Museum of Scotland; Sandwell Libraries; *The Scotsman*; Paul Screeton; Mary D. Shuker; Paul Sieveking; Dr Lala A.K. Singh; Malcolm Smith; Society For Psychical Research; Dora Stokes; *Strange Magazine*; *The Sun*; *Sunday Express*; *Sunday Independent*; *Sunday Mirror*; *Swara*; Syndication International; Lars Thomas; Dr Warren D. Thomas; Ernest and Gertrude Timmins; Dr Frank Turk; University of Birmingham; University of Leeds; *The Unknown*; John Valentini Jr; Dr G.R. Cunningham van Someren; Walsall Libraries; Sebastian Wang; Wednesbury Typewriter Service; *Western Morning News*; Prof Ron Westrum; Jan Williams; Ben Willis; Wolverhampton Libraries; Gerald L. Wood; Zoological Society of Glasgow & West of Scotland; Zoological Society of London.

And in deep gratitude for their encouragement and faith in my book, I wish to express my especial thanks to my 1989 edition's publisher Robert Hale Limited and their Managing Director John Hale, to my Editors Carmel Elwell, Robin Frampton, and Rachel Wright, and to my Assistant Editor Damian Thompson; and to this new 2020 edition's publisher Anomalist Books, and to my Editor Patrick Huyghe.

DISCLAIMER

The author has sought permission for the use of all illustrations and lengthy quotes known by him to be still in copyright. Any omissions brought to his attention will be rectified in future editions of this book.

ABOUT THE AUTHOR

Born and still living in the West Midlands, England, Dr. Karl P.N. Shuker graduated from the University of Leeds with a Bachelor of Science (Honors) degree in pure zoology, and from the University of Birmingham with a Doctor of Philosophy degree in zoology and comparative physiology. Ever since then, he has worked full-time as a freelance zoological consultant to the media, and as a prolific published writer.

Dr. Shuker is currently the author of 32 books and hundreds of articles, principally on animal-related subjects, with an especial interest in cryptozoology and animal mythology, on which he is an internationally-recognized authority, but also including two published collections of poetry. In addition, he has acted as consultant for several major multi-contributor volumes as well as for the world-renowned *Guinness Book of Records/Guinness World Records* (he has been its Life Sciences Consultant and a contributor for 23 years); and he has compiled questions for the BBC's long-running cerebral quiz *Mastermind*. He is also the editor of the *Journal of Cryptozoology*, the world's only existing peer-reviewed scientific journal devoted to mystery animals.

Dr. Shuker has travelled the world in the course of his researches and writings, and has appeared regularly on television and radio. Aside from work, his diverse range of interests include motorbikes, quizzes, philately, poetry, travel, world mythology, the history of animation, the life and career of James Dean, collecting masquerade and carnival masks, and anything relating to Sherlock Holmes (novels, short stories, literary pastiches and parodies, TV shows, movies, collectabilia).

He is a Scientific Fellow of the prestigious Zoological Society of London, a Fellow of the Royal Entomological Society, and a Member of various other wildlife-related organizations. He is also Cryptozoology Consultant to the Centre for Fortean Zoology, and a Member of the Society of Authors.

Dr. Shuker's personal website can be accessed at http://www.karlshuker.com and his mystery animals blog, *ShukerNature*, can be accessed at http://karlshuker.blogspot.com

His *Star Steeds* poetry blog can be accessed at http://starsteeds.blogspot.com and his *Eclectarium of Doctor Shuke*r blog can be accessed at http://eclectariumshuker.blogspot.com

His *Shuker In MovieLand* film review blog can be accessed at https://shukerinmovieland.blogspot.com and his *Shuker's Literary Likings* blog showcasing highly-recommended works of fiction can be accessed at https://shukersliterarylikings.blogspot.com.

There is also an entry for Dr. Shuker in the online encyclopedia Wikipedia at http://en.wikipedia.org/wiki/Karl_Shuker and a Like (fan) page on Facebook.

AUTHOR BIBLIOGRAPHY

Mystery Cats of the World: From Blue Tigers to Exmoor Beasts (Robert Hale: London, 1989)

Extraordinary Animals Worldwide (Robert Hale: London, 1991)

The Lost Ark: New and Rediscovered Animals of the 20th Century (HarperCollins: London, 1993)

Dragons: A Natural History (Aurum: London/Simon and Schuster: New York, 1995; republished Taschen: Cologne, 2006)

In Search of Prehistoric Survivors: Do Giant 'Extinct' Creatures Still Exist? (Blandford: London, 1995)

The Unexplained: An Illustrated Guide to the World's Natural and Paranormal Mysteries (Carlton: London/JG Press: North Dighton, 1996; republished Carlton: London, 2002)

From Flying Toads To Snakes With Wings: From the Pages of FATE Magazine (Llewellyn: St Paul, 1997; republished Bounty: London, 2005)

Mysteries of Planet Earth: An Encyclopedia of the Inexplicable (Carlton: London, 1999)

The Hidden Powers of Animals: Uncovering the Secrets of Nature (Reader's Digest: Pleasantville/Marshall Editions: London, 2001)

Historic Realms of Marvels and Miracles: Between Myth and Materiality (Chelsea House: New York, 2001)

Ancient Worlds, Ancient Mysteries: Legends of Many Millennia (Chelsea House: New York, 2001)

Lost Worlds and Forgotten Secrets: Riddles of Earth and Beyond (Chelsea House: New York, 2001)

The New Zoo: New and Rediscovered Animals of the Twentieth Century (House of Stratus Ltd: Thirsk, UK/House of Stratus Inc: Poughkeepsie, USA, 2002)

The Beasts That Hide From Man: Seeking the World's Last Undiscovered Animals (Paraview: New York, 2003)

Extraordinary Animals Revisited: From Singing Dogs To Serpent Kings (CFZ Press: Bideford, 2007)

Dr Shuker's Casebook: In Pursuit of Marvels and Mysteries (CFZ Press: Bideford, 2008)

Dinosaurs and Other Prehistoric Animals on Stamps: A Worldwide Catalogue (CFZ Press: Bideford, 2008)

Star Steeds and Other Dreams: The Collected Poems (CFZ Press: Bideford, 2009)

Karl Shuker's Alien Zoo: From the Pages of Fortean Times (CFZ Press: Bideford, 2010)

The Encyclopaedia of New and Rediscovered Animals: From The Lost Ark to The New Zoo – And Beyond (Coachwhip Publications: Landisville, 2012)

Cats of Magic, Mythology, and Mystery: A Feline Phantasmagoria (CFZ Press: Bideford, 2012)

Mirabilis: A Carnival of Cryptozoology and Unnatural History (Anomalist Books: San Antonio, 2013)

Dragons in Zoology, Cryptozoology, and Culture (Coachwhip Publications: Greenville, 2013)

The Menagerie of Marvels: A Third Compendium of Extraordinary Animals (CFZ Press: Bideford, 2014)

A Manifestation of Monsters: Examining the (Un)Usual Suspects (Anomalist Books: San Antonio, 2015)

More Star Steeds and Other Dreams: The Collected Poems – 2015 Expanded Edition (Fortean Words: Bideford, 2015).

Here's Nessie! A Monstrous Compendium From Loch Ness (CFZ Press: Bideford, 2016)

Still In Search Of Prehistoric Survivors: The Creatures That Time Forgot? (Coachwhip Publications: Greenville, 2016).

ShukerNature Book 1: Antlered Elephants, Locust Dragons, and Other Cryptic Blog Beasts (Coachwhip Publications: Greenville, 2019).

ShukerNature Book 2: Living Gorgons, Bottled Homunculi, and Other Monstrous Blog Beasts (Coachwhip Publications: Greenville, 2020).

This Cryptid World: A Global Survey of Undiscovered Beasts (Herb Lester Associates: London, 2020).

Mystery Cats of the World Revisited: Blue Tigers, King Cheetahs, Spotted Lions, Black Cougars, and More (Anomalist Books: San Antonio, 2020)

Consultant and also Contributor

Man and Beast (Reader's Digest: Pleasantville, 1993)

Secrets of the Natural World (Reader's Digest: Pleasantville, 1993)

Almanac of the Uncanny (Reader's Digest: Surry Hills, 1995)

The Guinness Book of Records/Guinness World Records 1998-present day (Guinness: London, 1997-present day)

GWR Amazing Animals (Guinness: London, 2017)

GWR Wild Things (Guinness: London, 2018)

Consultant

Monsters (Lorenz: London, 2001)

Contributor

Of Monsters and Miracles CD-ROM (Croydon Museum/Interactive Designs: Oxton, 1995)

Fortean Times Weird Year 1996 (John Brown Publishing: London, 1996)

Mysteries of the Deep (Llewellyn: St Paul, 1998)

Guinness Amazing Future (Guinness: London, 1999)

The Earth (Channel 4 Books: London, 2000)

Mysteries and Monsters of the Sea (Gramercy: New York, 2001)

Chambers Dictionary of the Unexplained (Chambers: Edinburgh, 2007)

Chambers Myths and Mysteries (Chambers: Edinburgh, 2008)

The Fortean Times Paranormal Handbook (Dennis Publishing: London, 2009)

Death Worm: Metamorphosis of the Allghoi Khorkhoi [in Russian] (Salamandra P.V.V.: Moscow, 2014)

Folk Horror Revival: Field Studies (Wyrd Harvest Press/Lulu, 2015)

Tales of the Damned: An Anthology of Fortean Horror (Fortean Fiction: Bideford, 2016)

Plus numerous contributions to the annual *CFZ Yearbook* series of volumes, *Fortean Studies*, and various other annual publications.

Editor

The *Journal of Cryptozoology* (CFZ Press: Bideford, 2012-present day)

INDEX OF ANIMAL NAMES